Perspectives in Neural Computing

Springer
London
Berlin
Heidelberg
New York
Barcelona
Budapest
Hong Kong
Milan
Paris
Santa Clara
Singapore
Tokyo

Also in this series:

J.G. Taylor
The Promise of Neural Networks
3-540-19773-7

Maria Marinaro and Roberto Tagliaferri (Eds)
Neural Nets - WIRN VIETRI-96
3-540-76099-7

Adrian Shepherd
Second-Order Methods for Neural Networks: Fast and Reliable Training Methods for Multi-Layer Perceptrons
3-540-76100-4

Jason Kingdon
Intelligent Systems and Financial Forecasting
3-540-76098-9

Dimitris C. Dracopoulos
Evolutionary Learning Algorithms for Neural Adaptive Control
3-540-76161-6

M. Kárný, K. Warwick and V. Kůrková (Eds)
Dealing with Complexity: A Neural Networks Approach
3-540-76160-8

Maria Marinaro and Roberto Tagliaferri (Eds)
Neural Nets - WIRN VIETRI-97
3-540-76157-8

John A. Bullinaria, David W. Glasspool and George Houghton (Eds)
4th Neural Computation and Psychology Workshop, London, 9-11 April 1997: Connectionist Representations
3-540-76208-6

L.J. Landau and J.G. Taylor (Eds)

Concepts for Neural Networks

A Survey

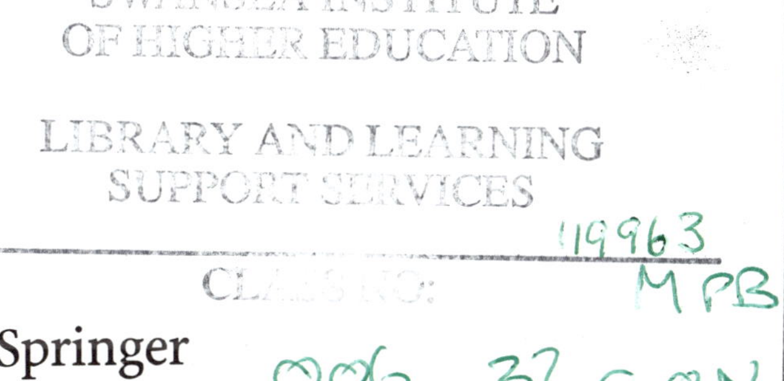

Springer

L.J. Landau, BSc, MA, PhD
Department of Mathematics,
Kings College, Strand, London WC2R 2LS, UK

J.G. Taylor, BA, BSc, MA, PhD, FlnstP
Centre for Neural Networks,
Department of Mathematics, Kings College,
Strand, London WC2R 2LS, UK

Series Editor

J.G. Taylor, BA, BSc, MA, PhD, FlnstP
Centre for Neural Networks,
Department of Mathematics, Kings College,
Strand, London WC2R 2LS, UK

ISBN 3-540-76163-2 Springer-Verlag Berlin Heidelberg New York

British Library Cataloguing in Publication Data
A catalogue record for this book is available from the British Library

Library of Congress Cataloging-in-Publication Data
concepts for neural networks : a survey / L.J. Landau and J.G. Taylor, eds.
p. cm. - - (Perspectives in neural computing)
Includes bibliographical references and index.
ISBN 3-540-76163-2 (pbk.)
1. Neural networks (copmuter science) I. Landau, Lawrence J.
(Lawrence Jay) II. Taylor, John Gerald, 1931- . III. Series.
QA76.87.C6656 1997 97-23815
006.3'2- -dc21 CIP

Printed in Great Britain

Typesetting: Camera ready by editors
Printed and bound at the Athenæum Press Ltd., Gateshead, Tyne and Wear
34/3830-543210 Printed on acid-free paper

Preface

Neural Networks is a subject which is now a well established discipline. It has learned journals ('Neural Networks', 'Neural Computation', 'Network', 'Neural Computing', 'Connection Science', 'IEEE Transactions on Neural Networks', 'Journal of Computational Neuroscience', to name but a few); international conferences are dedicated to it (at least three in each of Europe, the United States and Asia each year); and entrepreneurs have had great success exploiting its powers in commerce and industry. There are specialised University degrees in the subject, such as the M.Sc. in Information Processing and Neural Networks at King's College London, under the Centre for Neural Networks, which also offers Ph.D. degrees.

The discipline of Neural Networks is very broad, involving at one extreme sophisticated concepts and tools from mathematics and theoretical physics and, at the other, designs of VLSI chips to enable fast and light-weight computational systems to be constructed for a range of applications. In this volume it is only possible to scratch the surface of this vast area, but not only is the applications side considered, but also that associated with the brain.

It appears that Neural Networks is about to take off yet again in its explosive career over the last 50 years, to begin to attack the last bastion of the mind - consciousness. Being propelled forward on the new results now flooding in from brain imaging, neural networks offers the natural approach to begin to correlate this wealth of data. In this manner it would be constructing a theoretical framework in terms of which consciousness should find a scientific explanation. This exciting possibility provides one of the areas dealt with in this volume.

The book comprises contributions from authors working in diverse areas associated with neural networks. A summary of each chapter and details of its author(s) is found in the Chapter Summaries, after the table of contents. References for each chapter are placed in the bibliography at the end of the chapter. Topics include introductory presentations of basic ideas of neural networks and associated mathematics, robotics, philosophical issues, modelling of brain function, and a round table discussion on consciousness.

The authors were brought together by their participation in a mathematics summer school at King's College London on neural networks. The goal of the summer school was to make many of the areas touched on by neural networks accessible to a wide audience, among whom would be those who might contemplate further study or research in this field.

The editors would like to thank the staff of the Department of Mathematics, King's College London, for help in the organisation of the summer school, for their participation in the school, and for assistance in converting some Word documents into LaTeX. Above all, we thank the contributors to this volume for their time and effort in carefully preparing the individual chapters.

L.J. Landau and J.G. Taylor, *Editors*

Contents

Chapter Summaries

Chapter 1. Neural Networks: An Overview

J.G. Taylor
Centre for Neural Networks
Department of Mathematics, King's College London, UK

This is an introductory overview of neural networks. The subject is now divided into two tracks, artificial neural networks (ANNs) and computational neuroscience (CNS). Artificial neural networks are being applied in finance, industry, and commerce. Computational neuroscience attempts to model the brain, from the analysis of a single living nerve cell to the study of consciousness and self-consciousness of the whole brain. There is no doubt that neural networks will enter into a new regime of application and explanatory power in the next millenium.

Chapter 2. A Beginner's Guide to the Mathematics of Neural Networks

A.C.C. Coolen
Department of Mathematics
King's College London, UK

A description is given of the role of mathematics in shaping our understanding of how neural networks operate, and the curious new mathematical concepts generated by our attempts to capture neural networks in equations. A selection of relatively simple examples of neural network tasks, models and calculations, is presented.
Acknowledgement. It is my pleasure to thank Charles Mace for helpful comments and suggestions.

Chapter 3. Neurobiological Modelling

J.G. Taylor
Centre for Neural Networks,
Department of Mathematics, King's College London, UK

Neural networks are the modelling medium par excellence in attempting to understand the brain and central nervous system. Firstly, the single cell is considered, a system which in its own right has a very large amount of complexity. Then the retina is studied, a very accessible but complex part of the brain. Visual processing is then the natural consequent of retinal analysis.

Chapter 4. Neural Network Control of a Simple Mobile Robot

R.J. Mitchell and D.A. Keating
Department of Cybernetics
The University of Reading, UK

In recent years researchers in the Department of Cybernetics have been developing simple mobile robots capable of exploring their environment on the basis of the information obtained from a few simple sensors. These robots are used as the test bed for exploring various behaviours of single and multiple organisms: the work is inspired by considerations of natural systems. That part of the work which involves neural networks and related techniques is discussed. These neural networks are used both to process the sensor information and to develop the strategy used to control the robot. Here the robots, their sensors, and the neural networks used are all described.

Chapter 5. A Connectionist Approach to Spatial Memory and Planning

G. Bugmann
Centre for Neural and Adaptive Systems
School of Computing, University of Plymouth, UK

This chapter describes the design and testing of a biologically inspired vision-based model of spatial memory. Three theories of biological spatial memory are discussed. The Topological Network-map theory is translated into general principles, and two forms of connectionist implementation of these principles are discussed. This is followed by a discussion of planning and map-learning experiments performed with a robot. These experiments reveal problems with the implementation of the view-graph principle. The causes of these problems are discussed and solutions proposed.

Acknowledgements. This work has greatly benefited from discussions with Phillipe Gaussier. I gratefully acknowledge comments by Mike Denham, Philippe Gaussier and Loukas Michalis on earlier versions of the manuscript. I am grateful to Davi Vann Bugmann who provided pointers to psychological work on spatial memory. Earlier discussions with John G. Taylor, Chris Hindle and the Neural and Adaptive Systems Group have set the scene for this investigation. Many thanks to Alan Simpson for his support.

Chapter 6. Turing's Philosophical Error?

W. Hodges
Department of Mathematics
Queen Mary and Westfield College, London, UK

In 1936 Alan Turing described a 'universal machine' which he argued could compute the same things as any human 'computer'. Kurt Gödel praised Turing's work but spent much of his career looking for gaps in Turing's argument. In 1972 Gödel wrote a short note consisting of three Remarks. The third Remark was headed *A philosophical error in Turing's work.* He suggested that human beings might be able to compute by 'mental procedures' some things which can't be computed by 'mechanical procedures'. The background to this question is developed and a discussion is given as to what Gödel may have meant by his note.

Chapter 7. Penrose's Philosophical Error

L.J. Landau
Department of Mathematics
King's College London, UK

Computer learning techniques, as embodied in neural networks, are modelled on the way we learn and make decisions. With the ever greater power of computers, one may wonder if a point might be reached when it could be said that the computer has become conscious, that it behaves in the way that a thinking human being behaves. In his book *Shadows of the Mind*, Roger Penrose argues that our conscious mentality cannot be fully understood in terms of computational models, using mathematical theorems due to Gödel and Turing. The mathematics of computation and formal systems is developed in some detail, and it is shown that Penrose's argument does not work.
Acknowledgement. I thank Wilfrid Hodges for helpful comments and criticisms of various versions of this chapter.

Chapter 8.
Attentional Modulation in Visual Pathways

C. Büchel and K.J. Friston
The Wellcome Department of Cognitive Neurology
Institute of Neurology, London, UK

In the past decade functional neuroimaging has been extremely successful in establishing functional segregation as a principle of organisation in the human brain. The nature of the functional specialisation is attributed to the sensorimotor or cognitive process that has been manipulated experimentally. Newer approaches have introduced a number of concepts into neuroimaging. An experiment that studies the effect of attention on visual pathways is described. The analysis reveals a change in the connectivity of two cortical areas involved in visual motion analysis. A nonlinear modulation of this connectivity by the prefrontal regions is shown.
Acknowledgement. C.B. and K.J.F. were supported by the Wellcome Trust.

Chapter 9.
Neural Networks and the Mind

J.G. Taylor
Centre for Neural Networks, King's College London, UK
and
Institut für Medizin
Forschungzentrum Jülich GmbH, Germany

The brain still presents many enigmas to the working neurophysiologist. In order to make progress it seems necessary to take a high level view of the information processing performed by the brain and search for the general organisational plan. The Relational Mind model is used to begin to construct a detailed model of the Mind. A set of guiding Principles is developed, and the two stage Relational Mind model explored to understand the emergence of consciousness at the level of phenomenal experience.
Acknowledgements. I would like to thank Dr A Ioannides and Prof H Muller-Gartner for excellent discussions on non-invasive brain imaging, and the latter for arranging the possibility of my working at IME , KFA-Jülich, where this chapter was written up.

Chapter 10. Confusions about Consciousness

D. Papineau
Philosophy Department
King's College London

I am suspicious about the current enthusiasm for 'consciousness studies'. Many people are now touting consciousness as the last unconquered region of science, and thinkers from many different disciplines are racing to find a 'theory of consciousness' which will unlock this final secret of nature. But I fear that much of this brouhaha is generated by philosophical confusion. In the end there is no special secret, and no special key is needed to unlock it.

Chapter 11. Round Table Discussion

This is the transcription, by L.J. Landau from audio tape, of a round table discussion on the modelling and understanding of consciousness. The session was chaired by John Taylor and included questions and comments from the audience. The panelists were:

Igor Aleksander	Head of Department and Dennis Gabor Chair, Department of Electrical and Electronic Engineering, Imperial College, London, UK
Christian Büchel	Leopold Muller Functional Imaging Laboratory, Wellcome Department of Cognitive Neurology, Institute of Neurology, London, UK
Peter Fenwick	Institute of Psychiatry, Maudsley Hospital, London, UK
Susan Greenfield	Gresham Professor of Physic, Department of Pharmacology, Oxford University, UK
Brian Josephson	Department of Physics, Cambridge University, UK
David Papineau	Head, Department of Philosophy, King's College London, UK
John Taylor	Centre for Neural Networks, King's College London, UK

Acknowledgement. L.J. Landau wishes to thank Jonathan Landau for assistance with the audio transcription.

Chapter 1

Neural Networks: An Overview

1.1 Introduction

There is great excitement in the air these days about the subject of Neural Networks. These devices now mean many things: (1) a model of parts (or even the whole) of the brain; (2) an information processing system which can be taught to classify patterns such as fingerprints or images taken from a satellite of parts of the earth's surface; (3) a mechanism which can be trained to predict tomorrow's foreign exchange rate of the pound or dollar; (4) a fast learning device on a chip which could (but not yet) be sent into space to make satellite control systems more intelligent; (5) last but not least, although far more controversial, a machine which could be built so as to possess a modicum of consciousness similar to that which we all experience.

The first and last of the applications described above are in the domain of neuroscience (and the last even in that of neurophilosophy). The others may have been inspired originally by the powers of the brain but are now more properly termed 'artificial neural networks'. The latest developments in neural networks make it abundantly clear that there are at least two branches of the subject:

1. artificial neural networks (ANNs for short),
2. neuroscientific neural networks. This branch of neural networks, involved with the attempt to model the brain, is termed 'computational neuroscience' (CNS for short).

The task domains in which artificial neural networks are presently in operation include finance (as in application (3) above), industry (in (2) above) and commerce (again using (2) above). More generally these task domains cover the areas of:

- finance:
  credit rating and risk assessment; prediction; variable selection (as in choosing financial variables to build a model of the stock exchange or the bond markets);

- industry:
  plant modelling and control; energy demand prediction; fault maintenance; assignment and scheduling as constrained optimisation;

- commerce:
  credit risk assessment; resource allocation % prediction; personnel selection; quality control.

All of these areas, and many more, present very difficult problems. These are so partly because they may involve many variables or long computing times to solve the relevant problems. Thus the allocation of thousands of radio frequencies between transmitters on a battlefield, when there may be tens of thousands of constraints which must be satisfied to avoid interference between the radio signals, is a very hard problem involving heavy computational costs. When a small change in the allowed frequencies then occurs, due possibly to changes in the weather conditions or the loss of transmitters due to enemy action, then the computational costs can become enormous to find a new set of allowed frequencies as rapidly as possible.

The tasks to which computational neuroscientists have addressed themselves are also as difficult. The brain is composed of billions of living nerve cells, and the job of creating models of even local parts of this most complex of all objects in the universe is enormous. At one end of the scale of the problems is that of modelling and analysing the single living nerve cell itself, which is in its own right a very hard problem. At the other end of the scale is the whole brain, with its remarkable information processing powers leading to consciousness and self-consciousness. The race is on to crack the problem as to how these latter features of the brain can arise from its action. At the same time more straightforward aspects of brain processing - vision, audition, memory, thinking, motor action, and so on - are all being modelled by neural networks.

In all, then, there are exciting challenges to the subject of neural networks from industry and from neuroscience. That these challenges are now being faced up to is related to:

- increased computer facilities,

- the achievement of real applications in the above two fields,

- the interdisciplinary character of the subject, bringing new techniques from the various fields together to enable more efficient solutions to old problems.

The disciplines involved include (but are not restricted to):

- neuroscience

- cognitive science
- artificial intelligence
- computing
- engineering
- statistics
- physics
- mathematics
- neurophilosophy

Alongside the intellectual challenges described above for neural networks has been the associated establishment of the subject as a discipline in its own right, with learned societies, refereed journals, Masters Degree programs (such as the Master of Science program in Information Processing and Neural Networks at King's College London).

In this chapter the very simplest aspects of the subject will be explained. These are then extended into the mathematical domain in Chapter 2 and into the neurobiological area in Chapter 3. The manner in which neural networks may explain consciousness is considered in Chapter 9.

1.2 What is a Neural Network?

A neural network (NN) is composed of a set of very simple processing elements. As such, it may be regarded as the simplest possible parallel computer. Each element is called a neuron. It is often termed an artificial neuron in the literature, but it is useful to drop the epithet "artificial" both for reasons of space and also because the elements are being modelled in terms of the most essential features of living neurons. These are in any case so complex that any attempt to model them would always have to use the adjectives "artificial" or "model", and so these epithets would lose all information value.

Each of the neurons can send a signal, usually normalised to have the value one, to the other neurons in the network of which they are constituents. These signals are sent along connection "wires" similar to the wires in an electric circuit. Each neuron then receives the signals from its companions. These signals are rescaled by the so-called "connection weights", one for each wire coming to a neuron from another one. Thus if the ith neuron receives a signal from the jth, and the connection weight for this wire from neuron j to neuron i has value w_{ij} (as shown in Figure 1.1), then the activity received by the ith neuron will have the amount w_{ij}.

The total activity received by the ith neuron will thus be

$$A_i = \sum_i w_{ij} u_j \tag{1.1}$$

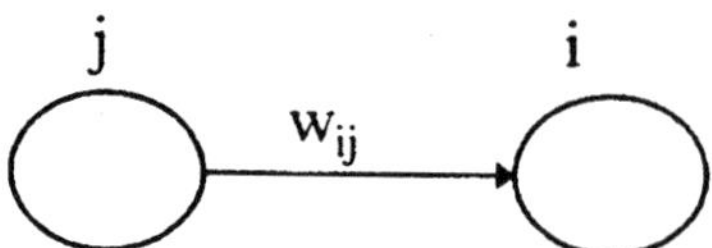

Figure 1.1: The connection from neuron j to neuron i, with connection weight w_{ij} used to renormalise the strength with which the neuron j can affect the neuron i.

where u_j is the activity of the jth neuron, being 1 if the jth neuron is active and 0 if it is inactive. The ith neuron responds with a signal which depends on the value of its activity at that time. This response may be a purely binary decision: give a response (a one) if the activity A_i is positive, otherwise remain silent. The neuron and its companions may, on the other hand, respond in a probabilistic manner by giving out a one with the probability $f(A_i)$, where $f(x)$ is some sigmoidal function (see below) of its variable x. The former net are termed deterministic neurons, since their response is certain, whilst the latter are called probabilistic and give a 'noisy' output closer to the response of living nerve cells in the brain. It is also possible to have neurons which pass round real values, and not just zeros or ones; in that case the real-valued output would be $u_i(t+1) = f(A_i(t))$, where the activity at time t is calculated from the activities $u_j(t)$ at time t, and these quantities are now real valued.

Neuron type	Output	Output function $f(A_i)$
1) deterministic	1 if $A_i > 0$, 0 if $A_i < 0$	Step function $\theta(A_i)$
2) probabilistic	1 with probability $f(A_i)$	Sigmoidal function (such as $[1 + \exp(-A_i)]^{-1}$)
3) real-valued	the real number $f(A_i)$	ditto

Table 1.1: Table of neuron response types.

These possibilities are summarised in Table 1.1, in which the three classes of neurons are delineated in terms of their response functions. The first class has as output function the step function $\theta(activity)$, which is sometimes called the 'hard-limiting response function' since there are only the response values of 0 or 1, with the first if the activity of the neuron is too low and 1 if it is large enough; there are no response values in between and no values above 1 (thus the epithet 'limiting' noted earlier). The second class of neurons are noisy ones, in which the neuron output is only ever 0 or 1 but may be so only with a certain probability. This is determined, for the neuron considered in the table, by the sigmoidal function $[1+\exp(-activity)]^{-1}$, which is very small for very negative values of the activity of the neuron and then grows gradually as this activity increases until it ever more closely approaches unity at very high neuron activity values. Finally the third class of neurons uses the same sigmoidal output response function as the previous class but outputs the real value of the sigmoidal function directly instead of using it to govern the neuron response indirectly. This latter response function is sometimes considered as corresponding to the average, over time, of the output of the noisy neuron of class 2.

Each of these neuron classes has been used in the construction and training of both models of living neural networks and for artificial neural networks. They all have their advantages and disadvantages which help solve some problems but not others. Thus the noisy neurons were originally suggested by J.G.Taylor in 1972 as helping to resolve problems of training nets to classify a set of patterns so that when they meet similar patterns, but not quite the same as the original ones they were trained on, they can still give a useful classification. This property is called generalisation and is very important to incorporate into any neural system; otherwise the system will be very vulnerable to any small environmental changes in the patterns it is tested on. The use of training in noise is now part of the standard practice of training artificial neural networks.

1.3 Neural Network Dynamics

The neural net dynamics specified in the above cases allow the temporal development of the activities of all of the neurons of the net to be calculated in terms of their initial activities (at time 0, say) and the values of the connection weights. These activities may be used for various processing tasks, by means of various types of connectivities or network architectures. There are two extreme nets:

1. feedforward, in which the neurons are laid out in layers labelled 1, 2, 3, etc, and activity feeds through from the input at layer 1 to layer 2, then to layer 3, and so on (as shown in Figure 1.2). Such a feedforward net can set up transforms of one pattern into another:

$$\mathbf{x} \rightarrow \mathbf{y} \tag{1.2}$$

   where $\mathbf{x}$ is the input pattern and $\mathbf{y}$ the output (not necessarily of the same dimension). Such a transformation can be used in classification (when $\mathbf{y}$ may be a binary vector as classifier) or to achieve more generally an associative memory.

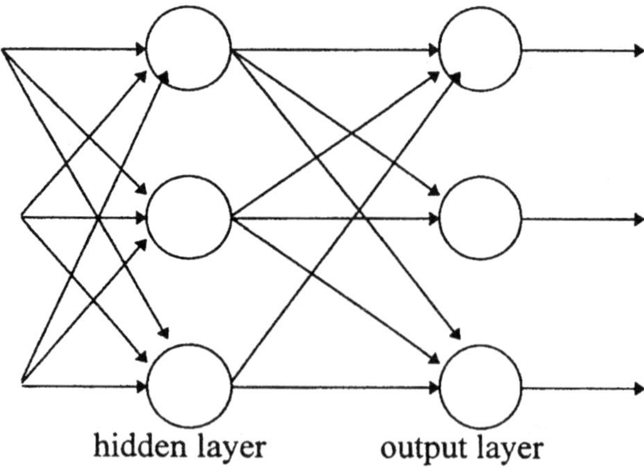

Figure 1.2: The architecture of a feed-forward network, in which input enters from the left and is transformed successively by the neurons of the hidden layer and then by those of the output layer. The hidden layer is so-called because it has no direct output to the observer, so that its contribution to the net error of the net is difficult to assess; the method of back-propagation mentioned in the text gives a solution to this error-assignment problem.

2. recurrent, where the output of any layer may be fed back to itself and to earlier (or later) layers (as shown in Figure 1.3). The activity of such a net is then one of settling or relaxing of the activity of the net to a stable or fixed point, sometimes called an attractor. The most well- known of such recurrent nets is the single layer Hopfield net, with all-to-all connectivity, with a symmetric connection matrix, $w_{ij} = w_{ji}$, with $w_{ii} = 0$. In this case it is possible to obtain a lot of information about the activity of the net, and even to specify how the connection matrix must be chosen in order to obtain a given set of attractors of the net. The net then acts as a 'content addressable memory', so as to complete or fill out noisy or degraded versions of patterns purely from their content, with no need of further special address labels.

1.4 Training a Neural Network

The most important question that must then be answered about a neural net is how the connection matrix is chosen to be effective in the solution of a given task. A specification had been indicated above for the Hopfield net, but the most popular net, the feedforward one, has to attempt to solve the following problem :

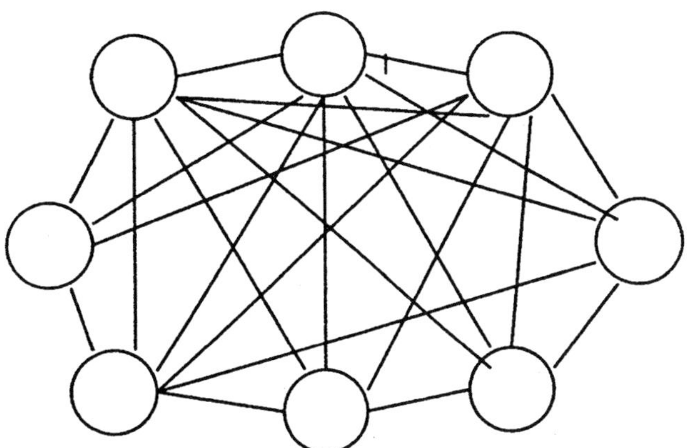

Figure 1.3: The architecture of a recurrent network; all neurons are connected to each other in the example shown, although this is an extreme case and there can be reduced connectivity to save on the total number of parameters (the weights) which need to be determined to help solve specific problems. Input from outside can be to any of the neurons of the network.

> Choose the best set of connection weights so that the network approximates most effectively (say in the mean square sense) the mapping of the given data set:
>
> $$\mathbf{x}_i \rightarrow \mathbf{y}_i \tag{1.3}$$
>
> for a given training set of pairs $\{\mathbf{x}_i, \mathbf{y}_i\}$, for $1 \leq i \leq N$, for some N.

That such a choice is possible, for a suitable large net composed of nodes emitting and receiving real values, is the result of the important Universal Approximation Theorem, which states that a single layer net, with a suitably large number of hidden nodes, can approximate any suitably smooth function:

$$\mathbf{y} = \mathbf{F}(\mathbf{x}) \tag{1.4}$$

The 'universal' approximating function will have the form:

$$\mathbf{y} = \mathbf{F}_{\mathbf{approx}}(\mathbf{x}) = \sum_i \mathbf{a}_i f(\mathbf{w}_i \cdot \mathbf{x} - t_i) \tag{1.5}$$

where $\mathbf{a}_i$ and $\mathbf{w}_i$ are weights for the output layer and hidden layer neurons respectively (as shown in Figure 1.4) and t_i are the corresponding thresholds. The approximating function $\mathbf{F}_{\mathbf{approx}}$ will only give an approximation to the original function $\mathbf{F}$. This approximation will become ever better as the number of hidden nodes (see Figure 1.4) is increased.

Such results allow for confidence that a neural net can always be found to get a good approximation to the function $\mathbf{F}$ of equation (1.4). Since the form of equation (1.4) is at the basis of most real-world tasks then the future looks good for neural networks. That is especially true now that ever more efficient learning methods are being developed to train the weights.

The most basic technique is that of back-error propagation, which is based on the mean square error function:

$$E = \sum_i [\mathbf{y}_i - \mathbf{F}_{\mathbf{approx}}(\mathbf{x}_i)]^2 \tag{1.6}$$

This error function assesses how good an approximation to the real function $\mathbf{F}$ is the approximating function $\mathbf{F}_{\mathbf{approx}}$. The error depends on the unknown parameters which are included in the function $\mathbf{F}_{\mathbf{approx}}$ through the equation (1.5). Thus the error (equation (1.6)) can be regarded as a surface over these parameters. The best choice for these parameters is that which will reduce the error to its minimum.

One way to attempt to find those parameter values which give a minimum of the error surface (equation (1.6)) is to perform what is called 'gradient descent'. That corresponds to starting at some arbitrary value of the parameters $\mathbf{a}_i \cdot \mathbf{w}_i$, t_i and changing them by a small amount so as to move along the line of steepest descent on the error surface. Such a direction may be shown to be along the gradient of the error function with respect to these parameters; that is why such an approach is called 'gradient descent'. The derivatives of the total function (equation (1.5)) with respect to the parameters are complicated, but may be reduced to a very compact formula so as to depend mainly on the response of the neuron whose parameters are being changed. This formula may be shown to correspond to allowing the inputs to propagate forward through the net (from left to right in Figure 1.2). The error function is then calculated as in equation (1.6) and the errors assigned to each output neuron. These are then 'back- propagated' by a simple formula to the hidden layer neurons of the network so as to allow the parameters to be so changed as to reduce the error assignment reached for them by back propagation.

There may be a number of choices of these parameters at which the error surface is locally lower than at surrounding parameter values; these are called local minima. The problem is to choose those parameter values for which the error surface is lower than all other possible choices of the parameters; this is called a global minimum. It is a very difficult problem to find a global minimum for a general network architecture like the one in Figure 1.4 and for a general pattern set. Even for very simple problems there can be numerous local minima which can trap the unwary neural network researcher and produce a non-optimal solution to the problem at hand. However there are now techniques which can help, such as starting a number of times from different initial weight values or modifying the inputs so as to make the problem of calculating the global minimum easier and then returning 'gently' to the original difficult problem (this is called performing a homotopy).

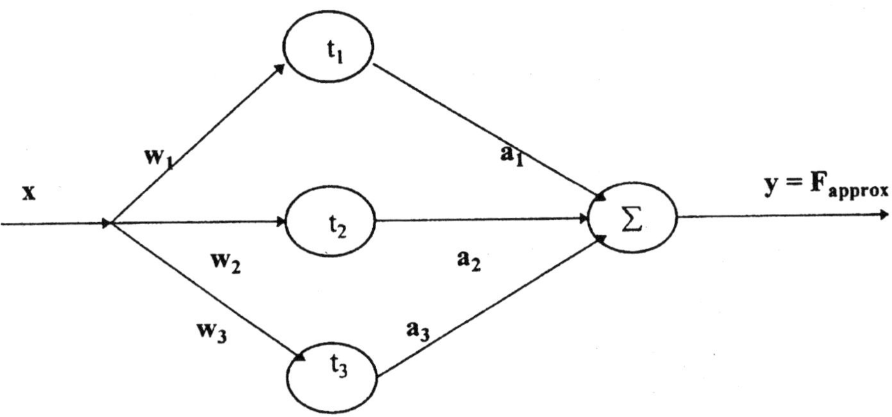

Figure 1.4: The simplest form of network used to implement the universal approximation formula (5). The input variable $\mathbf{x}$ is shown only connected to a single line, although it may have any number of components, as may the output $\mathbf{y}$ (each component then requiring a separate input line). The hidden unit weights $\mathbf{w}_i$ are only shown for three neurons, as are the threshold values t_i and the ouptut weights $\mathbf{a}_i$. The output neuron only needs to be a linear one, as the symbol Σ in the ouptut node denotes.

The above back-propagation method, and its variants, allows for a successive assignment of the error caused by an incorrect choice of connection weights to be reduced gradually so as to ultimately bring it below any criterial value. There are other neural network training techniques, involving reinforcement methods (where learning is achieved by criticism and not teaching, as in the supervised method associated with the error (equation (1.5)) above), or unsupervised methods. In the latter the net tries to determine some general statistical features of the input pattern set.

In summary, a neural network is a system of very simple input-output nodes whose strengths of connections can be trained so that the network produces an effective response to a set of inputs. That response may be to classify the inputs into a well-defined set of categories (as in speech recognition), to attempt to learn the pattern that may be presented next in a time series (as in financial forecasting, say of the exchange rate) or to give a response of a specific form for a given input (such as is needed in a control problem).

1.5 Problems for Artificial Neural Networks

One can give a list of criteria which indicate the suitability of those tasks for which neural networks can be considered as a possible tool. In particular if the problem possesses one of the following characteristics then it should be especially amenable to a neural approach:

- noise
- poorly defined characteristics
- changing environment

One or more of these characteristics are possessed by the problems being addressed by ANNs listed in the introduction. Neural networks are becoming more successful in such tasks; training allows them to be constructed to be ever more effective.

1.6 Problems for Neurobiologically Realistic Neural Networks

The applications of neural networks to biological systems is more constrained. Most generally the behaviour of a given animal is to be modelled. However beside the set of behaviours which will be performed by a given animal under a given set of conditions the animal in question has its own neural systems which may be very different from that of other animals. However there is also the bonus that there are many animals whose behaviour can be modelled, so that the range of possible experimental test beds for neural networks is vast.

The class of problems to which neural networks have been applied involve:

- small networks, such as ganglia in insects and arthropods,
- regularly arranged networks, such as in the retina, where the translation invariance allows simplification of the model of a large number of neurons, or in its fabrication if a chip design is suggested based on the model,
- realistic but scaled down versions of parts of the cortex, such as in modelling early vision and especially the orientation sensitivity of primary visual cortex,
- 'toy' models of large neural networks performing high level cognitive processes, such as in the holding of activity in frontal cortex to solve tasks over time,
- global models based on very general principles in an attempt to explain general features of memory or of awareness.

There is considerable progress in all of these areas, and neural networks are now an integral part of the neuroscientific endeavour.

1.7 Conclusions

A very simple introduction has been given to the subject of neural networks. It is now divided into two tracks, that of artificial neural networks (ANNs) and that of computational neuroscience (CNS). Both of these approaches use the training of neural networks in order to make them effective in their tasks, although the latter imposes the additional constraint that the architectures and learning rules should be biologically realistic. Even inside these constraints there is considerable progress being made in modelling and explaining the brain processes, many of which have so far not been transferred to artificial neural network technology. However the implementation of even the highest level control system in the brain, that of consciousness, is presently being tackled. There is no doubt that neural networks will enter into a new regime of application and explanatory power in the next millennium. For a subject which is only just over 50 years old this is a remarkable testament to its powers.

1.8 Further Reading

New books on Neural Networks are appearing every week, so the list of appropriate texts would be very long. However the true beginner should have a look at the two volume set which opened up the field to many people [1]. Further introductions are [2] and [3]. Two recent and helpful books are [4] and [5]. More advanced and very recent texts are [6] and [7]. Furthermore technical references may be obtained by consulting the specialised journals, such as Neural Networks, Neural Computation, Network, Neural Computing, Transactions on Neural Networks; these are in most University Libraries.

1.9 Bibliography

[1] Rumelhart DE, McClelland JL and the PDP Research Group (1986) Parallel Distributed Processing, Cambridge MA: MIT Press.

[2] Beale R and Jackson (1990) Introduction to Neural Networks, Bristol: Institute of Physics Publ.

[3] Taylor JG (1993) The Promise of Neural Networks, London: Springer Verlag

[4] Rojas R (1995) Neural Networks, A Systematic Introduction, Berlin: Springer Verlag

[5] Haykin S (1994) Neural Networks, A Comprehensive Foundation, New York: Macmillan

[6] Bishop CM (1995) Neural Networks for Pattern Recognition, Oxford: Clarendon Press.

[7] Golden RM (1996) Mathematical Methods for Neural Network Analysis and Design, Cambridge Mass: Bradford Books, MIT Press.

Chapter 2

A Beginner's Guide to the Mathematics of Neural Networks

2.1 Introduction: Neural Information Processing

Our brains perform sophisticated information processing tasks, using hardware and operation rules which are quite different from the ones on which conventional computers are based. The processors in the brain, the neurons (see Figure 2.1), are rather noisy elements[1] which operate in parallel. They are organised in dense networks, the structure of which can vary from very regular to almost amorphous (see Figure 2.2), and they communicate signals through a huge number of inter-neuron connections (the so-called synapses). These connections represent the 'program' of a network. By continuously updating the strengths of the connections, a network as a whole can modify and optimise its 'program', 'learn' from experience and adapt to changing circumstances.

From an engineering point of view neurons are in fact rather poor processors, they are slow and unreliable (see the table below). In the brain this is overcome by ensuring that a very large number of neurons are always involved in any task, and by having them operate in parallel, with many connections. This is in sharp contrast to conventional computers, where operations are as a rule performed sequentially, so that failure of any part of the chain of operations is usually fatal. Furthermore, conventional computers execute a detailed specification of orders, requiring the programmer to know exactly which data can be expected and how to respond. Subsequent changes in the actual situation, not foreseen by the programmer, lead to trouble. Neural networks, on the other hand,

[1] By this we mean that their output signals are to some degree subject to random variation; they exhibit so-called spontaneous activity which appears not to be related to the information processing task they are involved in.

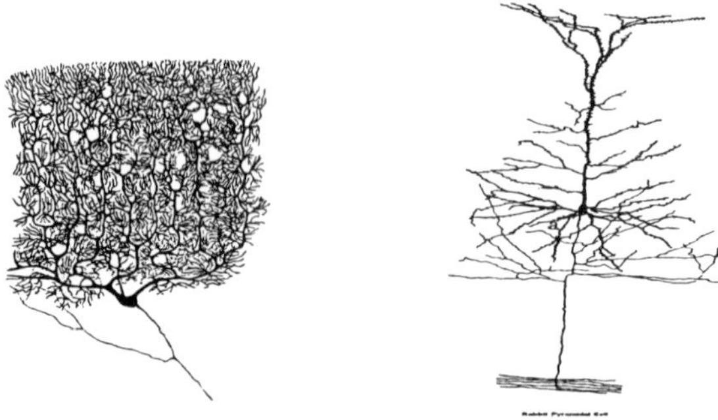

Figure 2.1: Left: a Purkinje neuron in the human cerebellum. Right: a pyramidal neuron of the rabbit cortex. The black blobs are the neurons, the trees of wires fanning out constitute the input channels (or dendrites) through which signals are received which are sent off by other firing neurons. The lines at the bottom, bifurcating only modestly, are the output channels (or axons).

can adapt to changing circumstances. Finally, in our brain large numbers of neurons end their careers each day unnoticed. Compare this to what happens if we randomly cut a few wires in our workstation.

conventional computers	biological neural networks
processors *operation speed* $\sim 10^8 Hz$ *signal/noise* $\sim \infty$ *signal velocity* $\sim 10^8 m/sec$ *connections* ~ 10	neurons *operation speed* $\sim 10^2 Hz$ *signal/noise* ~ 1 *signal velocity* $\sim 1 m/sec$ *connections* $\sim 10^4$
sequential operation program & data external programming	parallel operation connections, neuron thresholds self-programming & adaptation
hardware failure: fatal no unforseen data	robust against hardware failure messy, unforseen data

Roughly speaking, conventional computers can be seen as the appropriate tools for performing well-defined and rule-based information processing tasks, in stable and safe environments, where all possible situations, as well as how to respond in every situation, are known beforehand. Typical tasks fitting these criteria are, for example, brute-force chess playing, word processing, keeping accounts and rule-based (civil servant) decision making. Neural information processing systems, on the other hand, are superior to conventional computers in dealing with real-world tasks, such as communication (vision, speech recognition), movement coordination (robotics) and experience-based decision making (classification, prediction, system control), where data are often messy, uncer-

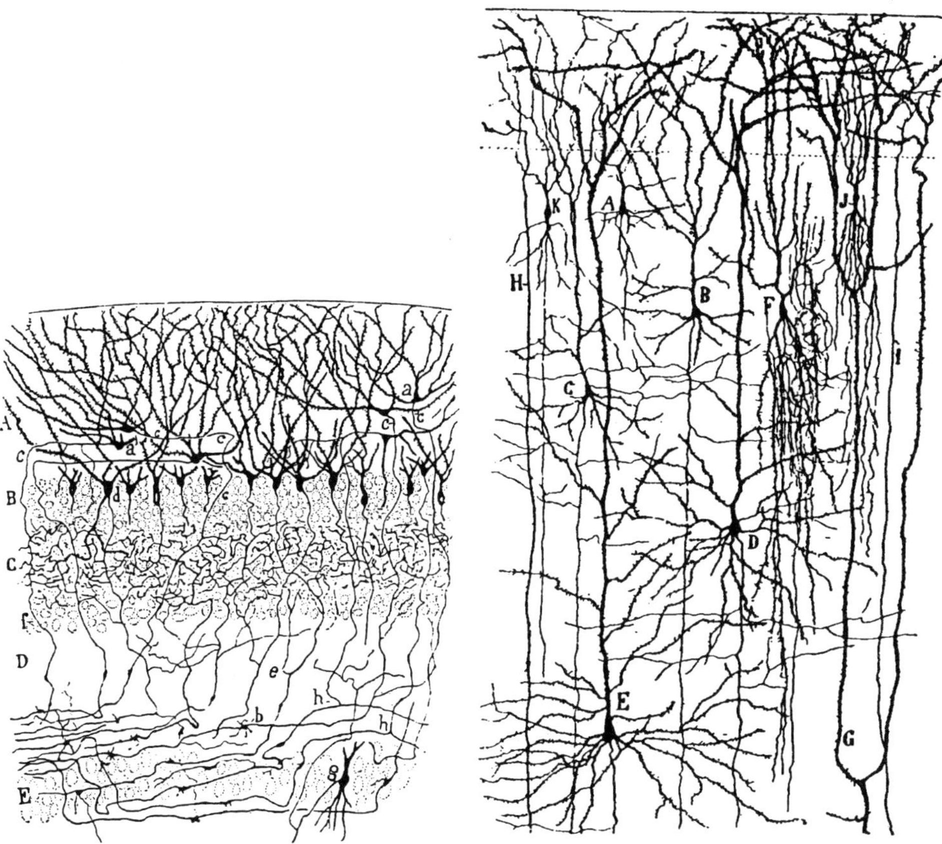

Figure 2.2: Left: a section of the human cerebellum. Right: a section of the human cortex. Note that the staining method used to produce such pictures colours only a reasonably modest fraction of the neurons present, so in reality these networks are far more dense.

tain or even inconsistent, where the number of possible situations is infinite and where perfect solutions are for all practical purposes non-existent.

One can distinguish three types of motivation for studying neural networks. Biologists, physiologists, psychologists and, to some degree also, philosophers aim to understand information processing in real biological nervous tissue. They study models, mathematically and through computer simulations, which are preferably close to what is being observed experimentally, and try to understand the global properties and functioning of brain regions.

Engineers and computer scientists would like to understand the principles behind neural information processing in order to use these for designing adaptive software and artificial information processing systems which can also 'learn'. They use highly simplified neuron models, which are again arranged in networks. As their biological counterparts, these artificial systems are not programmed, their inter-neuron connections are not prescribed, but they are 'trained'. They gradually 'learn' to perform tasks by being presented with examples of what they are supposed to do. The key question then is to understand the relationships between the network performance for a given type of task, the choice of 'learning rule' (the recipe for the modification of the connections) and the network architecture. Secondly, engineers and computer scientists exploit the emerging insight into the way real (biological) neural networks manage to process information efficiently in parallel, by building artificial neural networks in hardware, which also operate in parallel. These systems, in principle, have the potential to be incredibly fast information processing machines.

Finally, it will be clear that, due to their complex structure, the large numbers of elements involved, and their dynamic nature, neural network models exhibit a highly non-trivial and rich behaviour. This is why theoretical physicists and mathematicians have also become involved, challenged as they are by the many fundamental new mathematical problems posed by neural network models. Studying neural networks as a mathematician is rewarding in two ways. The first reward is to find nice applications for one's tools in biology and engineering. It is fairly easy to come up with ideas about how certain information processing tasks could be performed by (either natural or synthetic) neural networks; by working out the mathematics, however, one can actually quantify the potential and restrictions of such ideas. Mathematical analysis further allows for a systematic design of new networks, and the discovery of new mechanisms. The second reward is to discover that one's tools, when applied to neural network models, create quite novel and funny mathematical puzzles. The reason for this is the 'messy' nature of these systems. Neurons are not at all well-behaved: they are microscopic elements which do not live on a regular lattice, they are noisy, they change their mutual interactions all the time, etc.

Since this chapter aims at no more than sketching a biased impression of a research field, I will not give references to research papers along the way, but instead will mention textbooks and review papers in the final section, for those who are interested.

2.2 From Biology to Mathematical Models

We cannot expect to solve mathematical models of neural networks in which all electro-chemical details are taken into account (even if we knew all such details perfectly). Instead we start by playing with simple networks of model neurons, and trying to understand their basic properties (i.e. we study elementary electronic circuitry before we volunteer to repair the video recorder).

2.2.1 From Biological Neurons to Model Neurons

Neurons operate more or less in the following way. The cell membrane of a neuron maintains concentration differences between inside and outside the cell, of various ions (the main ones are Na^+, K^+ and Cl^-), by a combination of the action of active ion pumps and controllable ion channels. When the neuron is at rest, the channels are closed, and due to the activity of the pumps and the resultant concentration differences, the inside of the neuron has a net negative electric potential of around –70 mV, compared to the fluid outside. A sufficiently strong local electric excitation, however, making the cell potential temporarily less negative, leads to the opening of specific ion channels, which in turn causes a chain reaction of other channels opening and/or closing, with the net result that an electrical peak of height around +40 mV is generated, with a duration of about 1 msec, which will propagate along the membrane at a speed of about 5 m/sec: the so-called action potential. After this electro-chemical avalanche it takes a few milliseconds to restore peace and order. During this period, the so-called refractory period, the membrane can only be forced to generate an action potential by extremely strong excitation. The action potential serves as an electric communication signal, propagating and bifurcating along the output channel of the neuron, the axon, to other neurons. Since the propagation of an action potential along an axon is the result of an active electro/chemical process, the signal will retain shape and strength, even after bifurcation, much like a chain of tumbling dominos.

typical time-scales	
action potential:	$\sim 1msec$
reset time:	$\sim 3msec$
synapses:	$\sim 1msec$
pulse transport:	$\sim 5m/sec$

typical sizes	
cell body:	$\sim 50\mu m$
axon diameter:	$\sim 1\mu m$
synapse size:	$\sim 1\mu m$
synaptic cleft:	$\sim 0.05\mu m$

The junction between an output channel (axon) of one neuron and an input channel (dendrite) of another neuron, is called synapse (see Figure 2.3). The arrival at a synapse of an action potential can trigger the release of a chemical, the neurotransmitter, into the so-called synaptic cleft which separates the cell membranes of the two neurons. The neurotransmitter in turn acts to selectively open ion channels in the membrane of the dendrite of the receiving neuron. If these happen to be Na^+ channels, the result is a local increase of the potential at the receiving end of the synapse, if these are Cl^- channels the result is a decrease. In the first case the arriving signal will increase the probability of the receiving neuron starting to fire itself, therefore such a synapse is called excitatory. In the second case the arriving signal will decrease the probability of the receiving neuron being triggered, and the synapse is called inhibitory. However, there is also the possibility that the arriving action potential will not succeed in releasing the neurotransmitter; neurons are not perfect. This introduces an element of uncertainty, or noise, into the operation of the machinery.

Whether or not the receiving neuron will actually be triggered into firing itself will depend on the cumulative effect of all excitatory and inhibitory signals arriving, a detailed analysis of which also requires the electrical details of the dendrites to be taken into account. The region of the neuron membrane most

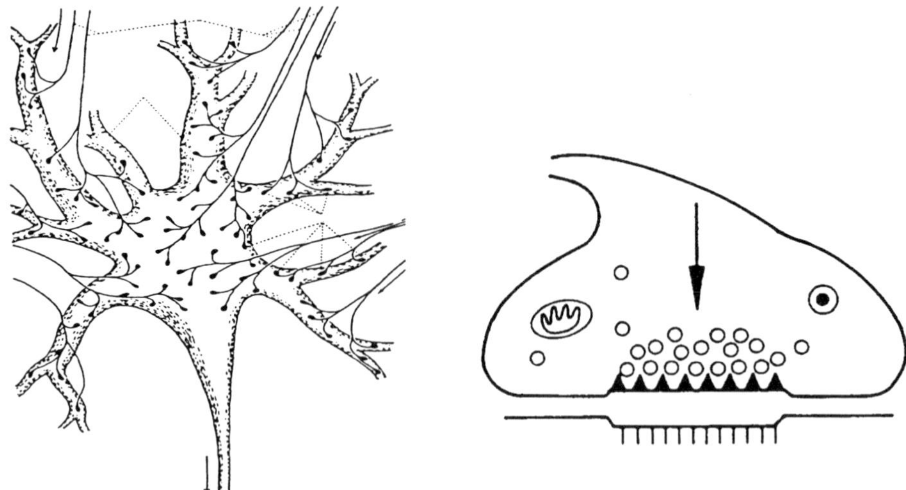

Figure 2.3: Left: drawing of a neuron. The black blobs attached to the cell body and the dendrites (input channels) represent the synapses (adjustable terminals which determine the effect communicating neurons will have on one another's membrane potential and firing state). Right: close-up of a typical synapse.

sensitive to be triggered into sending an action potential is the so-called hillock zone, near the root of the axon. If the potential in this region, the post-synaptic potential, exceeds some neuron-specific threshold (of the order of –30 mV), the neuron will fire an action potential. However, the firing threshold is not a strict constant, but can vary randomly around some average value (so that there will always be some non-zero probability of a neuron not doing what we would expect it to do with a given post-synaptic potential), which constitutes the second main source of uncertainty in the operation.

The key to the adaptive and self-programming properties of neural tissue and to being able to store information, is that the synapses and firing thresholds are not fixed, but are being updated all the time. It is not entirely clear, however, how this is realised at a chemical/electrical level. Most likely the amount of neurotransmitter in a synapse, available for release, and the effective contact surface of a synapse are modified.

The simplest caricature of a neuron is one where its possible firing states are reduced to just a single binary variable S, indicating whether it fires ($S = 1$) or is at rest ($S = 0$). See Figure 2.4. Which of the two states the neuron will be in is dictated by whether or not the total input it receives (i.e. the post-synaptic potential) does ($S \to 1$) or does not ($S \to 0$) exceed the neuron's firing threshold, denoted by θ (if we forget about the noise). As a bonus this allows us to illustrate the collective firing state of networks by colouring the constituent neurons: firing = $\bullet$, rest = $\circ$. We further assume the individual

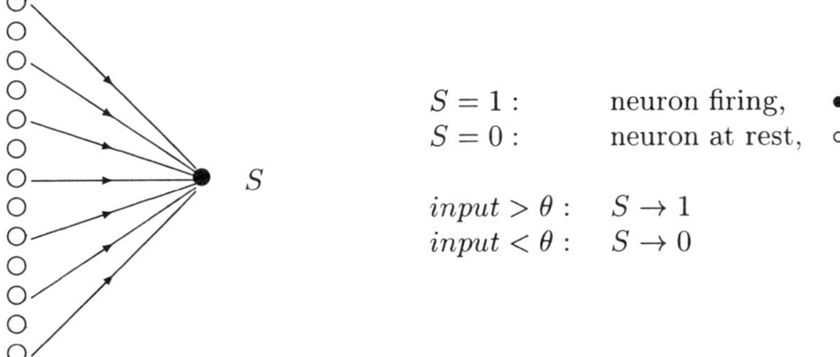

Figure 2.4: The simplest model neuron: a neuron's firing state is represented by a single instantaneous binary state variable S, whose value is solely determined by whether or not its input exceeds a firing threshold.

input signals to add up linearly, weighted by the strengths of the associated synapses. The latter are represented by real variables w_ℓ, whose sign denotes the type of interaction ($w_\ell > 0$: excitation, $w_\ell < 0$: inhibition) and whose absolute value $|w_\ell|$ denotes the magnitude of the interaction:

$$input = w_1 S_1 + \ldots + w_N S_N$$

Here the various neurons present are labelled by subscripts $\ell = 1, \ldots, N$. This rule indeed appears to capture the characteristics of neural communication. Imagine, for instance, the effect on the input of a quiescent neuron ℓ suddenly starting to fire:

$$S_\ell \to 1: \qquad input \to input + w_\ell \qquad \begin{cases} w_\ell > 0: & input \uparrow, \quad excitation \\ w_\ell < 0: & input \downarrow, \quad inhibition \end{cases}$$

We now adapt these rules for each of our neurons. We indicate explicitly at which time t (for simplicity to be measured in units of one) the various neuron states are observed, we denote the synaptic strength at a junction $j \to i$ (where j denotes the 'sender' and i the 'receiver') by w_{ij}, and the threshold of a neuron i by θ_i. This brings us to the following set of microscopic operation rules:

$$\begin{aligned} w_{i1} S_1(t) + \ldots + w_{iN} S_N(t) > \theta_i: \quad S_i(t+1) = 1 \\ w_{i1} S_1(t) + \ldots + w_{iN} S_N(t) < \theta_i: \quad S_i(t+1) = 0 \end{aligned} \tag{2.1}$$

These rules could either be applied to all neurons *at the same time*, giving so-called parallel dynamics, or to one neuron at a time (drawn randomly or according to a fixed order), giving so-called sequential dynamics.[2] Upon specifying the values of the synapses $\{w_{ij}\}$ and the thresholds $\{\theta_i\}$, as well as the

[2] Strictly speaking, we also need to specify a rule for determining $S_i(t+1)$ for the marginal case, where $w_{i1} S_1(t) + \ldots + w_{iN} S_N(t) = \theta_i$. Two common ways of dealing with this situation are to either draw $S_i(t+1)$ at random from $\{0, 1\}$, or to simply leave $S_i(t+1) = S_i(t)$.

initial network state $\{S_i(0)\}$, the system will evolve in time in a deterministic manner, and the operation of our network can be characterised by giving the states $\{S_i(t)\}$ of the N neurons at subsequent times, e.g.:

	S_1	S_2	S_3	S_4	S_5	S_6	S_7	S_8	S_9
$t=0:$	1	1	0	1	0	0	1	0	0
$t=1:$	1	0	0	1	0	1	1	1	1
$t=2:$	0	0	1	1	1	0	0	0	1
$t=3:$	1	0	0	1	1	1	1	0	1
$t=4:$	0	1	1	1	0	0	1	0	1

or, equivalently, by drawing the neuron states at different times as a collection of coloured circles, according to the convention 'firing' = $\bullet$, 'rest' = $\circ$, e.g.:

$t=0$	$t=1$	$t=2$	$t=3$	$t=4$
●●○	●○○	○○●	●○○	○●●
●○○	●○●	●●○	●●●	●○○
●○○	●●●	○○●	●○●	●○●

We have thus achieved a reduction of the operation of neural networks to a well-defined manipulation of a set of (binary) numbers, whose rules (2.1) can be seen as an extremely simplified version of biological reality. The binary numbers represent the states of the information processors (the neurons), and therefore describe the system *operation*. The details of the operation to be be performed depend on a set of control parameters (synapses and thresholds), which must be interpreted accordingly as representing the *program*. Moreover, manipulating numbers brings us into the realm of mathematics; the formulation (2.1) describes a non-linear discrete-time dynamical system.

2.2.2 Universality of Model Neurons

Although it is not a priori clear that our equations (2.1) are not an oversimplification of biological reality, there are at least two reasons for not making things more complicated yet. First of all, solving (2.1) for arbitrary control parameters and nontrivial system sizes is already impossible, in spite of its apparent simplicity. Secondly, networks of the type (2.1) are found to be *universal* information processing systems, in that (roughly speaking) they can perform any computation that can be performed by conventional digital computers, provided one chooses the synapses and thresholds appropriately.

The simplest way to show this is by demonstrating that the basic logical units of digital computers, the operations AND: $(x,y) \to x \wedge y$, OR: $(x,y) \to x \vee y$ and NOT: $x \to \neg x$ (with $x,y \in \{0,1\}$), can be built with our model neurons. Each logical unit (or 'gate') is defined by a so-called truth table, specifying its output for each possible input. All we need to do is to define for each of the above gates a model neuron of the type

$$
\begin{aligned}
w_1x+w_2y-\theta > 0: &\quad S=1 \\
w_1x+w_2y-\theta < 0: &\quad S=0
\end{aligned}
$$

by choosing appropriate values of the control parameters $\{w_1, w_2, \theta\}$, which has the same truth table. This turns out to be fairly easy:

AND:

x	y	$x \wedge y$	$x+y-\frac{3}{2}$	S
0	0	0	−3/2	0
0	1	0	−1/2	0
1	0	0	−1/2	0
1	1	1	1/2	1

x, y, S

$w_1 = w_2 = 1$
$\theta = \frac{3}{2}$

OR:

x	y	$x \vee y$	$x+y-\frac{1}{2}$	S
0	0	0	−1/2	0
0	1	1	1/2	1
1	0	1	1/2	1
1	1	1	3/2	1

x, y, S

$w_1 = w_2 = 1$
$\theta = \frac{1}{2}$

NOT:

x	$\neg x$	$-x+\frac{1}{2}$	S
0	1	1/2	1
1	0	−1/2	0

x, S

$w_1 = -1$
$\theta = -\frac{1}{2}$

This shows that we need not worry about a priori restrictions on the types of tasks our simplified model networks (2.1) can handle.

Furthermore, one can also make statements about the architecture required. Provided we employ model neurons with potentially large numbers of input channels, it turns out that every operation involving binary numbers can in fact be performed with a feed-forward network of at most two layers. Again this is proved by construction. Every binary operation $\{0,1\}^N \rightarrow \{0,1\}^K$ can be reduced (split-up) into specific sub-operations M, each performing a separation of the input signals $\boldsymbol{x}$ (given by N binary numbers) into two classes:

$$M : \{0,1\}^N \rightarrow \{0,1\}$$

(described by a truth table with 2^N rows). Each such M can be built as a neural realisation of a look-up exercise, where the aim is simply to check whether an $\boldsymbol{x} \in \{0,1\}^N$ is in the set for which $M(\boldsymbol{x}) = 1$. This set is denoted by Ω, with $L \leq 2^N$ elements which we label as follows: $\Omega = \{\boldsymbol{y}_1, \ldots, \boldsymbol{y}_L\}$. The basic tools of our construction are the so-called 'grandmother-neurons'[3] G_ℓ, whose sole task is to be on the look-out for one of the input signals $\boldsymbol{y}_\ell \in \Omega$:

$$\begin{aligned} w_1x_1 + \ldots + w_Nx_N > \theta : \quad & G_\ell = 1 \\ w_1x_1 + \ldots + w_Nx_N < \theta : \quad & G_\ell = 0 \end{aligned}$$

[3]This name was coined to denote neurons which only become active upon presentation of some unique and specific sensory pattern (visual or otherwise), e.g. an image of one's grandmother. Such neurons were at some stage claimed to have been observed experimentally.

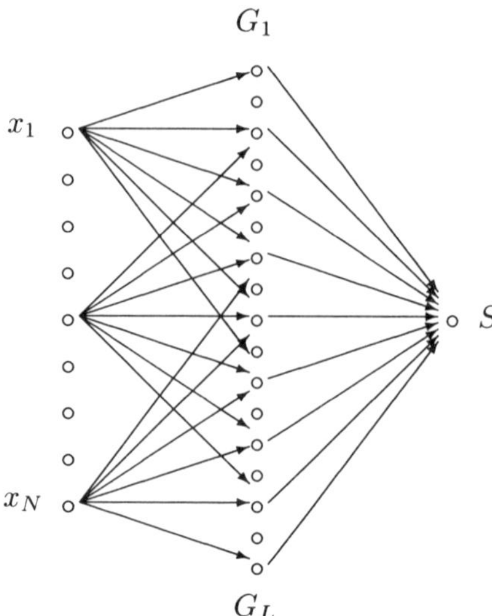

Figure 2.5: Universal architecture, capable of performing any classification $M : \{0,1\}^N \to \{0,1\}$, provided synapses and thresholds are choosen adequately.

with $w_\ell = 2(2y_\ell - 1)$ and $\theta = 2(y_1 + \ldots + y_N) - 1$. Inspection shows that with these definitions the output G_ℓ, upon presentation of input $\boldsymbol{x}$, is indeed (as required) given by:

$$\boldsymbol{x} = \boldsymbol{y}_\ell : \quad G_\ell = 1$$
$$\boldsymbol{x} \neq \boldsymbol{y}_\ell : \quad G_\ell = 0$$

Finally the outputs of the grandmother neurons are fed into a model neuron S, which is to determine whether or not one of the grandmother neurons is active:

$$y_1 + \ldots + y_L > 1/2 : \quad S = 1$$
$$y_1 + \ldots + y_L < 1/2 : \quad S = 0$$

The resulting feed-forward network is shown in Figure 2.5. For any input $\boldsymbol{x}$, the number of active neurons G_ℓ in the first layer is either 0 (leading to the final output $S = 0$) or 1 (leading to the final output $S = 1$). In the first case the input vector $\boldsymbol{x}$ is apparently not in the set Ω, in the second case it apparently is. This shows that the network thus constructed performs the separation M.

2.2.3 Directions and Strategies

Here the field effectively splits in two. One route leading away from equation (2.1) aims at solving it with respect to the evolution of the neuron states, for increasingly complicated but prescribed choices of synapses and thresholds. Here the key phenomenon is *operation*, the central dynamical variables are the neurons, whereas synapses and thresholds play the role of parameters. The alternative route is to concentrate on the complementary problem: which are the possible modes of operation equation (2.1) would allow for, if we were to vary synapses and thresholds in a given architecture, and how can one find learning rules (rules for the modification of synapses and thresholds) that will generate values such that the resulting network will meet some specified performance criterion. Here the key phenomenon is *learning*, the central dynamical variables are the synapses and thresholds, whereas neuron states (or, more often, their statistics) induce constraints and operation targets.

Operation		Learning	
variables:	neurons	variables:	synapses, thresholds
parameters:	synapses, thresholds	parameters:	required neuron states

Although quite prominent, in reality this separation is, of course, not perfect; in the field of learning theory one often specifies neuron states only in part of the system, and solves for the remaining neuron states, and there even exist non-trivial but solvable models in which both neurons and synapses/thresholds evolve in time. In the following sections examples from both main problem classes will be described.

A general rule in dealing with mathematical models, whether they describe phenomena in biology, physics, economics or any other discipline, is that one usually finds that the equations involved are most easily solved in extreme limits for the control parameters. This is also true for neural network models, in particular with respect to the system size N and the spatial distance over which the neurons are allowed to interact. Analysing models with just two or three neurons (on one end of the scale of sizes) is not much of a problem, but realistic systems happen to scale differently, both in biology (where even small brain regions are at least of size $N \sim 10^6$) and in engineering (where at least $N \sim 10^3$). Therefore one usually considers the opposite limit $N \to \infty$. In turn, one can only solve the equations descibing infinitely large systems when either interactions are restricted to occur only between neighbouring neurons (which is quite unrealistic), or when a large number (if not all) of the neurons are allowed to interact (which is a better approximation of reality).

The strategy of the model solver is then to identify global observables which characterise the system state at a macroscopic level (this is often the most difficult bit), and to calculate their values. For instance, in statistical mechanics one is not interested in knowing the positions and velocities of individual molecules in a gas, but rather in knowing the values of global obervables like pressure; in modelling (and predicting) exchange rates we do not care about which individuals buy certain amounts of a currency, but rather in the sum over all such buyers. Which macroscopic observables constitute the natural language for describing the operation of neural networks turns out to depend strongly on their function or task (as might have been expected). If the exercise is carried out properly, and if the model at hand is sufficiently friendly, one will observe that in the $N \to \infty$ limit clean and transparent analytical relations emerge. This happens for various reasons. If there is an element of randomness involved (noise) it is clear that in finite systems we can only speak about the probability of certain averages occurring, whereas in the $N \to \infty$ limit one would find averages being replaced by exact expressions. Secondly, as soon as spatial structure of a network is involved, the limit $N \to \infty$ allows us to take continuum limits and to replace discrete systems by continuous ones.

The operation a neural network performs depends on its program: the choice made for architecture, synaptic interactions and thresholds (equivalently, on the learning rule used to generate these parameters). Several examples will now be given involving different types of information processing tasks and, consequently, different types of analysis (although all share the reductionist strategy of calculating global properties from underlying microscopic laws).

2.3 Neural Networks as Associative Memories

Our definition of model neurons has led to a relatively simple scenario, where a global network state is described by specifying for each neuron the value of its associated binary variable. It can be conveniently drawn in a picture with black circles denoting active neurons and white circles denoting neurons at rest. If we choose the neural thresholds such that a disconnected neuron would be precisely critical (with a potential *at* threshold), we can simplify our equations further by choosing $\{-1, 1\}$ as the two neuron states (rest/firing), instead of $\{0, 1\}$ (see below), giving the rules:

$$\begin{array}{ll} w_{i1}S_1(t) + \ldots + w_{iN}S_N(t) > 0: & S_i(t+1) = 1 \\ w_{i1}S_1(t) + \ldots + w_{iN}S_N(t) < 0: & S_i(t+1) = -1 \end{array} \quad (2.2)$$

to be depicted as:

$\bullet: \quad S_i = 1$ (neuron i firing)
$\circ: \quad S_i = -1$ (neuron i at rest)

$input_i > 0: \quad S_i \to 1$
$input_i < 0: \quad S_i \to -1$

$input_i = w_{i1}S_1 + \ldots + w_{iN}S_N$

If this network is to operate as a memory, for storing and retrieving patterns (pictures, words, sounds, etc.), we must assume that the information is (physically) stored in the synapses, and that pattern retrieval must correspond to a dynamical process of neuron states. We are thus led to representing patterns to be stored as global network states, i.e. each pattern corresponds to a specific set of binary numbers $\{S_1, \ldots, S_N\}$, or, equivalently, to a specific way of colouring circles in the picture above. This is similar to what happens in conventional computers. However, in computers one retrieves such information by specifying the label of the pattern in question, which codes for the address of its physical memory location. This will be quite different here. Let us introduce the principles behind the neural way of storing and retrieving information, by working out the details for a very simple model example.

Biologically realistic learning rules for synapses are required to meet the constraint that the way a given synapse w_{ij} is modified can depend only on information locally available: the electro-chemical state properties of the neurons i and j.[4] One of the simplest such rules is the following: increase w_{ij} if the neurons i and j are in the same state, decrease w_{ij} otherwise. With our definition of the allowed neuron states being $\{-1, 1\}$, this can be written as:

$$\begin{array}{ll} S_i = S_j: & w_{ij} \uparrow \\ S_i \neq S_j: & w_{ij} \downarrow \end{array} \qquad w_{ij} \to w_{ij} + S_i S_j \tag{2.3}$$

If we apply this rule to just one specific pattern, denoted by $\{\xi_1, \ldots, \xi_N\}$ (each component $\xi_i \in \{-1, 1\}$ represents a specific state of a single neuron), we obtain the following recipe for the synapses: $w_{ij} = \xi_i \xi_j$. How would a network with such synapses behave? Note, firstly, that:

$$input_i = w_{i1} S_1 + \ldots + w_{iN} S_N = \xi_i \left[\xi_1 S_1 + \ldots + \xi_N S_N\right]$$

so the dynamical rules (2.2) for the neurons become:

$$\begin{array}{ll} \xi_i \left[\xi_1 S_1(t) + \ldots + \xi_N S_N(t)\right] > 0: & S_i(t+1) = 1 \\ \xi_i \left[\xi_1 S_1(t) + \ldots + \xi_N S_N(t)\right] < 0: & S_i(t+1) = -1 \end{array} \tag{2.4}$$

Note also that $\xi_i S_i(t) = 1$ if $\xi_i = S_i(t)$, and that $\xi_i S_i(t) = -1$ if $\xi_i \neq S_i(t)$. Therefore, if at time t more than half of the neurons are in the state $S_i(t) = \xi_i$ then $\xi_1 S_1(t) + \ldots + \xi_N S_N(t) > 0$. It subsequently follows from (2.4) that:

$$\text{for all } i: \quad \text{sign}(input_i) = \xi_i \quad \text{so} \quad S_i(t+1) = \xi_i$$

If the dynamics (2.4) is of the parallel type (all neurons change their state at the same time), this convergence $(S_1, \ldots, S_N) \to (\xi_1, \ldots, \xi_N)$ is completed in a single iteration step. For sequential dynamics (neurons change states one after the other), the convergence is a gradual process. In both cases, the choice $w_{ij} = \xi_i \xi_j$ achieves the following: the state $(S_1, \ldots, S_N) = (\xi_1, \ldots, \xi_N)$ has become a *stable state* of the network dynamics. The network dynamically reconstructs the full pattern $(\xi_1, \ldots, \xi_N)$ if it is prepared in an initial state which bears sufficient resemblance to the state corresponding to this pattern.

2.3.1 Recipes for Storing Patterns and Pattern Sequences

If the operation described above for the case of a single stored pattern turns out to carry over to the more general case of an arbitrary number p of patterns, we arrive at the following recipe for information storage and retrieval:

[4] This constraint is to be modified if in addition we wish to take into account the more global effects of modulatory chemicals like hormones and drugs.

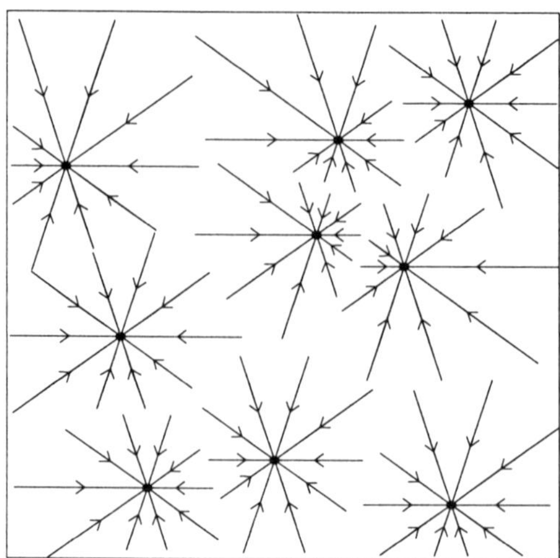

Figure 2.6: Information storage through the creation of attractors in the space of states. The state vector $(S_1, \ldots, S_N)$ evolves towards the nearest stable state. If the stable states are the patterns stored, and the initial state is an input pattern to be recognised, this system performs *associative* pattern recall.

- Represent each pattern as a specific network state $(\xi_1^\mu, \ldots, \xi_N^\mu)$.
- Construct synapses $\{w_{ij}\}$ such that these patterns become *stable states* (fixed-point attractors) for the network dynamics.
- An input to be recognised will serve as the initial state $\{S_i(t=0)\}$.
- The final state reached $\{S_i(t=\infty)\}$ can be interpreted as the pattern recognised by the network from the input $\{S_i(t=0)\}$.

From any given initial state, the system will, by construction, evolve towards the 'nearest'[5] stable state, i.e. towards the stored pattern which most closely resembles the initial state (see Figure 2.6). The system performs so-called *associative pattern recall*: patterns are not retrieved from memory by giving an address label (as in computers), but by an association process. By construction, this system will also recognise corrupted or incomplete patterns.

If we apply the learning rule (2.3) to a collection of p patterns, $(\xi_1^\mu, \ldots, \xi_N^\mu)$, where the superscript $\mu \in \{1, \ldots, p\}$ labels the patterns, we obtain:

$$w_{ij} = \xi_i^1 \xi_j^1 + \ldots + \xi_i^p \xi_j^p \tag{2.5}$$

Due to their simplicity it is easy and entertaining to write a computer program which simulates equations (2.2) for the choice (2.5), in order to verify that the

[5]The distance between two system states $(S_1, \ldots, S_N)$ and $(S_1', \ldots, S_N')$ is defined in terms of the number of neurons for which $S_i \neq S_i'$.

Figure 2.7: Ten patterns represented as specific microscopic states of an $N = 841$ attractor network. Individual pixels represent neuron states: $\{\bullet, \circ\} = \{1,-1\}$.

recipe described above works. An example is shown in Figures 2.7 and 2.8. We choose a set of ten patterns, each represented by $N = 841$ binary variables (pixels) (see Figure 2.7), and calculate synapses according to (2.5). For the initial state of the network equipped with these synapses we choose a corrupted version of one of the patterns. Following this initialisation the system is left to itself, and the dynamical rules (2.2) then generate processes such as those shown in Figure 2.8. If the corruption of the state to be recognised is modest, the system indeed evolves towards (i.e. 'recognises') the desired pattern. Note, however, that the particular recipe (2.5) en passant creates additional attractors, in the form of mixtures of the p stored patterns, to which the system is found to evolve if started from a completely random initial state (see Figure 2.8). Such 'mixture' states can be removed easily, either by adding noise to the dynamical rules, by introducing a non-zero threshold in the equations (2.2) or by using more sophisticated learning rules.

It turns out that the game described so far can be generalised to the situation where one wants to store not just individual (static) patterns, but sequences of patterns (films rather than individual pictures, sentences rather than words, or even an arbitrary set of required state transitions). To see this, let us make a small modification in the simple learning rule (2.3):

$$w_{ij} \to w_{ij} + S_i' S_j \tag{2.6}$$

Now *two* microscopic configurations $(S_1, \ldots, S_N)$ and $(S_1', \ldots, S_N')$ play a role. Rules like (2.6) emerge naturally if one takes transmission delays into account. If we apply (2.6) to a single pair of specific patterns, $(\xi_1, \ldots, \xi_N)$ and $(\xi_1', \ldots, \xi_N')$, we obtain $w_{ij} = \xi_i' \xi_j$, giving:

$$input_i = w_{i1} S_1 + \ldots + w_{iN} S_N = \xi_i' \left[\xi_1 S_1 + \ldots + \xi_N S_N \right]$$

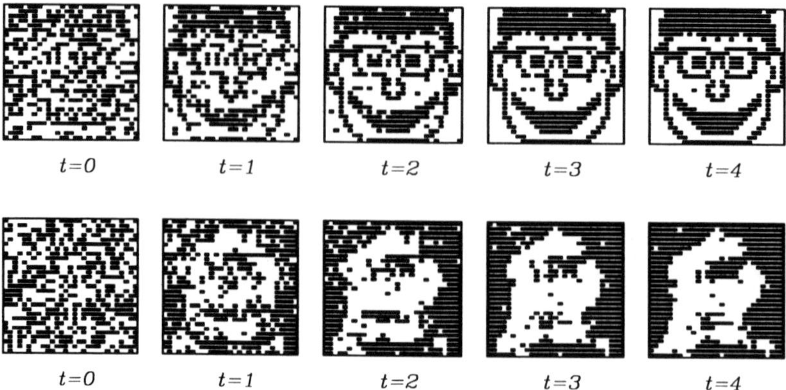

Figure 2.8: Two simulation examples: snapshots of the microscopic system state $\{S_1, \ldots, S_N\}$ at times $t = 0, 1, 2, 3, 4$ iteration steps per neuron. Dynamics: sequential. Top row: associative recall of a stored pattern from an initial state which is a corrupted version thereof. Bottom row: evolution towards a spurious (mixture) state from a randomly drawn initial state.

so that the dynamical rules (2.2) for the neuron states become:

$$\begin{array}{ll} \xi_1 S_1(t) + \ldots + \xi_N S_N(t) > 0: & S_i(t+1) = \xi_i' \\ \xi_1 S_1(t) + \ldots + \xi_N S_N(t) < 0: & S_i(t+1) = -\xi_i' \end{array} \qquad (2.7)$$

If at time t more than half of the neurons are in the state $S_i(t) = \xi_i$ then for all i: $S_i(t+1) = \xi_i'$. In other words: if the system is in a state sufficiently 'close' to state $(\xi_1, \ldots, \xi_N)$ it will tend to evolve towards state $(\xi_1', \ldots, \xi_N')$.[6]

This simple example shows several interesting things. Firstly, we can apparently store pattern sequences with the following rule:

- Represent each pattern sequence as a sequence of network state.
- Construct synapses $\{w_{ij}\}$ such that these sequences become attractors for the network dynamics.
- An input to trigger a squence will be the initial network state $\{S_i(t=0)\}$.
- The final attractor reached $\{S_i(t)\}$ (large t) can be interpreted as the sequence recalled by presenting input $\{S_i(t=0)\}$.

For example, we can achieve the storage of a given sequence of p states by applying the rule (2.7) to each of the individual constituent state transitions $(\xi_1^\mu, \ldots, \xi_N^\mu) \to (\xi_1^{\mu+1}, \ldots, \xi_N^{\mu+1})$ that we want to build in, giving the recipe:

$$w_{ij} = \xi_i^2 \xi_j^1 + \ldots + \xi_i^p \xi_j^{p-1} \qquad (2.8)$$

[6]The original recipe (2.3) just corresponds to the special case $(\xi_1, \ldots, \xi_N) = (\xi_1', \ldots, \xi_N')$.

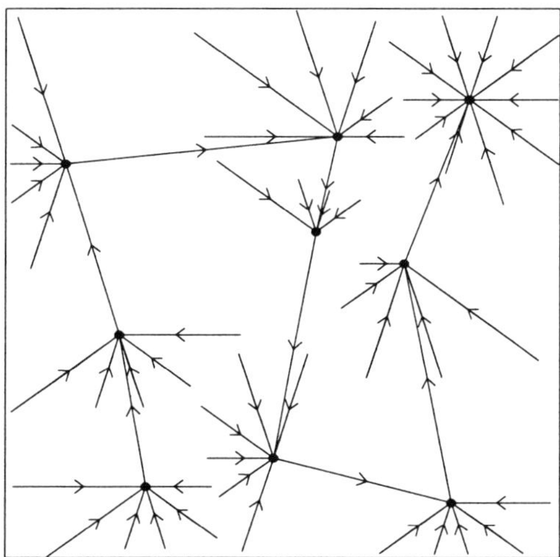

Figure 2.9: Information storage through the creation of attractors in the space of states. The state vector $(S_1, \ldots, S_N)$ evolves towards the nearest attractor. If the attractors are the pattern sequences stored, and the initial state is a constituent pattern, this system performs *associative* sequence recall.

It turns out that (2.8) indeed leads to the required operation described above, provided that the dynamics is of the parallel type. If the neurons change their states sequentially, an additional mechanism is found to be needed to stabilise the sequences (such as delayed interactions between the neurons).

Secondly, at least for parallel dynamics we now see how one might 'teach' these networks any arbitrary set of instructions, since it appears that the following interpretation holds:

$$\begin{array}{ll} \textbf{synaptic change}: & w_{ij} \to w_{ij} + \xi_i'\xi_j \\ \textbf{rule learned}: & \text{if in state } (\xi_1, \ldots, \xi_N) \text{ go to state } (\xi_1', \ldots, \xi_N') \end{array} \tag{2.9}$$

The resulting synapses are the sum over all individual stored instructions (2.9). Note the invariance of the synaptic change under $(\xi_i, \xi_i') \to (-\xi_i, -\xi_i')$ for all i. Although thinking in terms of instruction sets is reminiscent of conventional computers, the way these instructions are executed and combined is different. Let me just note a few points, some of which are immediately obvious, some of which require some more analysis which I will not discuss here:

- The procedure works best for orthogonal or random patterns.
- The microscopic realisation of the patterns is irrelevant. They define the language in terms of which instructions are written; if the 'words' of the language are sufficiently different from one another, any language will do.

- The system still operates by association: if it finds itself in a state not identical to any of the patterns in the instruction set, it will do the operation(s) corresponding to the pattern(s) it resembles most.
- Contradictory instructions just annihilate one another: $\xi_i'\xi_j+(-\xi_i')\xi_j = 0$.

2.3.2 Symmetric Networks: The Energy Picture

The simple learning rule (2.3) for storing (static) patterns will give rise to symmetric synapses, i.e. $w_{ij} = w_{ji}$ for all (ij), whereas the more general presciption (2.6) for storing attractors which are not fixed-points will generate predominantly non-symmetric synapses. It turns out that there is a deeper reason for this difference. For symmetric networks without self-interactions, i.e. $w_{ij} = w_{ji}$ for:

$$S_i \rightleftarrows S_j$$

all (ij) and $w_{ii} = 0$ for all i, one can easily show that the evolution of the neuron states is such that a certain quantity (termed the 'energy') is always decreasing. For sequential dynamics this energy is found to be:

$$E = -\frac{1}{2}\left[S_1.input_1 + \ldots + S_N.input_N\right] \tag{2.10}$$

For parallel dynamics one finds a different but related quantity.[7] Since the energy (2.10) is bounded from below, and since with each state change the energy decreases by at least some minimum amount (which depends on the values of the synapses), this process will have to stop at some point. We conclude: whatever the synapses (provided they are symmetric), the state dynamics will always end up in a fixed-point. The converse statement is not true: although in most cases this will not happen, the states of non-symmetric networks could also evolve towards a fixed-point (depending on the details of the synapses).

During the march for the lowest energy state, each individual transition $(S_1,\ldots,S_N) \to (S_1',\ldots,S_N')$ must decrease the energy, i.e. $E' < E$, which implies that one need not end up in the state with the lowest energy. Just imagine a downhill walk in a hilly landscape; in order to arrive at the lowest point one will occasionally have to cross a ridge to go from one valley to another. The network cannot do this, and can consequently end up in a local minimum of E, different from the global one.

Although quite natural in the context of neural networks, having asymmetry in the interactions of pairs of elements is in fact an unusual situation for the modeller. In physics the equivalent would be, for instance, a pair of two molecules A and B, with A exerting an attractive force on B and at the same time B repelling A. Or, likewise, a pair of magnets A and B such that magnet A prefers to have its poles (north and south) opposite to the poles of B, whereas B is keen on configurations where similar poles point in the same direction. If

[7] For parallel dynamics the condition that self-interactions must be absent can be dropped.

we add noise to symmetric systems we find that interaction symmetry implies *detailed balance*, which guarantees an evolution towards equilibrium. Since this is what all physical systems do, most analytical techniques developed to study interaction particle systems are based on this property. This makes neural networks the more interesting: since they are mostly non-symmetric, they will in general not evolve to equilibrium (compared to the possible modes of operation of non-symmetric networks, the symmetric ones are in fact quite boring), and they require novel intuition and techniques for analysis.

2.3.3 Solving Models of Noisy Attractor Networks

So far we have been concerned with qualitative properties of attractor networks. Let us turn to analysis now, and show how one proceeds to solve such models. I will skip details and only discuss the solution for sequential dynamics (for parallel dynamics one proceeds in a similar way). I will also introduce noise into the dynamics; this simplifies many calculations and often turns out to be beneficial to the operation of the system.

Stage 1: define the dynamical rules

The simplest way to add noise to the dynamics is to add to the each of the neural inputs at each time-step t an independent zero-average random number $z_i(t)$. This changes the noise-free dynamical laws (2.2) to:

$$\begin{array}{ll} w_{i1}S_1(t)+\ldots+w_{iN}S_N(t)+Tz_i(t)>0: & S_i(t+1)=1 \\ w_{i1}S_1(t)+\ldots+w_{iN}S_N(t)+Tz_i(t)<0: & S_i(t+1)=-1 \end{array} \tag{2.11}$$

Here T is an overall parameter to control the amount of noise ($T=0$: no noise, $T=\infty$: noise only). We store various operations of the type (2.9), defined using a set of p patterns:

$$w_{ij}=\frac{1}{N}\sum_{\mu,\nu=1}^{p}A_{\mu\nu}\xi_i^\mu\xi_j^\nu \qquad \overbrace{(\xi_1^1,\ldots,\xi_N^1)}^{\text{pattern 1}}\;\ldots\;\overbrace{(\xi_1^p,\ldots,\xi_N^p)}^{\text{pattern p}} \tag{2.12}$$

The prefactor $\frac{1}{N}$ is inserted to ensure that the inputs will not diverge in the limit $N\to\infty$ which we will eventually take. The associative memory rule (2.5) corresponds to $A_{\mu\nu}=\delta_{\mu\nu}$ (i.e. $A_{\mu\mu}=1$ for all μ and $A_{\mu\nu}=0$ for $\mu\neq\nu$).

Stage 2: rewrite dynamical rules in terms of probabilities

To suppress notation I will abbreviate $\boldsymbol{S}=(S_1,\ldots,S_N)$. Due to the noise we can only speak about the *probability* $p_t(\boldsymbol{S})$ to find a given state $\boldsymbol{S}$ at a given time t. In order to arrive at a description where time is a continuous variable, we choose the individual durations of the update steps at random from a Poisson distribution[8], with an average step duration of $\frac{1}{N}$, i.e. the probability $\pi_\ell(t)$ that at time t exactly ℓ neurons states have been updated is defined as:

$$\pi_\ell(t) = \frac{1}{\ell!}(Nt)^\ell e^{-Nt}$$

In one unit of time each neuron will on average have had one state update (as in the simulations of Figure 2.8). What remains is to do the bookkeeping of the possible sequential transitions properly, which results in an equation for the rate of change of the microscopic probability distribution:

$$\frac{d}{dt}p_t(\boldsymbol{S}) = \sum_{i=1}^{N} \{w_i(F_i\boldsymbol{S})p_t(F_i\boldsymbol{S}) - w_i(\boldsymbol{S})p_t(\boldsymbol{S})\} \tag{2.13}$$

Here $w_i(\boldsymbol{S})$ denotes the rate at which the transition $\boldsymbol{S} \to F_i\boldsymbol{S}$ occurs if the system is in state $\boldsymbol{S}$, and F_i is the operation 'change the state of neuron i: $F_i\boldsymbol{S} = (S_1, \ldots, S_{i-1}, -S_i, S_{i+1}, \ldots, S_N)$'. Equation (2.13) is quite transparent: the probability to find the state $\boldsymbol{S}$ increases due to transitions of the type $F_i\boldsymbol{S} \to \boldsymbol{S}$, and decreases due to transitions of the type $\boldsymbol{S} \to F_i\boldsymbol{S}$. The values of the transition rates $w_i(\boldsymbol{S})$ depend on the choice made for the distribution $P(z)$ of the noise variables $z_i(t)$. One convenient choice is:

$$P(z) = \frac{1}{2}[1-\tanh^2(z)] : \qquad w_i(\boldsymbol{S}) = \frac{1}{2}\left[1-\tanh[S_i\sum_{j=1}^{N} w_{ij}S_j/T]\right] \tag{2.14}$$

We have now translated our problem into solving a well-defined linear differential equation (2.13). Unfortunately this is still too difficult (except for networks obtained by making trivial choices for the synapses $\{w_{ij}\}$ or the noise level T).

Stage 3: find the relevant macroscopic features

We now try to find out which are the key macroscopic quantities (if any) that characterise the dynamical process. Unfortunately there is no general method to do this; one must rely on intuition, experience and common sense. Combining equations (2.11,2.12) shows that the neural inputs depend on the instantaneous state $\boldsymbol{S}$ only through the values of p specific macroscopic quantities $m_\mu(\boldsymbol{S})$:

$$input_i = \sum_{\mu\nu=1}^{p} A_{\mu\nu}\xi_i^\mu m_\nu(\boldsymbol{S}) + Tz_i$$

$$m_\nu(\boldsymbol{S}) = \frac{1}{N}[S_1\xi_1^\nu + \ldots + S_N\xi_N^\nu] \tag{2.15}$$

[8]This particular choice turns out to generate the simplest equations later.

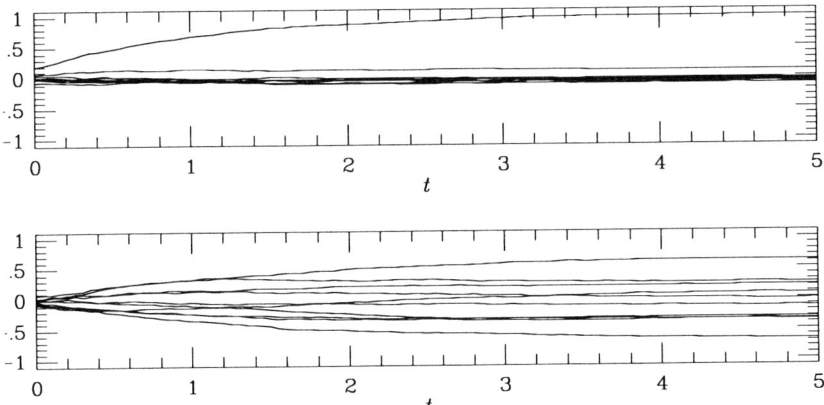

Figure 2.10: The two simulation examples of Figure 2.8: here we show the values of the p pattern overlaps $m_\mu(\boldsymbol{S})$, as measured at times $t = 0, 1, 2, 3, 4$ iteration steps per neuron. Top row: associative recall of a stored pattern from an initial state which is a corrupted version thereof. Bottom row: evolution towards a spurious (mixture) state from a randomly drawn initial state.

Note that these so-called 'overlaps' $m_\nu(\boldsymbol{S})$ measure the similarity between the state $\boldsymbol{S}$ and the stored patterns,[9] e.g.:

$$m_\mu(\boldsymbol{S}) = 1: \quad (S_1, \ldots, S_N) = (\xi_1^\mu, \ldots, \xi_N^\mu)$$
$$m_\mu(\boldsymbol{S}) = -1: \quad (S_1, \ldots, S_N) = (-\xi_1^\mu, \ldots, -\xi_N^\mu)$$

Further evidence of their status as our macroscopic level of description is provided by measuring their values during simulations, see Figure 2.10, for example. Since a description in terms of the observables $\{m_1(\boldsymbol{S}), \ldots, m_p(\boldsymbol{S})\}$ will only be simpler than the microscopic one in terms of $(S_1, \ldots, S_N)$ for modest numbers of patterns, i.e. for $p \ll N$, we will assume p to be finite (if $p \sim N$ we will simply have to think of something else). Having identified our macroscopic description, we can now define the macroscopic equivalent $P_t(m_1, \ldots, m_p)$ of the microscopic probability distribution $p_t(\boldsymbol{S})$, and calculate the macroscopic equivalent of the differential equation (2.13), which for $N \to \infty$ reduces to:

$$\frac{d}{dt} P_t(m_1, \ldots, m_p) = -\sum_{\mu=1}^{p} \frac{\partial}{\partial m_\mu} \{P_t(m_1, \ldots, m_p) F_\mu(m_1, \ldots, m_p)\} \quad (2.16)$$

$$F_\mu(m_1, \ldots, m_p) = \sum_{\boldsymbol{\xi} \in \{-1,1\}^p} p(\boldsymbol{\xi})\ \xi_\mu \tanh[\sum_{\lambda,\nu=1}^{p} A_{\lambda\nu} \xi_\lambda m_\nu / T] - m_\mu \quad (2.17)$$

[9]They are linearly related to the distance between the system state and the stored patterns.

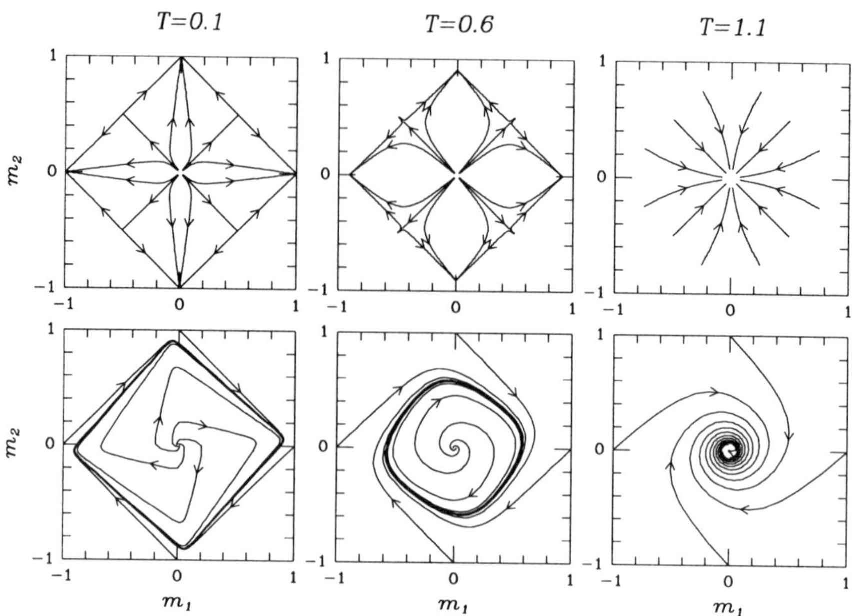

Figure 2.11: Solutions of the coupled equations (2.19) for the overlaps, with $p = 2$, obtained numerically and drawn as trajectories in the (m_1, m_2) plane. Row one: $A_{\mu\nu} = \delta_{\mu\nu}$, associative memory. Each of the four stable macroscopic states found for sufficiently low noise levels ($T < 1$) corresponds to the reconstruction of either a stored pattern $(\xi_1^\mu, \ldots, \xi_N^\mu)$ or its negative $(-\xi_1^\mu, \ldots, -\xi_N^\mu)$. Row two: $\boldsymbol{A} = \left(\begin{smallmatrix} 1 & 1 \\ -1 & 1 \end{smallmatrix}\right)$. For sufficiently low noise levels T this choice gives rise to the creation of a limit-cycle of the type $\boldsymbol{\xi}^1 \to (-\boldsymbol{\xi}^2) \to (-\boldsymbol{\xi}^1) \to \boldsymbol{\xi}^2 \to \boldsymbol{\xi}^1 \to \ldots$.

$$p(\boldsymbol{\xi}) = \lim_{N\to\infty} \frac{1}{N} \sum_{i=1}^{N} \delta_{\xi_i^1, \xi_1} \cdots \delta_{\xi_i^p, \xi_p} \tag{2.18}$$

Here $\boldsymbol{\xi} = (\xi_1, \ldots, \xi_p)$. For $N \to \infty$ the microsopic details of the pattern components are irrelevant; only the probability distribution (2.18) plays a role. For randomly drawn patterns one finds $p(\boldsymbol{\xi}) = 2^{-p}$ for all $\boldsymbol{\xi}$. Note that equation (2.16) is closed, i.e. the evolution of $P_t(m_1, \ldots, m_p)$ is given by a law in which knowledge of the microscopic realisations $\boldsymbol{S}$ or their distribution $p_t(\boldsymbol{S})$ is not required. The level of description of the overlaps is found to be autonomous.

Stage 4: solve the equation for $P_t(m_1, \ldots, m_p)$

The partial differential equation (2.16) has deterministic solutions: infinitely sharp probability distributions, which depend on time only through the location of the peak. In other words: in the limit $N \to \infty$ the fluctuations in the values of $(m_1, \ldots, m_p)$ become negligable, so that we can forget about probabilities and

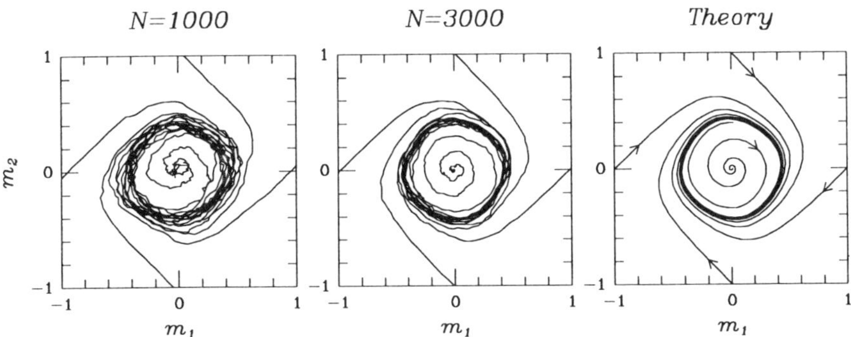

Figure 2.12: Comparison of the macroscopic dynamics in the (m_1, m_2) plane, as observed in finite-size numerical simulations, and the predictions of the $N = \infty$ theory, for the limit-cycle model with $\boldsymbol{A} = \begin{pmatrix} 1 & 1 \\ -1 & 1 \end{pmatrix}$ at noise level $T = 0.8$.

speak about the actual value of the macroscopic state $(m_1, \ldots, m_p)$. This value evolves in time according to the p coupled non-linear differential equations:

$$\frac{d}{dt}\begin{pmatrix} m_1 \\ \vdots \\ m_p \end{pmatrix} = \begin{pmatrix} F_1(m_1, \ldots, m_p) \\ \vdots \\ F_p(m_1, \ldots, m_p) \end{pmatrix} \tag{2.19}$$

with the functions F_μ given by (2.17). This is our final solution. One can now analyse these equations, calculate stationary states and their stability properties (if any), sizes of attraction domains, relaxation times, etc.

Equivalently we can solve the equations numerically, resulting in figures like 2.11 and 2.12. The first row of Figure 2.11 corresponds to $A_{\mu\nu} = \delta_{\mu\nu}$ and $p = 2$, representing a simple associative memory network of the type (2.5) with two stored patterns. The second row corresponds to a non-symmetric synaptic matrix, with $p = 2$, generating limit-cycle attractors. Finally, Figure 2.12 illustrates how the behaviour of finite networks (as observed in numerical simulations) for increasing values of the network size N approaches that described by the $N = \infty$ theory (described by the numerical solutions of (2.19)).

2.4 Creating Maps of the Outside World

Any flexible and robust autonomous system (whether living or robotic) will have to be able to create, or at least update, an internal 'map' or representation of its environment. Information on its environment, however, is usually obtained in an indirect manner, through a redundant set of sensors which each provide only partial and indirect information. The system responsible for forming this map needs to be adaptive, as both environment and sensors can change their characteristics during the system's life-time. Our brain performs recallibration of sensors all the time; for example simply because we grow, the neuronal information about limb positions (generated by sensors which measure the stretch of muscles) will have to be reinterpreted continually. Anatomic changes, and even learning new skills (like playing an instrument), are found to induce modifications of internal maps. At a more abstract level, one is confronted with a complicated non-linear mapping from a relatively low-dimensional and flat space (the 'physical world') into a high-dimensional one (the space of sensory signals), and the aim is to find the inverse of this operation. The key to achieving this is to exploit continuity and correlations in sensory signals, assuming similar sensory signals to represent similar positions in the environment, which therefore must correspond to similar positions in the internal map.

2.4.1 Map Formation Through Competitive Learning

Let us give a simple example. Image a system operating in a simple two-dimensional world, where positions are represented by two Cartesian coordinates (x, y), observed by sensors and fed into a neural network as input signals.

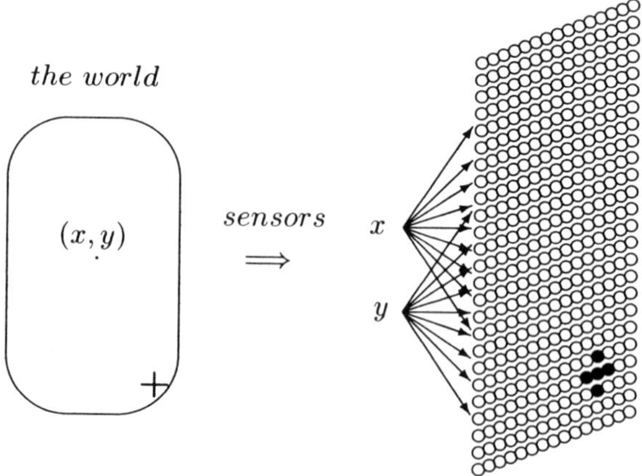

Each neuron i receives information on the input signals (x, y) in the usual way, through modifiable synaptic interaction strengths: $input_i = w_{ix}x + w_{iy}y$. If this network is to become an internal coordinate system, faithfully reflecting the events (x, y) observed in the outside world (in the present example its topology must accordingly be that of a two-dimensional array), the following objectives are to be met:

1. each neuron S_ℓ is more or less 'tuned' to a specific type of signal (x_ℓ, y_ℓ)
2. neighbouring neurons are tuned to similar signals
3. external 'distance' is monotonically related to internal 'distance'

Here the internal 'distance' between two signals (x_A, y_A) and (x_B, y_B) is defined as the physical distance between the two (groups of) neurons that would respond to these two signals.

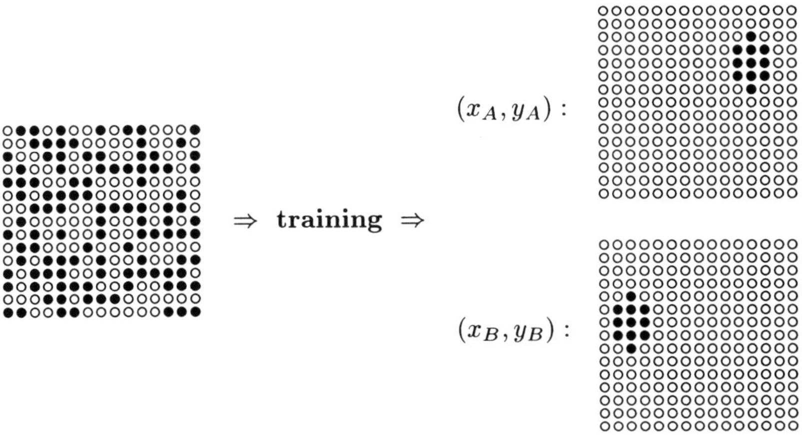

(x_A, y_A) :

$\Rightarrow$ **training** $\Rightarrow$

(x_B, y_B) :

It turns out that in order to achieve these objectives one needs learning rules where neurons effectively enter a competition for having signals 'allocated' to them, whereby neighbouring neurons stimulate one another to develop similar synaptic interactions and distant neurons are prevented from developing similar interactions. Let us try to construct the simplest such learning rule. Since our equations take their simplest form in the case where the input signals are normalised, we define $(x, y) \in [-1, 1]^2$ and add a dummy variable $z = \pm\sqrt{1-x^2-y^2}$ (together with an associated synaptic interaction w_z), so:

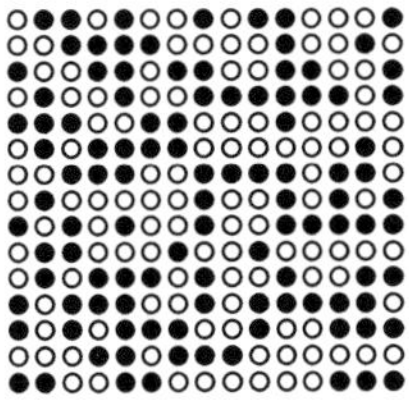

$$\begin{array}{lll} \bullet : & S_i = 1 & \text{(neuron } i \text{ firing)} \\ \circ : & S_i = -1 & \text{(neuron } i \text{ at rest)} \end{array}$$

$$\begin{array}{ll} input_i > 0 : & S_i \to 1 \\ input_i < 0 : & S_i \to -1 \end{array}$$

$$input_i = w_{ix}x + w_{iy}y + w_{iz}z$$

A learning rule with the desired effect is, starting from random synaptic interaction strengths, to iterate the following recipe until a (more or less) stable situation is reached:

$$\text{choose an input signal :} \quad (x, y, z)$$

$$\text{find most excited neuron :} \quad i, \quad input_i \geq input_k \quad (\text{for all } k)$$

$$\text{for } i \text{ and its neighbours :} \quad \left\{ \begin{array}{lll} w_{ix} & \to & (1-\epsilon)w_{ix} + \epsilon x \\ w_{iy} & \to & (1-\epsilon)w_{iy} + \epsilon y \\ w_{iz} & \to & (1-\epsilon)w_{iz} + \epsilon z \end{array} \right. \tag{2.20}$$

$$\text{for all others :} \quad \left\{ \begin{array}{lll} w_{ix} & \to & (1-\epsilon)w_{ix} - \epsilon x \\ w_{iy} & \to & (1-\epsilon)w_{iy} - \epsilon y \\ w_{iz} & \to & (1-\epsilon)w_{iz} - \epsilon z \end{array} \right.$$

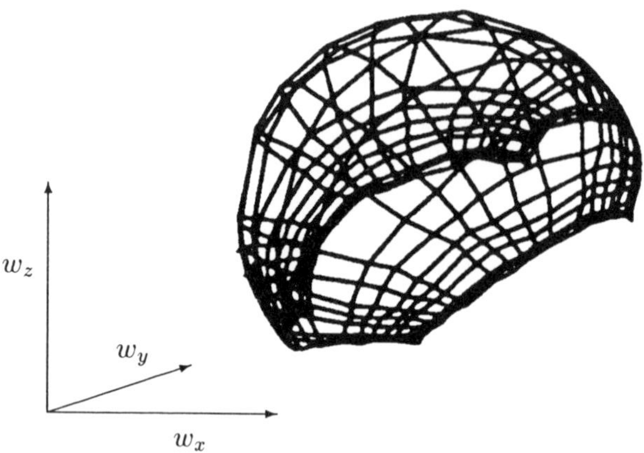

Figure 2.13: Graphical representation of the synaptic structure of a map forming network in the form of a 'fishing net'. The positions of the knots represent the signals in the world to which the neurons are 'tuned' and the cords connect the knots of neighbouring neurons.

In words: the neuron that was already the one most responsive to the signal (x, y, z) will be made even more so (together with its neighbours). The other neurons are made less responsive to (x, y, z). This is more obvious if we inspect the effect of the above learning rule on the actual neural inputs, using the built-in property $x^2+y^2+z^2 = 1$:

$$\begin{array}{ll} \text{for } i \text{ and its neighbours :} & input_i \to (1-\epsilon)input_i + \epsilon \\ \text{for all others :} & input_i \to (1-\epsilon)input_i - \epsilon \end{array}$$

In practice one often adds extra ingredients to this basic recipe, like explicit normalisation of synaptic interaction strengths to deal with non-uniform distributions of input signals (x, y, z), or a monotonically decreasing modification step size $\epsilon(t)$ to enforce and speed up convergence.

A nice way to illustrate what happens during the learning stage is based on exploiting the property that, apart from normalisation, one can interpret the synaptic strengths (w_{ix}, w_{iy}, w_{iz}) of a neuron as the signal (x, y, z) to which it is tuned. We can now draw each set of synaptic strengths (w_{ix}, w_{iy}, w_{iz}) as a point in space, and connect the points corresponding to neurons which are neighbours in the network. We end up with a graphical representation of the synaptic structure of a network in the form of a 'fishing net', with the positions of the knots representing the signals in the world to which the neurons are tuned and with the cords indicating neighbourship, see Figure 2.13. The three objectives of map formation set out at the beginning of this section thereby translate into:

1. all knots in the net are separated

2. all cords are similarly stretched

3. there are no regions with overlapping pieces of net

In Figure 2.13 all knots are more or less on the surface of the unit sphere, i.e. $w_{ix}^2 + w_{iy}^2 + w_{iz}^2 \approx 1$ for all i. This reflects the property that the length of the input vector (x, y, z) contains no information, due to $x^2 + y^2 + z^2 = 1$.

2.4.2 Solving Models of Map Formation

Let us now try to describe such learning processes analytically. The specific learning rules I will discuss here serve to illustrate only; they are by no means the most sophisticated or efficient ones, but they are sufficiently simple and transparent to allow for understanding and analysis. In addition they provide a nice example of how similarities between mathematical problems in remote scientific areas can be exploited, as will become clear shortly.

Stage 1: define the dynamical rules

One computationally nasty (and biologically unrealistic) feature of the learning rule described above is the need to find the neuron that is triggered most by a particular input signal (x, y, z) (to be given a special status, together with its neighbours). A more realistic but similar procedure is to base the decision about how synapses are to be modified only on the actual firing state of the neurons, and to realise the *neighbours-must-team-up* effect by a spatial smoothening of all neural inputs[10]. To be specific: before synaptic strengths are modified we replace

$$input_i \rightarrow Input_i = \langle input_j \rangle_{\text{near } i} - J \langle input \rangle_{\text{all}} \tag{2.21}$$

in which brackets denote taking the average over a group of neurons and J is a positive constant. This procedure has the combined effects that (i) neighbouring neurons will tend to have similar neural inputs (due to the first term in (2.21)), and (ii) the presence of a significant response somewhere in the network will evoke a global suppression of activity everywhere else, so that neurons are effectively encouraged to 'tune' to different signals (due to the second term in equation (2.21)).

Thus we arrive at the following recipe for the modification of synaptic strengths, to replace (2.20):

[10] In certain brain regions spatial smoothening is indeed known to take place, via diffusing chemicals and gases such as NO.

$$
\begin{array}{ll}
\text{choose an input signal :} & (x,y,z) \\
\text{smooth out all inputs :} & input_i \rightarrow Input_i \\
& Input_i = \langle input_j \rangle_{\text{near } i} - J\langle input \rangle_{\text{all}} \\
\text{for all } i \text{ with } S_i = 1 : & \left\{ \begin{array}{lcl} w_{ix} & \rightarrow & (1-\epsilon)w_{ix}+\epsilon x \\ w_{iy} & \rightarrow & (1-\epsilon)w_{iy}+\epsilon y \\ w_{iz} & \rightarrow & (1-\epsilon)w_{iz}+\epsilon z \end{array} \right. \\
\text{for all } i \text{ with } S_i = -1 : & \left\{ \begin{array}{lcl} w_{ix} & \rightarrow & (1-\epsilon)w_{ix}-\epsilon x \\ w_{iy} & \rightarrow & (1-\epsilon)w_{iy}-\epsilon y \\ w_{iz} & \rightarrow & (1-\epsilon)w_{iz}-\epsilon z \end{array} \right.
\end{array}
\tag{2.22}
$$

As before, the 'world' from which the input signals (x, y, z) are drawn is the surface of a sphere: $x^2+y^2+z^2 = C^2$.

Stage 2: consider small modifications $\epsilon \rightarrow 0$

The dynamical rules (2.22) define a stochastic process, in that at each time-step the actual synaptic modification depends on the (random) choice made for the input (x, y, z) at that particular instance. However, in the limit of infinitesimally small modification size ϵ one finds the procedure (2.22) being transformed into a deterministic differential equation (if we also choose ϵ as the duration of each modification step), which involves only *averages* over the distribution $p(x, y, z)$ of inputs signals:

$$
\begin{aligned}
\tfrac{d}{dt}w_{ix} &= \int dxdydz\ p(x,y,z)\ x\ \text{sgn}[Input_i(x,y,z)] - w_{ix} \\
\tfrac{d}{dt}w_{iy} &= \int dxdydz\ p(x,y,z)\ y\ \text{sgn}[Input_i(x,y,z)] - w_{iy} \\
\tfrac{d}{dt}w_{iz} &= \int dxdydz\ p(x,y,z)\ z\ \text{sgn}[Input_i(x,y,z)] - w_{iz}
\end{aligned}
\tag{2.23}
$$

in which the spatially smoothed out neural inputs $Input_i(x, y, z)$ are given by (2.21), and the function sgn[..] gives the sign of its argument (i.e. $\text{sgn}[u > 0] = 1$, $\text{sgn}[u < 0] = -1$). The spherical symmetry of the distribution $p(x, y, z)$ allows us to do the integrations in (2.23). The result of the integrations involves only the smoothed out synaptic weights $\{W_{ix}, W_{iy}, W_{iz}\}$, defined as:

$$
\begin{aligned}
W_{ix} &= \langle w_{jx} \rangle_{\text{near } i} - J\langle w_x \rangle_{\text{all}} \\
W_{iy} &= \langle w_{jy} \rangle_{\text{near } i} - J\langle w_y \rangle_{\text{all}} \\
W_{iz} &= \langle w_{jz} \rangle_{\text{near } i} - J\langle w_z \rangle_{\text{all}}
\end{aligned}
\tag{2.24}
$$

and takes the form:

$$
\begin{aligned}
\frac{d}{dt}w_{ix} &= \frac{1}{2}C\frac{W_{ix}}{\sqrt{W_{ix}^2+W_{iy}^2+W_{iz}^2}} - w_{ix} \\
\frac{d}{dt}w_{iy} &= \frac{1}{2}C\frac{W_{iy}}{\sqrt{W_{ix}^2+W_{iy}^2+W_{iz}^2}} - w_{iy}
\end{aligned}
\tag{2.25}
$$

$$\frac{d}{dt}w_{iz} = \frac{1}{2}C\frac{W_{iz}}{\sqrt{W_{ix}^2 + W_{iy}^2 + W_{iz}^2}} - w_{iz}$$

Stage 3: exploit equivalence with dynamics of magnetic systems

If the constant J in (2.24) (controlling the global competition between the neurons) is below some critical value J_c, one can show that the equations (2.25), with the smoothed out weights (2.24), evolve towards a stationary state. In stationary states, where $\frac{d}{dt}w_{ix} = \frac{d}{dt}w_{iy} = \frac{d}{dt}w_{iz} = 0$, all synaptic strengths will be normalised according to $w_{ix}^2+w_{iy}^2+w_{iz}^2 = \frac{1}{4}C^2$, according to (2.25). In terms of the graphical representation of Figure 2.13 this corresponds to the statement that in stationary states all knots must lie on the surface of a sphere. From now on we take $C = 2$, leading to stationary synaptic strengths on the surface of the unit sphere.

If one works out the details of the dynamical rules (2.25) for synaptic strengths which are normalised according to $w_{ix}^2+w_{iy}^2+w_{iz}^2 = 1$, one observes that they are suspiciously similar to the ones that describe a system of microscopic magnets, which interact in such a way that neighbouring magnets prefer to point in the same direction (NN and SS), whereas distant magnets prefer to point in opposite directions (NS and SN).

synapses to neuron i :	(w_{ix}, w_{iy}, w_{iz})
neighbouring neurons :	prefer similar synapses
distant neurons :	prefer different synapses

orientation of magnet i :	(w_{ix}, w_{iy}, w_{iz})
neighbouring magnets :	prefer $\uparrow\uparrow$, $\downarrow\downarrow$
distant magnets :	prefer $\uparrow\downarrow$, $\downarrow\uparrow$

This relation suggests that one can use physical concepts again. More specifically, such magnetic systems would evolve towards the minimum of their energy E, in the present language given by:

$$E = -\frac{1}{2}\left[w_{1x}W_{1x}+w_{1y}W_{1y}+w_{1z}W_{1z}\right] - \frac{1}{2}\left[w_{2x}W_{2x}+w_{2y}W_{2y}+w_{2z}W_{2z}\right]$$

$$\ldots \qquad -\frac{1}{2}\left[w_{Nx}W_{Nx}+w_{Ny}W_{Ny}\right] \qquad (2.26)$$

If we check this property for our equations (2.25), we indeed find that, provided $J < J_c$, from some stage onwards during the evolution towards the stationary state, the energy (2.26) will be decreasing monotonically. The situation thus becomes quite similar to the one with the dynamics of the attractor neural networks in a previous section, in that the dynamical process can ultimately be seen as a quest for a state with minimal energy. We now know that the *equilibrium state* of our map forming system is defined as the configuration of weights that satisfies:

$$\begin{aligned} I: &\quad w_{ix}^2 + w_{iy}^2 + w_{iz}^2 = 1 \quad \text{for all } i \\ II: &\quad E \ \text{ is minimal} \end{aligned} \tag{2.27}$$

with E given by (2.26). We now forget about the more complicated dynamic equations (2.25) and concentrate on solving (2.27).

Stage 4: switch to new coordinates, and take the limit $N \to \infty$

Our next step is to implement the conditions $w_{ix}^2 + w_{iy}^2 + w_{iz}^2 = 1$ (for all i) by writing for each neuron i the three synaptic strengths (w_{ix}, w_{iy}, w_{iz}) in terms of the two polar coordinates (θ_i, ϕ_i) (a natural step in the light of the representation of Figure 2.13):

$$w_{ix} = \cos\phi_i \sin\theta_i \qquad w_{iy} = \sin\phi_i \sin\theta_i \qquad w_{iz} = \cos\theta_i \tag{2.28}$$

Furthermore, for large systems we can replace the discrete neuron labels i by their position coordinates (x_1, x_2) in the network, i.e. $i \to (x_1, x_2)$, so that:

$$\theta_i \to \theta(x_1, x_2) \qquad \phi_i \to \phi(x_1, x_2)$$

In doing so we have to specify how in the limit $N \to \infty$ the average over neighbours in (2.21) is to be carried out. If one just expands any well-behaved local and normalised averaging distribution up to first non-trivial order in its width ϵ, and chooses the parameter $J > J_c$ (the critical value depends on ϵ), one finds, after some non-trivial bookkeeping, that the solution of (2.27) obeys the following coupled non-linear partial differential equations:

$$\sin\theta \left[\frac{\partial^2 \phi}{\partial x_1^2} + \frac{\partial^2 \phi}{\partial x_2^2} \right] + 2\cos\theta \left[\frac{\partial \phi}{\partial x_1} \frac{\partial \theta}{\partial x_1} + \frac{\partial \phi}{\partial x_2} \frac{\partial \theta}{\partial x_2} \right] = 0 \tag{2.29}$$

$$\frac{\partial^2 \theta}{\partial x_1^2} + \frac{\partial^2 \theta}{\partial x_2^2} - \sin\theta \cos\theta \left[\left(\frac{\partial \phi}{\partial x_1} \right)^2 + \left(\frac{\partial \phi}{\partial x_2} \right)^2 \right] = 0 \tag{2.30}$$

with the constraints:

$$\int dx_1 dx_2 \ \cos\phi \sin\theta = \int dx_1 dx_2 \ \sin\phi \sin\theta = \int dx_1 dx_2 \ \cos\theta = 0 \tag{2.31}$$

The corresponding value for the energy E is then given by the expression:

$$E = -\frac{1}{2} + \frac{\epsilon}{2} \int dx_1 dx_2 \left\{ \sin^2\theta \left[\left(\frac{\partial \phi}{\partial x_1} \right)^2 + \left(\frac{\partial \phi}{\partial x_1} \right)^2 \right] + \left(\frac{\partial \theta}{\partial x_1} \right)^2 + \left(\frac{\partial \theta}{\partial x_1} \right)^2 \right\}$$

Stage 5: use symmetries and pendulums ...

Finding the general solution to fierce equations like (2.29,2.30,2.31) is out of the question. However, we can find special solutions, namely those which are of the form suggested by the simulation results shown in Figure 2.13. These configurations appear to have many symmetries, which we can exploit. In particular we make an 'ansatz' for the angles $\phi(x_1, x_2)$, which states that if the array of neurons had been a circular disk, the configuration of synaptic strengths would have had rotational symmetry:

$$\cos\phi(x_1, x_2) = \frac{x_1}{\sqrt{x_1^2 + x_2^2}} \qquad \sin\phi(x_1, x_2) = \frac{x_2}{\sqrt{x_1^2 + x_2^2}} \tag{2.32}$$

Insertion of this ansatz into our equations (2.29,2.30,2.31) shows that solutions with this property do indeed exist, and that they imply a simple law for the remaining angle: $\theta(x_1, x_2) = \theta(r)$, with $r = \sqrt{x_1^2 + x_2^2})$ and:

$$r^2 \frac{d^2\theta}{dr^2} + r\frac{d\theta}{dr} - \sin\theta\cos\theta = 0 \tag{2.33}$$

This is already an enormous simplification, but we need not stop here. A simple transformation of variables turns this differential equation (2.33) into the one describing the motion of a pendulum!

$$u = \log r, \qquad \theta(r) = \frac{1}{2}\pi + \frac{1}{2}G(u), \qquad \frac{d^2}{du^2}G(u) + \sin G(u) = 0 \tag{2.34}$$

This means that we have basically cracked the problem, since the motion of a pendulum can be described analytically. There are two types of motion, the first one describes the familiar swinging pendulum and the second one describes a rotating one (the effect resulting from giving the pendulum significantly more than a gentle swing ...). If we calculate the synaptic energies E associated with the two types of motion (after translation back into the original synaptic strength variables) we find that the lowest energy is obtained upon choosing the solution corresponding to the pendulum motion which precisely separates rotation from swinging. Here the pendulum 'swings' just once, from one vertical position at $u = -\infty$ to another vertical position at $u = \infty$:

$$G(u) = \arcsin[\tanh(u + q)] \qquad (q \in \Re) \tag{2.35}$$

If we now translate all results back into the original synaptic variables, combining equations (2.32,2.34,2.32,2.28), we end up with a beautifully simple solution:

$$\begin{aligned} w_x(x_1, x_2) &= \frac{2kx_1}{k^2 + x_1^2 + x_2^2} \\ w_y(x_1, x_2) &= \frac{2kx_2}{k^2 + x_1^2 + x_2^2} \\ w_z(x_1, x_2) &= \frac{k^2 - x_1^2 - x_2^2}{k^2 + x_1^2 + x_2^2} \end{aligned} \tag{2.36}$$

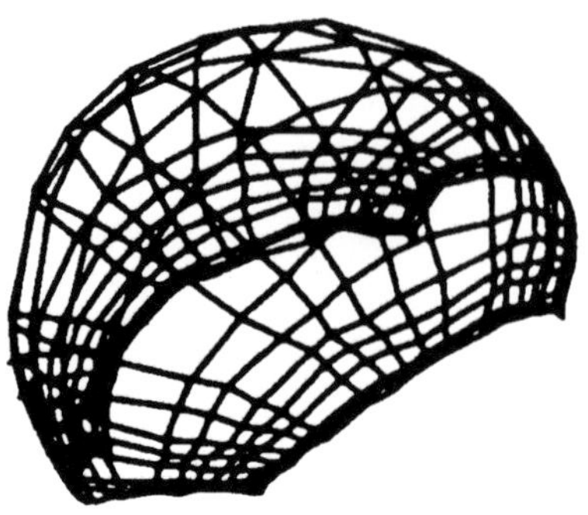

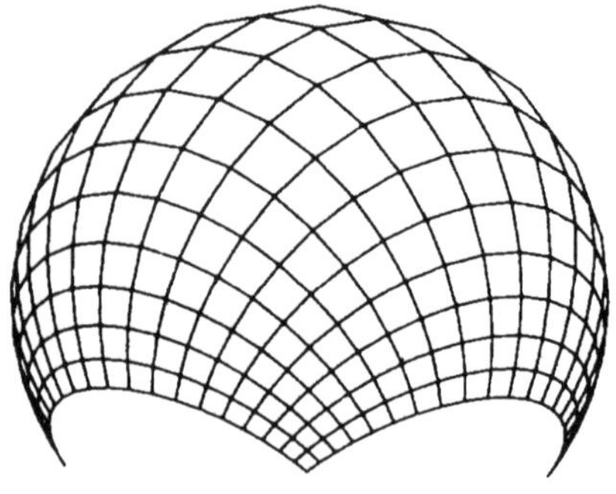

Figure 2.14: Graphical representation of the synaptic structure of a map forming network in the form of a 'fishing net'. Left: stationary state resulting from numerical simulations. Right: the analytical result (2.36).

in which the remaining constant k is uniquely defined as the solution of:

$$\int dx_1 dx_2 \, \frac{k^2 - x_1^2 - x_2^2}{k^2 + x_1^2 + x_2^2} = 0$$

(which is the translation of the constraints (2.31)). The final test, of course, is to draw a picture of this solution in the representation of Figure 2.13, which results in Figure 2.14. The agreement is quite satisfactory.

2.5 Learning a Rule From an Expert

Finally let us turn to the class of neural systems most popular among engineers: layered neural networks. Here the information flows in only one direction, so that calculating all neuron states at all times can be done iteratively (layer by layer), and has therefore become trivial. As a result one can concentrate on developing and studying non-trivial learning rules. The popular types of learning rules used in layered networks are the so-called 'supervised' ones, where the networks are trained by using examples of input signals ('questions') and the required output signals ('answers'). The latter are provided by a 'teacher'. The learning rules are based on comparing the network answers to the correct ones, and subsequently making adjustments to synaptic weights and thresholds to reduce the differences between the two answers to zero.

The most important property of neural networks in this context is their ability to generalise. In contrast to the situation where we would have just stored the question-answer pairs in memory, neural networks can, after having been confronted with a sufficient number of examples, generalise their 'knowledge' and provide reasonable, if not correct, answers even for new questions.

2.5.1 Perceptrons

The simplest feed-forward 'student' network one can think of is just a single binary neuron, with the standard operation rules:

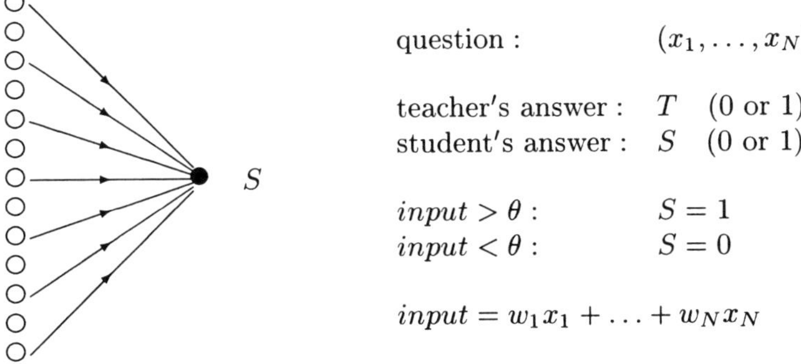

trying to adjust its synaptic strengths $\{w_\ell\}$ such that for each question $(x_1, \ldots, x_N)$ its answer S coincides with the answer T given by the teacher. We now define a very simple learning rule (the so-called perceptron learning rule) to achieve this, where changes are made only if the neuron makes a mistake, i.e. if $T \neq S$:

1. select a question $(x_1, \ldots, x_N)$

2. compare $S(x_1, \ldots, x_N)$ and $T(x_1, \ldots, x_N)$:

$S = T$:	do nothing
if $S = 1, T = 0$:	change $w_i \to w_i - x_i$ and $\theta \to \theta + 1$
if $S = 0, T = 1$:	change $w_i \to w_i + x_i$ and $\theta \to \theta - 1$

Binary neurons, equipped with the above learning rule, are called 'perceptrons' (reflecting their original use in the fifties to model perception). If a perceptron mistakenly produces an answer $S = 1$ for question $(x_1, \ldots, x_N)$ (i.e. *input* $> \theta$, whereas we would have preferred *input* $< \theta$), the modifications made ensure that the input produced upon presentation of this particular question will be reduced. If the perceptron mistakenly produces an answer $S = 0$ (i.e. *input* $< \theta$, whereas we would have preferred *input* $> \theta$), the changes made increase the input produced upon presentation of this particular question.

The perceptron learning rule appears to make sense, but it is as yet just one choice from an infinite number of possible rules. What makes the above rule special is that it comes with a convergence proof, in other words, it is guaranteed to work! More precisely, provided the input vectors are bounded:

If values for the parameters $\{w_\ell\}$ and θ exists, such that $S = T$ for each question $(x_1, \ldots, x_N)$, then the perceptron learning rule will find these, or equivalent ones, in a finite number of modification steps

This is a remarkable statement. It could, for instance, easily have happened that correcting the system's answer for a given question $(x_1, \ldots, x_N)$ would affect the performance of the system on those questions it had so far been answering correctly, thus preventing the system from ever arriving at a state without errors. Apparently this does not happen. What is even more remarkable, is that the proof of this powerful statement is quite simple.

The standard version of the convergence proof assumes the set of questions Ω to be finite and discrete (in the continuous case one can construct a similar proof)[11]. To simplify notation we use a trick: we introduce a 'dummy' input variable $x_0 = -1$ (a constant). Upon giving the threshold θ a new name, $\theta = w_0$, our equations can be written in a very compact way. We denote the vector of weights as $\boldsymbol{w} = (w_0, \ldots, w_N)$, the questions as $\boldsymbol{x} = (x_0, \ldots, x_N)$, inner products as $\boldsymbol{w} \cdot \boldsymbol{x} = w_0 x_0 + \ldots w_N x_N$, and the length of a vector as $|\boldsymbol{x}| = \sqrt{\boldsymbol{x} \cdot \boldsymbol{x}}$. For the operation of the perceptron we get:

$$\boldsymbol{w} \cdot \boldsymbol{x} > 0 : \; S = 1, \qquad \boldsymbol{w} \cdot \boldsymbol{x} < 0 : \; S = 0$$

whereas the learning rule becomes:

$$\boldsymbol{w} \rightarrow \boldsymbol{w} + [T(\boldsymbol{x}) - S(\boldsymbol{x})]\boldsymbol{x} \tag{2.37}$$

There exists an error-free system (by assumption), with as yet unknown parameters $\boldsymbol{w}^\star$ (these include the threshold w_0). This system by definition obeys:

$$T(\boldsymbol{x}) = 1 : \;\; \boldsymbol{w}^\star \cdot \boldsymbol{x} > 0, \qquad T(\boldsymbol{x}) = 0 : \;\; \boldsymbol{w}^\star \cdot \boldsymbol{x} < 0$$

Define $X = \max_\Omega |\boldsymbol{x}| > 0$ and $\delta = \min_\Omega |\boldsymbol{w}^* \cdot \boldsymbol{x}| > 0$. The convergence proof relies on the following inequalities:

$$\text{for all } \boldsymbol{x} \in \Omega : \qquad |\boldsymbol{x}| \leq X \qquad \text{and} \qquad |\boldsymbol{w}^* \cdot \boldsymbol{x}| \geq \delta \tag{2.38}$$

At each modification step $\boldsymbol{w} \rightarrow \boldsymbol{w}'$ (2.37), where $S = 1 - T$ (otherwise: no modification !), we can inspect what happens to the quantities $\boldsymbol{w} \cdot \boldsymbol{w}^\star$ and $|\boldsymbol{w}|^2$:

$$\begin{aligned} &\boldsymbol{w}' \cdot \boldsymbol{w}^\star = \boldsymbol{w} \cdot \boldsymbol{w}^\star + [2T(\boldsymbol{x}) - 1]\boldsymbol{x} \cdot \boldsymbol{w}^\star \\ &|\boldsymbol{w}'|^2 = |\boldsymbol{w}|^2 + 2[2T(\boldsymbol{x}) - 1]\boldsymbol{x} \cdot \boldsymbol{w} + [2T(\boldsymbol{x}) - 1]^2 |\boldsymbol{x}|^2 \end{aligned}$$

Note that $2T(\boldsymbol{x}) - 1 = -1$ if $\boldsymbol{w}^\star \cdot \boldsymbol{x} < 0$, and $2T(\boldsymbol{x}) - 1 = 1$ if $\boldsymbol{w}^\star \cdot \boldsymbol{x} > 0$, and that therefore $[2T(\boldsymbol{x}) - 1]\boldsymbol{x} \cdot \boldsymbol{w} < 0$ (otherwise one would have had $S = T$), so:

$$\begin{aligned} &\boldsymbol{w}' \cdot \boldsymbol{w}^\star = \boldsymbol{w} \cdot \boldsymbol{w}^\star + |\boldsymbol{x} \cdot \boldsymbol{w}^\star| \geq \boldsymbol{w} \cdot \boldsymbol{w}^\star + \delta \\ &|\boldsymbol{w}'|^2 < |\boldsymbol{w}|^2 + |\boldsymbol{x}|^2 \leq |\boldsymbol{w}|^2 + X^2 \end{aligned}$$

After n such modification steps we therefore find:

$$\begin{aligned} &\boldsymbol{w}(n) \cdot \boldsymbol{w}^\star \geq \boldsymbol{w}(0) \cdot \boldsymbol{w}^\star + n\delta \\ &|\boldsymbol{w}(n)|^2 \leq |\boldsymbol{w}(0)|^2 + nX^2 \end{aligned}$$

[11] One also assumes for simplicity that the borderline situation *input* $= \theta$ never occurs. This is not a big issue; such cases can be dealt with quite easily, should the need arise.

In combination this implies the following inequality:

$$\frac{\boldsymbol{w}(n)\cdot\boldsymbol{w}^\star}{|\boldsymbol{w}^\star||\boldsymbol{w}(n)|} \geq \frac{\boldsymbol{w}(0)\cdot\boldsymbol{w}^\star+n\delta}{|\boldsymbol{w}^\star|\sqrt{|\boldsymbol{w}(0)|^2+nX^2}}$$

giving:

$$\lim_{n\to\infty}\frac{1}{\sqrt{n}}\frac{\boldsymbol{w}(n)\cdot\boldsymbol{w}^\star}{|\boldsymbol{w}^\star||\boldsymbol{w}(n)|} \geq \frac{\delta}{|\boldsymbol{w}^\star|X} > 0 \tag{2.39}$$

We see that the number of modifications made *must* be bounded, since otherwise (2.39) leads to a contradiction with the Schwarz inequality $|\boldsymbol{w}\cdot\boldsymbol{w}^\star| \leq |\boldsymbol{w}||\boldsymbol{w}^\star|$. As soon as no more modifications are made, we must be in a situation where $S(\boldsymbol{x}) = T(\boldsymbol{x})$ for each question $\boldsymbol{x} \in \Omega$. This completes the proof.

Figure 2.15 gives an impression of the learning process described by the preceptron learning rule. In these simulation experiments the task T is defined by a teacher perceptron with a (randomly drawn) synaptic weight vector $\boldsymbol{w}^\star$:

$$\boldsymbol{w}^\star\cdot\boldsymbol{x} > 0: \quad T(\boldsymbol{x}) = 1 \qquad\qquad \boldsymbol{w}^\star\cdot\boldsymbol{x} < 0: \quad T(\boldsymbol{x}) = 0$$

The evolution of the student perceptron's synaptic vector $\boldsymbol{w}$ is monitored by calculating at each iteration step the quantity ω which already played a dominant role in the convergence proof for the perceptron learning rule, and which measures the resemblance between $\boldsymbol{w}$ and $\boldsymbol{w}^\star$:

$$\omega = \frac{\boldsymbol{w}\cdot\boldsymbol{w}^\star}{|\boldsymbol{w}||\boldsymbol{w}^\star|} \tag{2.40}$$

At each iteration step each component x_i of the questions $\boldsymbol{x}$ posed during the simulation experiments was drawn randomly from $\{-1,1\}$. Figure 2.15 suggests that for large perceptrons, $N\to\infty$, our general strategy of trying to derive exact analytical and deterministic dynamical laws might again be succesful. This turns out to be true, as we will show in a subsequent section.

Since we know that the perceptron learning rule will converge for each realisable task, we need only worry about which tasks are learnable by perceptrons and which are not. Tasks that can be performed by single binary neurons are called 'linearly separable'. Unfortunately, not all rules $(x_1,\ldots,x_N)\to T(x_1,\ldots,x_N)\in\{0,1\}$ can be performed with simple binary neurons. The simplest counter-example is the so-called XOR operation, XOR: $\{0,1\}^2\to\{0,1\}$, defined below. One can prove quite easily that there cannot exist a choice of parameters $\{w_1,w_2,\theta\}$ such that:

$$\begin{aligned}\text{XOR}(x_1,x_2)=1: &\quad w_1x_1+w_2x_2>\theta\\ \text{XOR}(x_1,x_2)=0: &\quad w_1x_1+w_2x_2<\theta\end{aligned}$$

by just checking explicitly the four possibilities for (x_1,x_2):

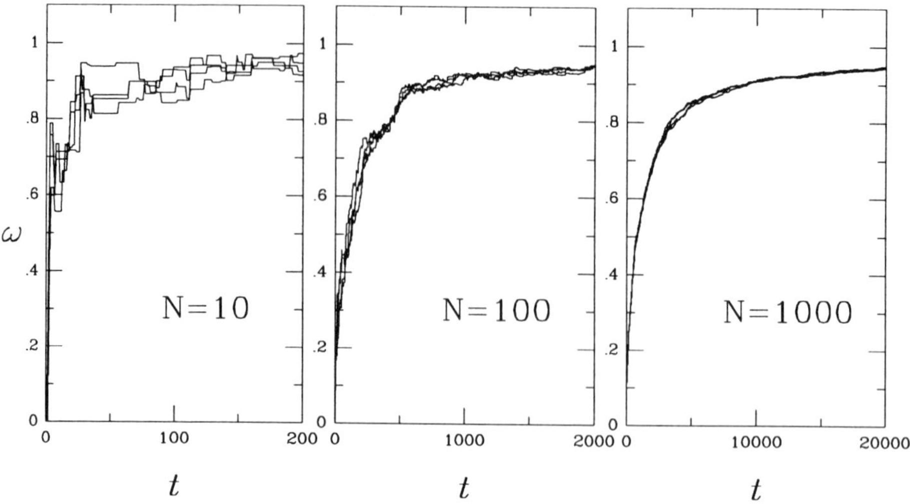

Figure 2.15: Evolution in time of the observable $\omega = \boldsymbol{w} \cdot \boldsymbol{w}^\star / |\boldsymbol{w}||\boldsymbol{w}^\star|$, obtained by numerical simulation of the perceptron learning rule, for a randomly drawn teacher vector $\boldsymbol{w}^\star$ and binary questions $(x_1, \ldots, x_N) \in \{-1, 1\}^N$. Each picture shows the results following four different random initialisations of the student vector $\boldsymbol{w}$.

x_1	x_2	$\mathrm{XOR}(x_1, x_2)$	requirement
0	0	0	$\theta > 0$
0	1	1	$w_2 > \theta$
1	0	1	$w_1 > \theta$
1	1	0	$w_1 + w_2 < \theta$

The four parameter requirements are clearly contradictory.

2.5.2 Multi-Layer Networks

If we want a neural network to perform an operation T that cannot be performed by a single binary neuron, like the XOR operation in the previous section, we need a more complicated network architecture. For the case where the question variables x_i are binary we know that the two-layer architecture shown in Figure 2.5 is in principle sufficiently complicated (see the proof in section 2.2), but as yet we have no learning rule available that will allow us to train such systems. For real-valued question variables one can prove that two-layer architectures can perform any sufficiently regular[12] task T with arbitrary accuracy, if the number of neurons in the 'hidden' layer is sufficiently large and

[12] $T(x_1, \ldots, x_N)$ should, for instance, be bounded for all $(x_1, \ldots, x_N)$

provided we turn our binary neurons into so-called 'graded-response' ones:

$$S = f(w_1 x_1 + \ldots w_N x_N - \theta) \tag{2.41}$$

in which the non-linear function $f(z)$ has the properties:

$$f'(z) \geq 0, \qquad \lim_{z \to \pm\infty} |f(z)| < \infty$$

Commonly made choices for f are $f(z) = \tanh(z)$ and $f(z) = \text{erf}(z)$. Definition (2.41) can be seen as a generalisation of the binary neurons considered so far, which correspond to choosing a step-function: $f(z > 0) = 1$, $f(z < 0) = 0$. From now on we will assume the task T and the nonlinear function f to be normalised according to $|T(x_1, \ldots, x_N)| \leq 1$ for all $(x_1, \ldots, x_N)$, and $|f(z)| \leq 1$ for all $z \in \Re$. Given a 'student' S in the form of a (universal) two-layer feed-forward architecture with neurons of the type (2.41), we can write the student answer $S(x_1, \ldots, x_N)$ to question $(x_1, \ldots, x_N)$ as:

$$\begin{aligned} S(x_1, \ldots, x_N) &= f(w_1 y_1 + \ldots + w_K y_K - \theta) \\ y_\ell(x_1, \ldots, x_N) &= f(w_{\ell 1} x_1 + \ldots + w_{\ell N} x_N - \theta_\ell) \end{aligned} \tag{2.42}$$

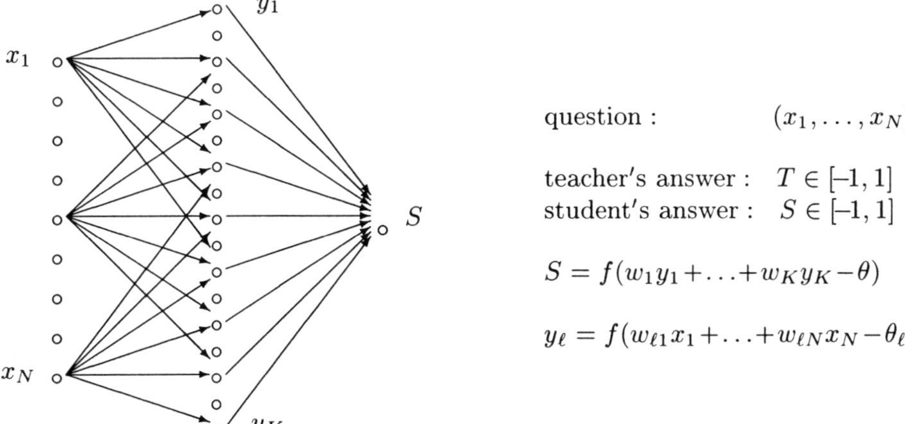

As before, we can simplify (2.42) and eliminate the thresholds θ and $\{\theta_\ell\}$ by introducing dummy neurons $x_0 = -1$ and $y_0 = -1$, with corresponding synaptic strengths w_0 and $\{w_{\ell 0}\}$.

Our goal is to find a learning rule. In this case it would be a recipe for the modification of the synaptic strengths w_i (connecting 'hidden' neurons to the output neuron) and $\{w_{ij}\}$ (connecting the input signals, or questions components, x_i to the hidden neurons), based on the observed differences between the teacher's answers $T(x_0, \ldots, x_N)$ and the student's answers $S(x_0, \ldots, x_N)$. If we denote the set of all possible questions by Ω, and the probability that a given question $\boldsymbol{x} = (x_0, \ldots, x_N)$ will be asked by $p(\boldsymbol{x})$, we can quantify the student's performance at any stage during training by its average quadratic error E[13]:

$$E = \frac{1}{2} \sum_{\boldsymbol{x} \in \Omega} p(\boldsymbol{x}) \left[T(\boldsymbol{x}) - S(\boldsymbol{x})\right]^2 \tag{2.43}$$

Training is supposed to minimise E, preferably to zero. This suggests a very simple way to modify the system's parameters, the so-called 'gradient descent' procedure. Here one inspects the change in E following infinitesimally small parameter changes, and then chooses those modifications for which E would decrease most (like a blind mountaineer in search of the valley). In terms of partial derivatives this implies the following 'motion' for the parameters:

$$\begin{aligned} &\text{for all } i = 1, \ldots, K: && \tfrac{d}{dt} w_i = -\tfrac{\partial}{\partial w_i} E \\ &\text{for all } i = 1, \ldots, K,\ j = 1, \ldots, N: && \tfrac{d}{dt} w_{ij} = -\tfrac{\partial}{\partial w_{ij}} E \end{aligned} \tag{2.44}$$

Although the equations one finds upon working out the derivatives in (2.44) are rather messy compared to the simple perceptron rule (2.37), the procedure (2.44) is guaranteed to decrease the value of the error E until a stationary state is reached, since from (2.44) we can deduce via the chain rule:

$$\frac{dE}{dt} = \sum_{i=1}^{K} \left[\frac{\partial E}{\partial w_i}\right] \left[\frac{dw_i}{dt}\right] + \sum_{i=1}^{K} \sum_{j=1}^{N} \left[\frac{\partial E}{\partial w_{ij}}\right] \left[\frac{dw_{ij}}{dt}\right]$$

$$= - \sum_{i=1}^{K} \left[\frac{dw_i}{dt}\right]^2 - \sum_{i=1}^{K} \sum_{j=1}^{N} \left[\frac{dw_{ij}}{dt}\right]^2 \leq 0$$

A stable stationary state is reached only when every small modification of the synaptic weights would lead to an increase in E, i.e. when we are at a *local minimum* of E. However, this local minimum need not be the desired *global minimum*; the blind mountaineer might find himself in some small valley high up in the mountains, rather than the one he is aiming for.

It will be clear that the principle behind the learning rule (2.44) can be applied to any feed-forward architecture, with arbitrary numbers of 'hidden' layers of arbitrary size, since it relies only on our being able to write down an explicit expression for the student's answers $S(\boldsymbol{x})$ in expression (2.43) for the error E. One just writes down equations of the form (2.44) for every parameter to be modified in the network. One can even generalise the construction to the case of multiple output variables (multiple answers):

[13] Note that this is an arbitrary choice; any monotonic function of $|T(\boldsymbol{x}) - S(\boldsymbol{x})|$ other than $|T(\boldsymbol{x}) - S(\boldsymbol{x})|^2$ would do.

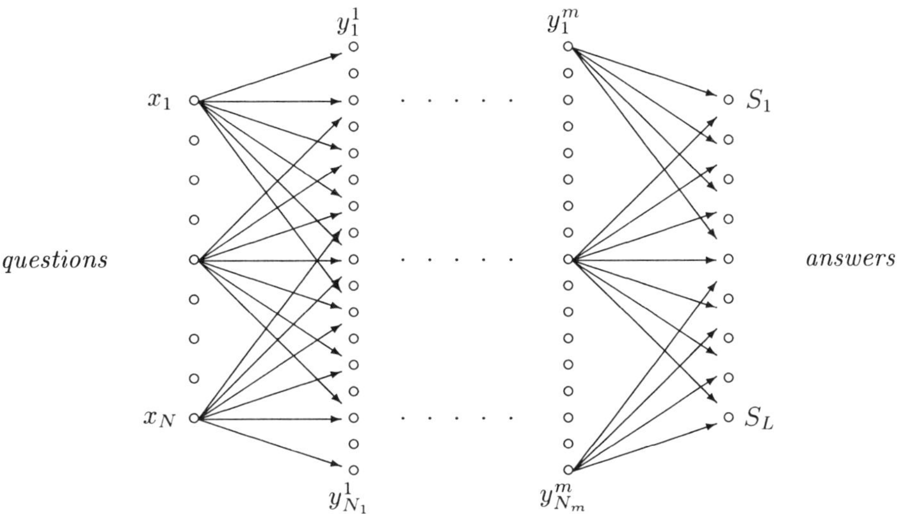

In this case there are L student answers $S_\ell(x_0, \ldots, x_N)$ and L teacher answers $T_\ell(x_0, \ldots, x_N)$ for each question $(x_0, \ldots, x_N)$, so that the error (2.43) to be minimised by gradient descent is to be generalised to:

$$E = \frac{1}{2} \sum_{\boldsymbol{x} \in \Omega} p(\boldsymbol{x}) \sum_{\ell=1}^{L} [T_\ell(\boldsymbol{x}) - S_\ell(\boldsymbol{x})]^2 \tag{2.45}$$

The strategy of tackling tasks which are not linearly separable (i.e. not executable by single perceptrons) using multi-layer networks of graded-response neurons which are being trained by gradient descent has several advantages, but also several drawbacks. Just to list a few:

+ the networks are in principle universal
+ we have a learning rule that minimises the student's error
+ the rule works for arbitrary numbers of layers and layer sizes

– we don't know the required network dimensions beforehand
– the rule cannot be implemented exactly (infinitely small modifications !)
– convergence is not guaranteed (we can end up in a local minimum of E)
– at each step we need to evaluate *all* student and teacher answers

The lack of a priori guidelines in choosing numbers of layers and layer sizes, and the need for discretisation and approximation of the differential equation (2.44) unfortunately generate quite a number of parameters to be tuned by hand. This problem can be solved only by having exact theories to describe the learning process; I will give a taste of recent analytical developments in a subsequent section.

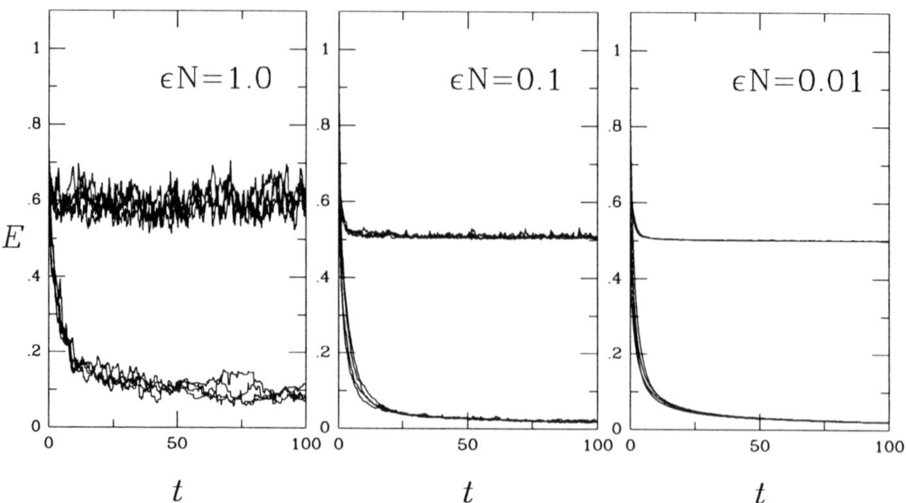

Figure 2.16: Evolution of the student's error E (2.43) in a two-layer feed-forward network with $N = 15$ input neurons and $K = 10$ 'hidden' neurons. The parameter ϵ gives the elementary time-steps used in the discretisation of the gradient descent learning rule. The upper graphs refer to task I (not linearly separably), the lower graphs to task II (which is linearly separable).

Figures 2.16 and 2.17 give an impression of the learning process described by the following discretisation/approximation of the original equation (2.44):

$$w_i(t+\epsilon) = w_i(t) - \epsilon \frac{\partial}{\partial w_i} E[\boldsymbol{x}(t)] \qquad w_{ij}(t+\epsilon) = w_{ij}(t) - \epsilon \frac{\partial}{\partial w_{ij}} E[\boldsymbol{x}(t)]$$

$$E[\boldsymbol{x}] = \frac{1}{2}[T(\boldsymbol{x}) - S(\boldsymbol{x})]^2$$

in a two-layer feed-forward network, with graded response neurons of the type (2.41), in which the non-linear function was choosen to be $f(z) = \tanh(z)$. Apart from having a discrete time, as opposed to the continuous one in (2.44), the second approximation made consists of replacing the *overall* error E (2.43) in (2.44) by the error $E[\boldsymbol{x}(t)]$ made in answering the current question $\boldsymbol{x}(t)$. The rationale is that for times $t \ll \epsilon^{-1}$ the above learning rule tends towards the original one (2.44) in the limit $\epsilon \to 0$. In the simulation examples the following two tasks T were considered:

$$\boldsymbol{x} \in \{-1, 1\}^N : \quad \begin{cases} \text{task I}: & T(\boldsymbol{x}) = \prod_{i=1}^N x_i \\ \text{task II}: & T(\boldsymbol{x}) = \text{sgn}[\boldsymbol{w}^\star \cdot \boldsymbol{x}] \end{cases} \tag{2.46}$$

with, in the case of task II, a randomly drawn teacher vector $\boldsymbol{w}^\star$. Task I is not linearly separable, task II is. At each iteration step each component x_i

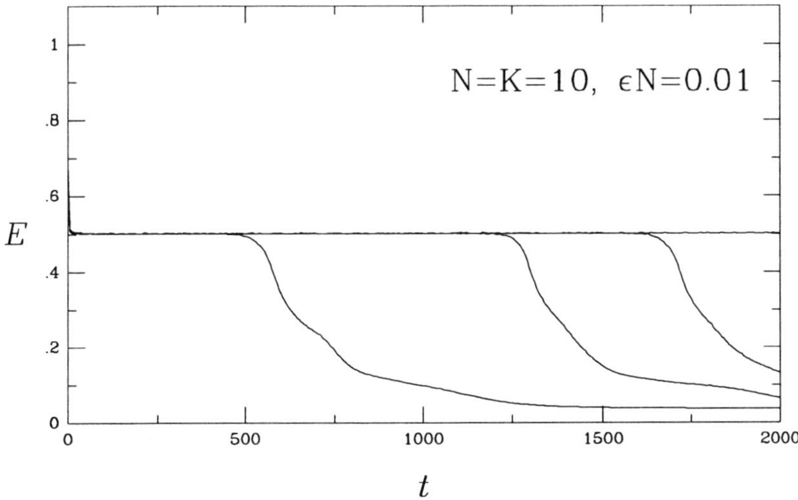

Figure 2.17: Evolution of the student's error E (2.43) in a two-layer feed-forward network with $N = 10$ input neurons and $K = 10$ 'hidden' neurons, being trained on task I (which is not linearly separably). The so-called 'plateau phase' is the intermediate (slow) stage in the learning process, where the error appears to have stabilised.

of the question $\boldsymbol{x}$ in the simulation experiments was drawn randomly from $\{-1, 1\}$. In spite of the fact that task I can be performed with a two-layer feed-forward network if $K \geq N$ (which can be demonstrated by construction), Figure 2.16 suggests that the learning procedure used does not converge to the desired configuration. This, however, is just a demonstration of one of the characteristic features of learning in multi-layer networks: the occurrence of so-called 'plateau phases'. These are phases in the learning process where the error appears to have stabilised (suggesting arrival at a local minimum), but where it is in in fact only going through an extremely slow intermediate stage. This is illustrated in Figure 2.17, where much larger observation times are choosen.

2.5.3 Calculating what is Achievable

There are several reasons why one would like to know beforehand for any given task T which is the minimal architecture necessary for a student network S to be able to 'learn' this task. Firstly, if we can get away with using a perceptron, as opposed to a multi-layer network, then we can use the perceptron learning rule, which is simpler and is guaranteed to converge. Secondly, the larger the number of layers and layer sizes, the larger the amount of computer time needed to carry out the training process. Thirdly, if the number of adjusted parameters is too large compared to the number actually required, the network might end up doing brute-force data-fitting which resembles the creation of a look-up table for the answers, rather than learning the underlying rule, leading to poor generalisation.

We imagine having a task (teacher), in the form of a rule T assigning an answer $T(\boldsymbol{x})$ to each question $\boldsymbol{x} = (x_0, \ldots, x_N)$, drawn from a set Ω with probability $p(\boldsymbol{x})$. We also have a student network with a given architecture, and a set of adjustable parameters $\boldsymbol{w} = (w_0, \ldots, w_N)$, assigning an answer $S(\boldsymbol{x}; \boldsymbol{w})$ to each question (the rule operated by the student depends on the adjustable paremeters $\boldsymbol{w}$, which we now emphasise in our notation). The answers could take discrete or continuous values. The degree to which a student with parameter values $\boldsymbol{w}$ has succesfully learned the rule operated by the teacher is measured by an error $E[\boldsymbol{w}]$, usually choosen to be of the form:

$$E[\boldsymbol{w}] = \frac{1}{2} \sum_{\boldsymbol{x} \in \Omega} p(\boldsymbol{x}) \left[T(\boldsymbol{x}) - S(\boldsymbol{x}; \boldsymbol{w})\right]^2 \tag{2.47}$$

What we would like to calculate is $E_0 = \min_{\boldsymbol{w} \in W} E[\boldsymbol{w}]$, since:

$$E_0 = \min_{\boldsymbol{w} \in W} E[\boldsymbol{w}] : \qquad \begin{cases} E_0 > 0: & \text{task not feasable} \\ E_0 = 0: & \text{task feasable} \end{cases} \tag{2.48}$$

W denotes the set of allowed values for $\boldsymbol{w}$. The freedom in choosing W allows us to address a wider range of feasibility questions; for instance, we might have constraints on the allowed values of $\boldsymbol{w}$ due to restrictions imposed by the (biological or electrical) hardware used. If $E_0 = 0$ we know that it is at least in principle possible for the present student architecture to arrive at a stage with zero error; if $E_0 > 0$, on the other hand, no learning process will ever lead to error-free performance. In the latter case, the actual magnitude of E_0 still contains valuable information, since allowing for a certain fraction of errors could be a price worth paying if it means a drastic reduction in the complexity of the architecture (and therefore in the amount of computing time).

Performing the minimisation in (2.48) explicitly is usually impossible, however, as long as we do not insist on knowing for which parameter setting(s) $\boldsymbol{w}^\star$ the minimum $E_0 = E[\boldsymbol{w}^\star]$ is actually obtained, we can use the following identity[14]:

$$E_0 = \lim_{\beta \to \infty} \frac{\int_W d\boldsymbol{w}\; E[\boldsymbol{w}] e^{-\beta E[\boldsymbol{w}]}}{\int_W d\boldsymbol{w}\; e^{-\beta E[\boldsymbol{w}]}} \tag{2.49}$$

The expression (2.49) can also be written in the following way, requiring us just to calculate a single integral:

$$E_0 = -\lim_{\beta \to \infty} \frac{\partial}{\partial \beta} \log \mathcal{Z}[\beta] \qquad \mathcal{Z}[\beta] = \int_W d\boldsymbol{w}\; e^{-\beta E[\boldsymbol{w}]} \tag{2.50}$$

[14]Strictly speaking this is no longer true if the minimum is obtained *only* for values of $\boldsymbol{w}$ in sets of measure zero. In practice this is not a serious restriction; in the latter case the system would be extremely sensitive to noise (numerical or otherwise) and thus of no practical use.

The integral in (2.50) need not (and usually will not) be trivial, but it can often be done. The results thus obtained can save us from running extensive (and expensive) computer simulations, only to find out the hard way that the architecture of the network at hand is too poor. Alternatively they can tell us what the minimal architecture is, given a task, and thus allow us to obtain networks with optimal generalisation properties.

Often we might wish to reduce our ambition further, in order to obtain exact statements. For instance, we could be interested in a certain *family* of tasks T, i.e. $T \in \mathcal{B}$, with $\mathcal{P}(T)$ denoting the probability of task T to be encountered, and try to find out about the feasibility of a generic task from this family, rather than the feasibility of each individual family member:

$$E[T;\boldsymbol{w}] = \frac{1}{2}\sum_{\boldsymbol{x}\in\Omega} p(\boldsymbol{x})\,[T(\boldsymbol{x}) - S(\boldsymbol{x};\boldsymbol{w})]^2 \tag{2.51}$$

$$\begin{aligned}\langle E_0[T]\rangle_{\mathcal{B}} &= \int_{\mathcal{B}} dT\ \mathcal{P}(T)\ E_0[T] \\ &= -\lim_{\beta\to\infty}\frac{\partial}{\partial\beta}\int_{\mathcal{B}} dT\ \mathcal{P}(T)\ \log\left\{\int_W d\boldsymbol{w}\ e^{-\beta E[T;\boldsymbol{w}]}\right\}\end{aligned} \tag{2.52}$$

In those cases where the original integral in (2.50) cannot be done analytically, one often finds that the quantity (2.52) can be calculated by doing the integration over the tasks before the integration over the students. Since (2.47) ensures $E_0 \geq 0$, we know that $\langle E_0[T]\rangle_{\mathcal{B}} = 0$ implies that if we randomly choose a task from the family $\mathcal{B}$, then the probability that a randomly drawn task from the family $\mathcal{B}$ is not feasible is zero.

Application of the above ideas to a single binary neuron leads to statements on which types of operations $T(x_1,\ldots,x_N)$ are linearly separable. For example, let us define a task by choosing at random p binary questions $(\xi_1^\mu,\ldots,\xi_N^\mu)$, with $\xi_i^\mu \in \{-1,1\}$ for all (i,μ) and $\mu = 1,\ldots,p$, and assign randomly an output value $T_\mu = T(\xi_1^\mu,\ldots,\xi_N^\mu) \in \{0,1\}$ to each. The synaptic strengths $\boldsymbol{w}$ of an error-free perceptron would then be the solution of the following problem:

$$\text{for all } \mu: \qquad \begin{cases} T_\mu = 1: & w_1\xi_1^\mu + \ldots + w_N\xi_N^\mu > 0 \\ T_\mu = 0: & w_1\xi_1^\mu + \ldots + w_N\xi_N^\mu < 0 \end{cases} \tag{2.53}$$

The larger p, the more complicated the task, and the smaller the set of solutions $\boldsymbol{w}$. For large N, the maximum number p of random questions that a perceptron can handle turns out to scale with N, i.e. $p = \alpha N$ for some $\alpha > 0$. Finding out for a specific task T, which is specified by the choice made for all question/answer pairs $\boldsymbol{\xi}^\mu = (\xi_1^\mu,\ldots,\xi_N^\mu;T_\mu)$, whether a solution of (2.53) exists, either directly or by calculating (2.50), is impossible (except for trivial and pathological cases). Our family $\mathcal{B}$ is the set of all such tasks T obtained by choosing different realisations of the question/answer pairs $(\xi_1^\mu,\ldots,\xi_N^\mu;T_\mu)$; there are 2^{p+1} possible choices of question/answer pairs, each equally likely.

The error (2.51) and the family-averaged mimimum error in (2.52) thus become:

$$E[T;\boldsymbol{w}] = \frac{1}{2p}\sum_{\mu=1}^{p}[T_\mu - S(\boldsymbol{\xi}^\mu;\boldsymbol{w})]^2 \tag{2.54}$$

$$\langle E_0[T]\rangle_{\mathcal{B}} = -\lim_{\beta\to\infty}\frac{\partial}{\partial\beta}\frac{1}{2^{p+1}}\sum_{\boldsymbol{\xi}^\mu}\log\left\{\int_W d\boldsymbol{w}\ e^{-\frac{\beta}{2}\sum_\mu[T_\mu-S(\boldsymbol{\xi}^\mu;\boldsymbol{w})]^2}\right\} \tag{2.55}$$

If one specifies the set W of allowed student vectors $\boldsymbol{w}$ by simply requiring $w_1^2+\ldots+w_N^2=1$, one finds that the average (2.55) can indeed be calculated[15], which results in the following statement on $\alpha = p/N$: for large N there exists a critical value α_c such that for $\alpha < \alpha_c$ the tasks in the family $\mathcal{B}$ are linearly separable, whereas for $\alpha > \alpha_c$ the tasks in $\mathcal{B}$ are not. The critical value turns out to be $\alpha_c = 2$.

We can play many interesting games with this procedure. Note that, since the associative memory networks of a previous section consist of binary neurons, this result also has immediate applications in terms of network storage capacities: for large networks the maximum number p_c of random patterns that can be stored in an associative memory network of binary neurons obeys $p_c/N \to 2$ for $N \to \infty$. We can also investigate the effect on the storage capacity of of the degree of symmetry of the synaptic strengths w_{ij}, which is measured by:

$$\eta(\boldsymbol{w}) = \frac{\sum_{ij} w_{ij}w_{ji}}{\sum_{ij} w_{ij}^2} \in [-1,1]$$

For $\eta(\boldsymbol{w}) = -1$ the synaptic matrix is anti-symmetric, i.e. $w_{ij} = -w_{ji}$ for all neuron pairs (i,j); for $\eta(\boldsymbol{w}) = 1$ it is symmetric, i.e. $w_{ij} = w_{ji}$ for all neuron pairs. If one now specifies in expression (2.55) the set W of allowed synaptic strengths by requiring both $w_1^2+\ldots+w_N^2 = 1$ and $\eta(\boldsymbol{w}) = \eta$ (for some fixed η), our calculation will give us the storage capacity as a function of the degree of symmetry: $\alpha_c(\eta)$. Somewhat unexpectedly, the optimal network turns out not to be symmetric:

$\eta = -1$:	fully anti−symmetric synapses,	$\alpha_c = \frac{1}{2}$
$\eta = 0$:	no symmetry preference,	$\alpha_c \sim 1.94$
$\eta = \frac{1}{\pi}$:	optimal synapses,	$\alpha_c = 2$
$\eta = 1$:	fully symmetric synapses,	$\alpha_c \sim 1.28$

[15]This involves a few technical subtleties which go beyond the scope of the present paper.

2.5.4 Solving the Dynamics of Learning for Perceptrons

As with our previous network classes, the associative memory networks and the networks responsible for creating topology conserving maps, we can for the present class of layered networks obtain analytical results on the dynamics of supervised learning, provided we restrict ourselves to (infinitely) large systems. In particular, we can find the system error as a function of time. Here I will only illustrate the route towards this result for perceptrons, and restrict myself to specific parameter limits. A similar approach can be followed for multi-layer systems; this is in fact one of the most active research areas at present.

Stage 1: define the rules

For simplicity we will not deal with thresholds, i.e. $\theta = 0$, and we will draw at each time-step t each bit $x_i(t)$ of the question $\boldsymbol{x}(t)$ asked at random from $\{-1, 1\}$. Furthermore, we will only consider the case where a perceptron S is being trained on a linearly separable (i.e. feasable) task, which means that the operation of the teacher T can itself be seen as that of a binary neuron, with synaptic strengths $\boldsymbol{w}^\star = (w_1^\star, \ldots, w_N^\star)$. Since the (constant) value of the length of the teacher vector $\boldsymbol{w}^\star$ has no effect on the process (2.37), we are free to choose the simplest normalisation $|\boldsymbol{w}^\star|^2 = w_1^{\star 2}+\ldots+w_N^{\star 2} = 1$.

In the original perceptron learning rule we can introduce a so-called learning rate $\epsilon > 0$, which defines the magnitude of the elementary modifications of the student's synaptic strengths $\boldsymbol{w} = (w_1, \ldots, w_N)$, by rescaling the modification term in equation (2.37). This does not affect the convergence proof. It just gauges the time-scale of the learning process, so we choose this ϵ to define the duration of individual iteration steps. For realisable tasks the teacher's answer to a question $\boldsymbol{x} = (x_1, \ldots, x_N)$ depends only on the sign of $\boldsymbol{w}^\star \cdot \boldsymbol{x} = w_1^\star x_1+\ldots+w_N x_N$. Upon combining our ingredients we can replace the learning rule (2.37) by the following expression:

$$\boldsymbol{w}(t+\epsilon) = \boldsymbol{w}(t) + \frac{1}{2}\epsilon\, \boldsymbol{x}(t) \left\{ \text{sgn}[\boldsymbol{w}^\star \cdot \boldsymbol{x}(t)] - \text{sgn}[\boldsymbol{w}(t) \cdot \boldsymbol{x}(t)] \right\} \tag{2.56}$$

with $\text{sgn}[z > 0] = 1$, $\text{sgn}[z < 0] = -1$ and $\text{sgn}[0] = 0$. If we now consider very small learning rates $\epsilon \to 0$, the following two pleasant simplifications occur[16]: (*i*) the discrete-time iterative map (2.56) will be replaced by a continuous-time differential equation, and (*ii*) the right-hand side of (2.56) will be converted into an expression involving only *averages* over the distribution of questions, to be denoted by $\langle \ldots \rangle_{\boldsymbol{x}}$:

$$\frac{d}{dt}\boldsymbol{w} = \frac{1}{2}\langle \boldsymbol{x} \left\{ \text{sgn}[\boldsymbol{w}^\star \cdot \boldsymbol{x}] - \text{sgn}[\boldsymbol{w} \cdot \boldsymbol{x}] \right\} \rangle_{\boldsymbol{x}} \tag{2.57}$$

[16] Note that this is the same procedure we followed to analyse the creation of topology conserving maps, in a previous section.

Stage 2: find the relevant macroscopic features

The next stage, as always, is to decide which are the quantities we set out to calculate. For the perceptron there is a clear guide in finding the relevant macroscopic features, namely the perceptron convergence proof. In this proof the following two observables played a key role:

$$J = \sqrt{w_1^2 + \ldots w_N^2} \qquad \omega = \frac{w_1 w_1^\star + \ldots w_N w_N^\star}{\sqrt{w_1^2 + \ldots w_N^2}} \tag{2.58}$$

Last but not least one would like to know at any time the accuracy with which the student has learned the task, as measured by the error E:

$$E = \frac{1}{2}\langle \left\{ 1 - \text{sgn}[(\boldsymbol{w}^\star \cdot \boldsymbol{x})(\boldsymbol{w} \cdot \boldsymbol{x})] \right\} \rangle_{\boldsymbol{x}} \tag{2.59}$$

which simply gives the fraction of questions which is wrongly answered by the student. We can use equation (2.57) to derive a differential equation describing the evolution in time of the two observables (2.58), which after some rearranging can be written in the form:

$$\frac{d}{dt} J = - \int_0^\infty \int_0^\infty dydz \; z \, [P(y,-z) + P(-y, z)] \tag{2.60}$$

$$\frac{d}{dt} \omega = \frac{1}{J} \int_0^\infty \int_0^\infty dydz \; [y + \omega z] \, [P(y,-z) + P(-y, z)] \tag{2.61}$$

in which the details of the student and teacher vectors $\boldsymbol{w}$ and $\boldsymbol{w}^\star$ enter only through the probability distribution $P(y, z)$ for the two local field sums:

$$y = w_1^\star x_1 + \ldots + w_N^\star x_N \qquad z = \frac{1}{J}(w_1 x_1 + \ldots + w_N x_N)$$

Similarly we can write the student's error E in (2.59) as:

$$E = \int_0^\infty \int_0^\infty dydz \; [P(y,-z) + P(-y, z)] \tag{2.62}$$

Note that the way everything has been defined so far guarantees a more or less smooth limit $N \to \infty$ when we eventually go to large systems, since with our present choice of question statistics we find for any N:

$$\int dydz \; P(y, z) y = \int dydz \; P(y, z) z = 0 \tag{2.63}$$

$$\int dydz \; P(y, z) y^2 = \langle [\boldsymbol{w}^\star \cdot \boldsymbol{x}]^2 \rangle_{\boldsymbol{x}} = \boldsymbol{w}^{\star 2} = 1 \tag{2.64}$$

$$\int dydz \; P(y, z) z^2 = J^{-2} \langle [\boldsymbol{w} \cdot \boldsymbol{x}]^2 \rangle_{\boldsymbol{x}} = J^{-2} \boldsymbol{w}^2 = 1 \tag{2.65}$$

$$\int dydz \; P(y, z) yz = J^{-1} \langle [\boldsymbol{w}^\star \cdot \boldsymbol{x}][\boldsymbol{w} \cdot \boldsymbol{x}] \rangle_{\boldsymbol{x}} = J^{-1} \boldsymbol{w}^\star \cdot \boldsymbol{w} = \omega \tag{2.66}$$

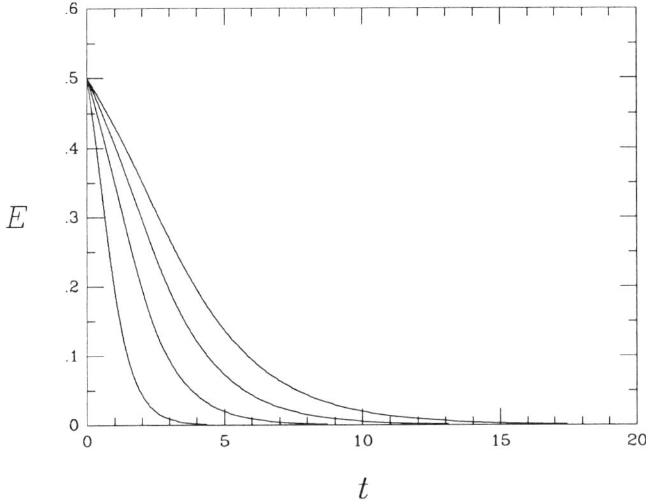

Figure 2.18: Evolution in time of the student's error E in an infinitely large perceptron in the limit of an infinitesimally small learning rate, according to (2.70). The four curves correspond to four random initialisations for the student vector $\boldsymbol{w}$, with lengths $J(0) = 2, \frac{3}{2}, 1, \frac{1}{2}$ (from top to bottom).

Stage 3: calculate $P(y, z)$ in the limit $N \to \infty$

For finite systems the shape of the distribution $P(y, z)$ depends in some complicated way on the details of the student and teacher vectors $\boldsymbol{w}$ and $\boldsymbol{w}^\star$, although the moments (2.63-2.66) will for any size N depend on the observable ω only. For large perceptrons ($N \to \infty$), however, a drastic simplification occurs: due to the statistical independence of our question components $x_i \in \{-1, 1\}$ the central limit theorem applies, which guarantees that the distribution $P(y, z)$ will become Gaussian[17]. This implies that it is characterised only by the moments (2.63-2.66), and therefore depends on the vectors $\boldsymbol{w}$ and $\boldsymbol{w}^\star$ *only* through the observable ω:

$$P(y, z) = \frac{1}{2\pi\sqrt{1-\omega^2}}\, e^{-\frac{1}{2}(y^2+z^2-2\omega yz)/(1-\omega^2)}$$

As a result the right-hand sides of both dynamic equations (2.60,2.61) as well as the error (2.62) can be expressed solely in terms of the two key observables J and ω. We can apparently forget about the details of $\boldsymbol{w}$ and $\boldsymbol{w}^\star$; the whole process can be described at a macroscopic level. Furthermore, due to the Gaussian shape of $P(y, z)$ one can even perform all remaining integrals analytically!

[17]Strictly speaking, for this to be true certain conditions on the vectors $\boldsymbol{w}$ and $\boldsymbol{w}^\star$ will have to be fulfilled, in order to guarantee that the random variables y and z are not effectively dominated by just a small number of their components.

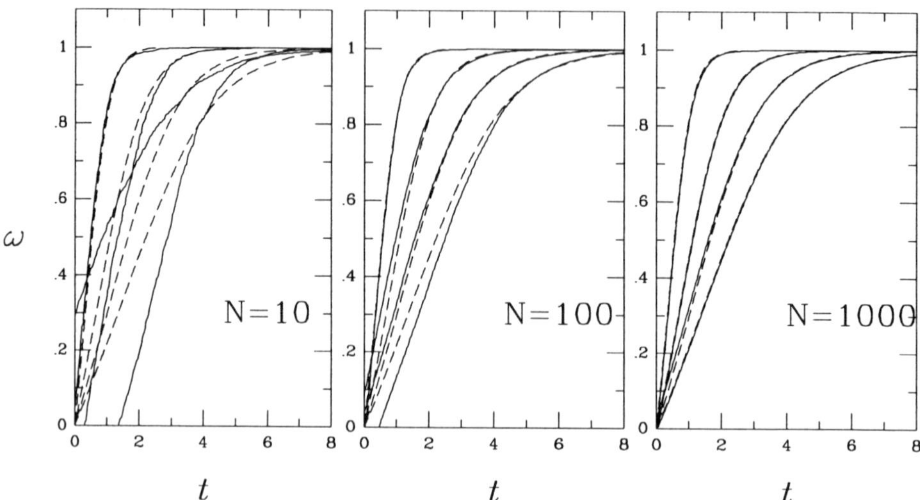

Figure 2.19: Evolution of the observable ω in a perceptron with learning rate $\epsilon = 0.01/N$ for various sizes N and a randomly drawn teacher vector $\boldsymbol{w}^{\star}$. In each picture the solid lines correspond to numerical simulations of the perceptron learning rule (for different values of the length $J(0)$ of the initial student vector $\boldsymbol{w}(0)$), whereas the dashed lines correspond to the theoretical predictions (2.69) for infinitely large perceptrons ($N \to \infty$).

Our main target, the student's error (2.59) turns out to become:

$$E = \frac{1}{\pi}\arccos(\omega) \tag{2.67}$$

whereas the dynamic equations (2.60,2.61) reduce to:

$$\frac{dJ}{dt} = -\frac{1-\omega}{\sqrt{2\pi}} \qquad \frac{d\omega}{dt} = \frac{1-\omega^2}{J\sqrt{2\pi}} \tag{2.68}$$

Stage 4: solve the remaining differential equations

In general one has to resort to numerical analysis of the macroscopic dynamic equations at this stage. For the present example, however, the dynamic equations (2.68) can actually be solved analytically, due to the fact that (2.68) describes evolution with a conserved quantity, namely the product $J(1+\omega)$ (as can be verified by substitution). This property allows us to eliminate the observable J altogether, and reduce (2.68) to just a single differential equation for ω only. This equation, in turn, can be solved. For initial conditions corresponding to a randomly chosen student vector $\boldsymbol{w}(0)$ with a given length $J(0)$, the solution takes its easiest form by writing t as a function of ω:

$$t = J(0)\sqrt{\frac{\pi}{2}}\left\{\log\left[\frac{1+\omega}{1-\omega}\right]^{\frac{1}{2}} + \frac{\omega}{1+\omega}\right\} \tag{2.69}$$

In terms of the error E, related to ω through (2.67), this result becomes:

$$t = J(0)\sqrt{\frac{\pi}{8}}\left\{1-\tan^2(\frac{1}{2}\pi E)-2\log\tan(\frac{1}{2}\pi E)\right\} \tag{2.70}$$

Examples of curves described by this equation are shown in Figure 2.18. What more can one ask for ? For any required student performance, relation (2.70) tells us exactly how long the student needs to be trained. Figure 2.19 illustrates how the learning process in finite perceptrons gradually approaches that described by our $N \to \infty$ theory, as the perceptron's size N increases.

2.6 Puzzling Mathematics

The models and model solutions described so far were reasonably simple. The mathematical tools involved where mostly quite standard and clear. Let us now open Pandora's box and see what happens if we move away from the nice and solvable region of the space of neural network models.

Most traditional models of systems of interacting elements (whether physical or otherwise) tend to be quite regular and 'clean'; the elements usually interact with one another in a more or less similar way and there is often a nice lattice-type translation invariance[18]. In the last few decennia, however, attention in science is moving away from these nice and clean systems to the 'messy ones', where there is no apparent spatial regularity and where at a microscopic level all interacting elements appear to operate different rules. The latter types, also called 'complex systems', play an increasingly important role in physics (glasses, plastics, spin-glasses), computer science (cellular automata) economics and trading (exchange rate and derivative markets) and biology (neural networks, ecological systems, genetic systems). One finds that such systems have much in common: firstly, many of the more familiar mathematical tools to describe interacting particle systems (usually based on microscopic regularity) no longer apply, and secondly, in analysing these systems one is very often led to so-called 'replica theories'.

2.6.1 Complexity due to Frustration, Disorder and Plasticity

I will now briefly discuss the basic mechanisms causing neural network models (and other related models of complex systems) to be structurally different from the more traditional models in the physical sciences. Let us return to the

[18] i.e. the systems looks exactly the same, even microscopically, if we shift our microscope in any direction over any distance.

relatively simple recurrent networks of binary neurons as studied in section 3, with the neural inputs:

$$input_i = w_{i1}S_1 + \ldots + w_{iN}S_N$$

Here excitatory interactions $w_{ij} > 0$ tend to promote configurations with $S_i = S_j$, whereas inhibitory interactions $w_{ij} < 0$ tend to promote configurations with $S_i \neq S_j$. A so-called 'unfrustrated' system is one where there exist configurations $\{S_1, \ldots, S_N\}$ with the property that each pair of neurons can realise its most favourable configuration, i.e. where:

$$\text{for all } (i,j): \quad \begin{cases} S_i = S_j & \text{if } w_{ij} > 0 \\ S_i \neq S_j & \text{if } w_{ij} < 0 \end{cases} \tag{2.71}$$

In a frustrated system, on the other hand, no configuration $\{S_1, \ldots, S_N\}$ exists for which (2.71) is true.

Let us consider at first only recurrent networks with symmetric interactions, i.e. $w_{ij} = w_{ji}$ for all (i,j) and without self-interactions (i.e. $w_{ii} = 0$ for all i). Depending on the actual choice made for the synaptic strengths such networks can be either frustrated or unfrustrated (see Figure 2.20).

In a frustrated network compromises will have to be made; some of the neuron pairs will have to accept that for them the goal in (2.71) cannot be achieved. However, since there are often many different compromises possible, with the same degree of frustration, frustrated systems usually have a large number of stable or meta-stable states, which generate non-trivial dynamics. Due to the symmetry $w_{ij} = w_{ji}$ of the synaptic strengths it takes at least three neurons to have frustration. In the examples of Figure 2.20 the neurons interact only with their nearest neighbours; in neural systems with a high connectivity and where there is a degree of randomness (or disorder) in the microscopic arrangement of all synaptic strengths, frustration will play an even more important role. The above situation is also encountered in certain complex physical systems like glasses and spin-glasses, and gives rise to relaxation times[19] which are measured in years rather than minutes (for example: ordinary window glass is in fact a liquid, which takes literally ages to relax towards its crystalline and non-transparent stationary state).

In non-symmetric networks, where $w_{ij} \neq w_{ji}$ is allowed, or in networks with self-interactions, where $w_{ii} \neq 0$ is allowed, the situation is even worse. In the case of non-symmetric interactions it takes just two neurons to have frustration; in the case of self-interactions even single neurons can be frustrated (see Figure 2.21). In the terminology associated with the theory of stochastic processes we would find that non-symmetric networks and networks with self-interacting neurons fail to have the property of 'detailed balance'. This implies that they will never evolve towards a microscopic equilibrium state (although the values of certain macroscopic observables might become stationary), and

[19]The relaxation time is the time needed for a system to relax towards its equilibrium state, where the values of all relevant macroscopic observables have become stationary.

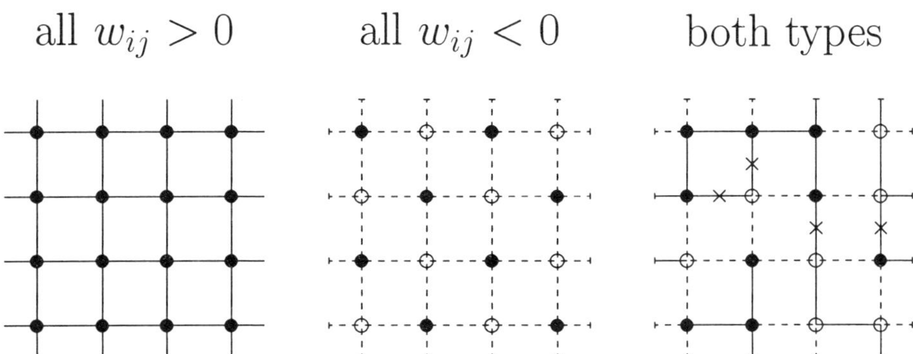

Figure 2.20: Frustration in symmetric networks. Each neuron is for simplicity assumed to interact with its four nearest neighbours only. Neuron states are drawn either as $\bullet$ (denoting $S_i = 1$) or as $\circ$ (denoting $S_i = 0$). Excitatory synapses $w_{ij} > 0$ are drawn as solid lines, inhibitory synapses $w_{ij} < 0$ as dashed lines. An unfrustrated configuration $\{S_1, \ldots, S_N\}$ now corresponds to a way of colouring the vertices such that solid lines connect identically coloured vertices ($\bullet\bullet$ or $\circ\circ$) whereas dashed lines connect differently coloured vertices ($\bullet\circ$ or $\circ\bullet$). In the left diagram (all synapses excitatory) and the middle diagram (all synapses inhibitory) the network is unfrustrated. The right diagram shows an example of a frustrated network: here there is no way to colour the vertices such that an unfrustrated configuration is achieved (in the present state the four 'frustrated' pairs are indicated with $\times$).

that consequently they will often show a remarkably rich phenomenology of dynamic behaviour. This situation never occurs in physical systems (although it does happen in cellular automata and ecological, genetic and economical systems), which, in contrast, always evolve towards an equilibrium state. As a result, many of the mathematical tools and much of the scientific intuition developed for studying interacting particle systems are based on the 'detailed balance' property, and therefore no longer apply.

Finally, and this is perhaps the most serious complication of all, the parameters in a real neural system, the synaptic strengths w_{ij} and the neural thresholds θ_i are not constants, but they evolve in time (albeit slowly) according to dynamical laws which, in turn, involve the states of the neurons and the values of the post-synaptic potentials (or inputs). The problem we face in trying to model and analyse this situation is equivalent to that of predicting what a computer will do when running a program that is continually being rewritten by the computer itself. A physical analogy would be that of a system of interacting molecules, where the formulae giving the strengths of the inter-molecular forces changed all the time, in a way that depends on the actual instantaneous

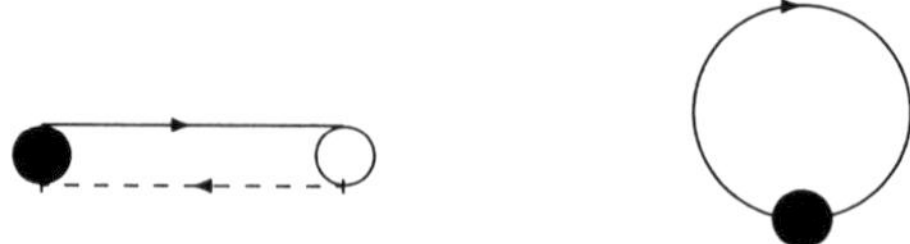

Figure 2.21: Frustration in non-symmetric networks and networks with self-interactions. Neuron states are drawn as • (denoting $S_i = 1$) or as ∘ (denoting $S_i = 0$). Solid lines denote excitatory synapses $w_{ij} > 0$, dashed lines denote inhibitory synapses $w_{ij} < 0$. An unfrustrated configuration corresponds to a way of colouring the vertices such that solid lines connect identically coloured vertices (•• or ∘∘) whereas dashed lines connect differently coloured vertices (•∘ or ∘•). The left diagram shows two interacting neurons S_1 and S_2; as soon as $w_{12} > 0$ and $w_{21} < 0$ this simple system is already frustrated. The right diagram shows a single self-interacting neuron. If the self-interaction is excitatory there is no problem, but if it is inhibitory we always have a frustrated state.

positions of the molecules. It will be clear why in all model examples discussed so far either the neuron states were the relevant dynamic quantities, or the synaptic strengths and thresholds; but never both at the same time.

In the models discussed so far we have taken care to steer away from the specific technical problems associated with frustration, disorder and simultaneously dynamic neurons and synapses. In the case of the associative memory models of section 3 this was achieved by restricting ourselves to situations where the number of patterns stored p was vanishingly small compared to the number of neurons N; the situation changes drastically if we try to analyse associative memories operating at $p = \alpha N$, with $\alpha > 0$. In the case of the topology conserving maps we will face the complexity problems as soon as the set of 'training examples' of input signals is no longer infinitely large, but finite (which is more realistic). In the case of the layered networks learning a rule we find that the naive analytical approach described so far breaks down (i) when we try to analyse multi-layer networks in which the number of neurons K in the 'hidden' layer is proportional to the number of input neurons N, and (ii) when we consider the case of having a restricted set of training examples. Although all these problems at first sight appear to have little in common, at a technical level they are quite similar. It turns out that all our analytical attempts in dealing with frustrated and disordered systems and systems with simultaneously dynamic neurons and synapses lead us to so-called replica theories.

2.6.2 The World of Replica Theory

It is beyond the scope of this chapter to explain replica theory in detail. In fact most researchers would agree that it is as yet only partly understood. Mathematicians are often quite hesitant in using replica theory because of this lack of understanding (and tend to call it 'replica trick'), whereas theoretical physicists are less scrupulous in applying such methods and are used to doing calculations with non-integer dimensions, imaginary times etc. (they call it 'replica method' or 'replica theory'). However, although it is still controversial, all agree that replica theory works.

The simplest route into the world of replica theory starts with the following representation of the logarithm:

$$\log z = \lim_{n\to 0} \frac{1}{n}\{z^n - 1\}$$

For systems with some sort of disorder (representing randomness in the choice of patterns in associative memories, or in the choice of input examples in layered networks, etc.) we usually find in calculating the observables of interest that we end up having to perform an average over a logarithm of an integral (or sum), which we can tackle using this representation:

$$\int dx\ p(x) \log \int dy\ z(x,y) = \lim_{n\to 0} \frac{1}{n}\left\{\int dx\ p(x)\left[\int dy\ z(x,y)\right]^n - 1\right\}$$

$$= \lim_{n\to 0} \frac{1}{n}\left\{\int dy_1 \cdots dy_n \int dx\ p(x) z(x,y_1)\cdots z(x,y_n) - 1\right]$$

The variable x represents the disorder, with $\int dx\ p(x) = 1$. We have managed to replace an average of a logarithm of a quantity (which is usually nasty) by the average of integer powers of this quantity. The last step, however, involved making the crucial replacement $[\int dy\ z(x,y)]^n \to \int dy_1 z(x,y_1)\cdots\int dy_n z(x,y_n)$. This is in fact only allowed for *integer* values of n, whereas we must take the limit $n \to 0$! Another route (which turns out to be equivalent to the previous one) starts with calculating averages:

$$\int dx\ p(x)\left\{\frac{\int dy\ f(x,y)z(x,y)}{\int dy\ z(x,y)}\right\} =$$

$$\int dx\ p(x)\left\{\frac{\int dy\ f(x,y)z(x,y)[\int dy\ z(x,y)]^{n-1}}{[\int dy\ z(x,y)]^n}\right\}$$

by taking the limit $n \to 0$ in both sides we get:

$$\int dx\ p(x)\left\{\frac{\int dy\ f(x,y)z(x,y)}{\int dy\ z(x,y)}\right\} =$$

$$\lim_{n\to 0} \int dy_1 \cdots dy_n \int dx\ p(x) f(x,y_1) z(x,y_1)\cdots z(x,y_n)$$

Now we appear to have succeeded in replacing the average of the ratio of two quantities (which is usually nasty) by the average of powers of these quantities, by performing a manipulation which is allowed only for integer values of n, followed by taking the thereby forbidden limit $n \to 0$[20].

If for now we just follow the route further, without as yet worrying too much about the steps we have taken so far, and apply the above identities to our calculations, we find in the limit of infinitely large systems and after a modest amount of algebra a problem of the following form. We have to calculate:

$$f = \lim_{n \to 0} \text{extr } \mathcal{F}[\boldsymbol{q}] \tag{2.72}$$

where $\boldsymbol{q}$ represents an $n \times n$ matrix with zero diagonal elements,

$$\boldsymbol{q} = \begin{pmatrix} 0 & q_{1,2} & \cdots & q_{1,n-1} & q_{1,n} \\ q_{2,1} & 0 & & & q_{2,n} \\ \vdots & & \ddots & & \vdots \\ q_{n-1,1} & & & 0 & q_{n-1,n} \\ q_{n,1} & q_{n,2} & \cdots & q_{n,n-1} & 0 \end{pmatrix}$$

$\mathcal{F}[\boldsymbol{q}]$ is some scalar function of $\boldsymbol{q}$, and the extremum is defined as the value of $\mathcal{F}[\boldsymbol{q}]$ in the saddle point which for $n \geq 1$ minimises $\mathcal{F}[\boldsymbol{q}]$[21]. This implies that for any given value of n we have to find the critical values of $n(n-1)$ quantities $q_{\alpha\beta}$ (the non-diagonal elements of the matrix $\boldsymbol{q}$). There would be no problem with this procedure if n were to be an integer, but here we have to take the limit $n \to 0$. This means, firstly, that the number $n(n-1)$ of variables $q_{\alpha\beta}$, i.e. the dimension of the space in which our extremisation problem is defined, will no longer be integer (which is somewhat strange, but not entirely uncommon). But secondly, the number of variables becomes *negative* as soon as $n < 1$! In other words, we will be exploring a space with a negative dimension. In such spaces life is quite different from what we are used to. For instance, let us calculate the sum of squares, as in $\sum_{\alpha\neq\beta=1}^{n} q_{\alpha\beta}^2$, which for integer $n \geq 1$ is always non-negative. For $n < 1$ this is no longer true. Just consider the example where $q_{\alpha\beta} = q \neq 0$ for all $\alpha \neq \beta$, which gives:

$$\sum_{\alpha\neq\beta=1}^{n} q_{\alpha\beta}^2 = q^2 n(n-1) < 0 \ !$$

To quote one of the founding fathers of replica theory: 'The whole program seems to be completely crazy'. Crazy or not, if we simply persist and perform all calculations required, accepting the rather strange objects we find along the way as they are, we end up with results which are, as far as the available evidence allows us to conclude, essentially correct.

[20] The name 'replica theory' refers to the fact that the resulting expressions, with their n-fold integrations over the variables y_α (with $\alpha = 1, \ldots, n$), are quite similar to what one would get if one were to study n identical copies (or replicas) of the original system.

[21] This version of the saddle-point problem is just the simplest one; in most calculations one finds several additional $n \times n$ matrices and n-vectors to be varied.

The key to success is not to try to calculate individual matrix elements $q_{\alpha\beta}$, but to concentrate wherever possible in the calculation on quantities that (at least formally) have a well-defined $n \to 0$ limit, such as $P(q)$, which is defined as the relative frequency with which the value $q_{\alpha\beta} = q$ occurs among the non-diagonal entries of the matrix $\boldsymbol{q}$. Since $q \in \Re$ this function becomes a probability density:

$$P(q)dq = \frac{\text{number of entries with } q - \frac{1}{2}dq < q_{\alpha\beta} < q + \frac{1}{2}dq}{\text{total number of entries}}$$

with $0 < dq \ll 1$. This quantity remains well-behaved. It will obey the relations $P(q) \geq 0$ and $\int dq\ P(q) = 1$, whatever the value of the dimension n; for $n < 1$ the minus sign generated by the denominator will be cancelled by a similar minus sign in the numerator. The problem in (2.72) of finding a saddle-point $\boldsymbol{q}$ can be translated into one which is formulated in terms of the $n \to 0$ limit of the function $P(q)$ only:

$$f = \text{extr } \mathcal{G}[\{P(q)\}] \tag{2.73}$$

Although at a technical level non-trivial, compared to (2.72) the problem (2.73) is conceptually quite sane; now one has to explore the (infinite-dimensional) space of all probability distributions, as opposed to a space with a negative dimension. Later it was discovered that the function $P(q)$ for which the extremum in (2.73) occurs can be interpreted in terms of the average probability of two identical copies A and B of the original system to be in a state with a given mutual overlap defined by q. For instance, for the associative memory networks of section 3, storing $p = \alpha N$ patterns (with $\alpha > 0$) we would find:

$$P(q)dq = \text{average probability of } q - \frac{1}{2}dq < \frac{1}{N}\sum_{i=1}^{N} S_i^A S_i^B < q + \frac{1}{2}dq \tag{2.74}$$

whereas for the perceptron feasability calculations of section 5.3 we would find:

$$P(q)dq = \text{average probability of } q - \frac{1}{2}dq < \sum_{i=1}^{N} \frac{w_i^A w_i^B}{|\boldsymbol{w}^A||\boldsymbol{w}^B|} < q + \frac{1}{2}dq \tag{2.75}$$

with $0 < dq \ll 1$. This leads to a convenient characterisation of complex systems. For non-complex systems, with just a few stable/metastable states, the quantity $P(q)$ would be just the sum of a small number of isolated peaks. On the other hand, as soon as our calculation generates a solution $P(q)$ with continuous pieces, it follows from (2.74,2.75) that the underlying system must have a huge number of stable/metastable states, which is the fingerprint of complexity.

Finally, even if we analyse certain classes of neural network models in which both neuron states and synaptic strengths evolve in time, described by coupled equations but with synapses changing on a much larger time-scale than the neuron states, we find a replica theory. This in spite of the fact that there is no disorder, no patterns have been stored, the network is just left to 'program' its synapses autonomously. However, in these calculations the parameter n in (2.72) (the replica dimension) does not necessarily go to zero, but turns out to be given by the ratio of the degrees of randomness (noise levels) in the two dynamic processes (neuronal dynamics and synaptic dynamics).

It appears that replica theory in a way constitutes the natural description of complex systems. Furthermore, replica theory clearly works, although we do not yet know why. This, I believe, is just a matter of time.

2.7 Further Reading

Since this chapter is just the result of a modest attempt to give a taste of the mathematical modelling and analysis of problems in neural network theory, by means of a biased selection of some characteristic solvable problems, and without distracting references, much has been left out. In this final section I want to try to remedy this situation, by briefly discussing research directions that have not been mentioned so far, and by giving references. Since I imagine the typical reader to be the novice, rather than the expert, I will only give references to textbooks and review papers (these will then hopefully serve as the entrance to more specialised research literature). Note, however, that due to the interdisciplinary nature of this subject, and the inherent fracturisation into sub-fields, each with their own preferred library of textbooks and papers, it is practically impossible to find textbooks with sketch a truly broad and impartial overview.

There are by now many books to serve as introductions to the field of neural computing, such as [1, 3, 2, 4, 5]. Most give a nice overview of the standard wisdom around the time of their appearance[22], with differences in emphasis depending on the background disciplines of the authors (mostly physics and engineering). A taste of the history of this field can be provided by one of the volumes with reprints of original articles, such as [6] (with a stronger emphasis on biology/psychology), and the influential book [7]. More specialised and/or advanced textbooks on associated memories and topology conserving maps are [8, 9, 10, 11]. Examples of books with review papers on more advanced topics in the analysis of neural network models are the trio [12, 13, 14], as well as [15]. A book containing review chapters and reprints of original articles, specifically on replica theory, is [16].

One of the subjects that I did not go into very much concerns the more accurate modelling of neurobiology. Many properties of neuronal and synaptic operation have been eliminated in order to arrive at simple models, such as Dale's law (the property that a neuron can have only one type of synapse attached to the branches of its axon; either exitatory ones or inhibitory ones, but never both), neuromodulators, transmission delays, genetic pre-structuring

[22] This could be somewhat critical: for instance, most of the models and solutions described in this chapter go back no further than around 1990, the analysis of the dynamics of on-line learning in perceptrons is even younger.

of brain regions, diffusive chemical messengers, etc. A lot of effort is presently being put into trying to take more of these biological ingredients into account in mathematical models, see for example [13]. More general references to such studies can be found by using the voluminous [17] as a starting point.

A second sub-field entirely missing in this chapter concerns the application of theoretical tools from the fields of computer science, information theory and applied statistics, in order to quantify the information processing properties of neural networks. Being able to quantify the information processed by neural systems allows for the efficient design of new learning rules, and for making comparisons with the more traditional information processing procedures. Here a couple of suitable introductions could be the textbooks [18] and [19], and the review paper [20], respectively.

Finally, there are an increasing number of applications of neural networks, or systems inspired by the operation of neural networks, in engineering. The aim here is to exploit the fact that neural information processing strategies are often complementary to the more traditional rule-based problem-solving algorithms. Examples of such applications can be found in books like [21, 22, 23].

2.8 Bibliography

[1] D. Amit, *Modeling Brain Function*, Cambridge U.P., 1989

[2] B. Müller and J. Reinhardt, *Neural Networks, an Introduction*, Springer (Berlin), 1990

[3] J. Hertz, A. Krogh and R.G. Palmer, *Introduction to the Theory of Neural Computation*, Addison-Wesley (Redwood City), 1991

[4] P. Peretto, *An Introduction to the Modeling of Neural Networks*, Cambridge U.P. 1992

[5] S. Haykin, *Neural Networks, A Comprehensive Foundation*, Macmillan (New York), 1994

[6] J.A. Anderson and E. Rosenfeld (eds.), *Neurocomputing, Foundations of Research*, MIT Press (Cambridge Mass.), 1988

[7] M.L. Minsky and S.A. Papert, *Perceptrons*, MIT Press (Cambridge Mass.), 1969

[8] T. Kohonen, *Self-organization and Associative Memory*, Springer (Berlin), 1984

[9] Y. Kamp and M. Hasler, *Recursive Neural Networks for Associative Memory*, Wiley (Chichester), 1990

[10] H. Ritter, T. Martinetz and K. Schulten, *Neural Computation and Self-organizing Maps*, Addison-Wesley (Reading Mass.), 1992

[11] T. Kohonen, *Self-organizing Maps*, Springer (Berlin), 1995

[12] E. Domany, J.L. van Hemmen and K.Schulten (eds.), *Models of Neural Networks I*, Springer (Berlin), 1991

[13] E. Domany, J.L. van Hemmen and K.Schulten (eds.), *Models of Neural Networks II*, Springer (Berlin), 1994

[14] E. Domany, J.L. van Hemmen and K.Schulten (eds.), *Models of Neural Networks III*, Springer (Berlin), 1995

[15] J.G. Taylor, *Mathematical Approaches to Neural Networks*, North-Holland (Amsterdam), 1993

[16] M. Mezard, G. Parisi and M.A. Virasoro, *Spin-Glass Theory and Beyond*, World-Scientific (Singapore), 1987

[17] M.A. Arbib (ed.), *Handbook of Brain Theory and Neural Networks*, MIT Press (Cambridge Mass.), 1995

[18] M. Anthony and N. Biggs, *Computational Learning Theory*, Cambridge U.P., 1992

[19] G. Deco and D. Obradovic, *An Information-theoretic Approach to Neural Computing*, Springer (New-York), 1996

[20] D.J.C. MacKay, *Probably Networks and Plausible Predictions - a Review of Practical Bayesian Methods for Supervised Neural Networks*, Network **6**, 1995, p. 469

[21] A.F. Murray (ed.), *Applications of Neural Networks*, Kluwer (Dordrecht), 1995

[22] C.M. Bishop, *Neural Networks for Pattern Recognition*, Oxford U.P., 1995

[23] G.W. Irwin, K. Warwick and K.J. Hunt (eds.), *Neural Network Applications in Control*, IEE London, 1995

Chapter 3

Neurobiological Modelling

3.1 Introduction

For a considerable period of time, and indeed even until the last few years, there has been hostility expressed by neuroscientists against the use of theorising in their subject. Even today it is as if, for some, neuroscience is still only at the stage of gathering data, which should be unsullied by theorists who would like to put such results into a larger framework. The use of mathematics to help explain neurobiological phenomena was especially regarded as a complete waste of time, and considerable hostility was displayed against mathematicians who ventured into the murky waters of neuroscience. That has now all changed, and more specifically the ideas of neural network modelling have become of increasing importance in the attempt to understand the nature of the control systems, exemplified by nervous tissue, as animals, including ourselves, carry out their tasks of survival. Such control indeed involves natural processes, so that ever more complex models of living tissue are required to give a deeper understanding of the manner in which such tissue is used effectively. The example of the success of theoretical physics in its attack on the nature of matter is an example which shows the effectiveness of modelling the natural world.

Neural networks are the modelling medium par excellence for use in attempting to understand the subtleties of the brain and central nervous system. This medium launches into new areas of mathematics, involving statistical techniques, non- linear analysis, discrete mathematics and numerous other branches of mathematics which are not usually regarded as relevant to the problem of understanding nerve cells and tissue. This chapter will discuss several areas in which different branches of mathematics have been used to help understand the nature of these systems more effectively.

The single cell is a system which in its own right has a very large amount of complexity so that there are still many problems that are unanswered in single cell analysis. After considering the single cell, the discussion then turns to the retina, a very accessible part of the brain but one with its own complexity to amaze and perplexe us. Visual processing is then the natural consequence of retinal analysis, with its two streams of 'what' and 'where', and will continue the theme of displaying the complexity of the mathematical problems facing us in analysing the brain. In all of this analysis no prior knowledge is assumed, although some understanding of the principles of biology would be of value.

3.2 The Single Nerve Cell

The single nerve cell is a very complex machine in its own right. A typical cell from the part of the cortex devoted to smell in the cat (the so-called pyriform cortex) is shown in Figure 3.1. This cell has a central cell body from which a number of protuberances extend. Most of these are covered in little 'spines' where it may be shown that other cells send their activity to affect the cell under consideration. One of these protuberances, however, arising from the base of the cell body, is smooth and has no inputs from other cells onto it. That is the so-called 'axon' of the cell and carries its 'nerve impulses', brief millisecond sudden changes of membrane potential of the interior of the axon, which move along the axon like the fuse of a firework burning gradually; the nerve impulses travel at about 10 metres per second. It is these nerve impulses which are the signals sent by one cell to the next and which are carried by the axon and its ramifications to many other nearby (or distant) cells to arrive at the spines on them and then transfer their potential change down their long protuberances, called 'dendrites', and so to the cell body. The total activity arriving at the cell body is then assessed and if it is above a certain threshold potential level there will be the sudden response of the cell and the signal will 'wing' its way to the other cells to which it is connnected along the axon and its collaterals.

Such is the general story of how a cell acts as a transformer of information on its input to send a modified output. What that modification might achieve on its own is something we will consider shortly. The more general network processing will be considered in the context of vision later. In the meantime let us return to the single cell and consider how it might be modelled mathematically. We will then develop more complexity in the cell corresponding to the observed features of living cells and note how such complexity might be modelled by extending our original model somewhat.

The first question to answer is as to how the cell is to be described. We could work at the very simplest level and consider only if the cell is active at the present moment, sending out a nerve impulse, or not. In the former case we could assign a value +1 to its 'state', in the latter a value 0 (or -1, as discussed in Chapter 2; the value 0 is more realistic from a neurobiological point of view, although there is no loss of generality to assume either value). We would thus arrive at the very simplest neuron with state either 'on' or 'off' (so with state variable +1 or 0). The updating of the state variables would then be in terms of the passing round, in a network of such neurons, 0's or 1's between the neurons. If we let the ith neuron have the state variable $u_i(t)$ at time t, then we might

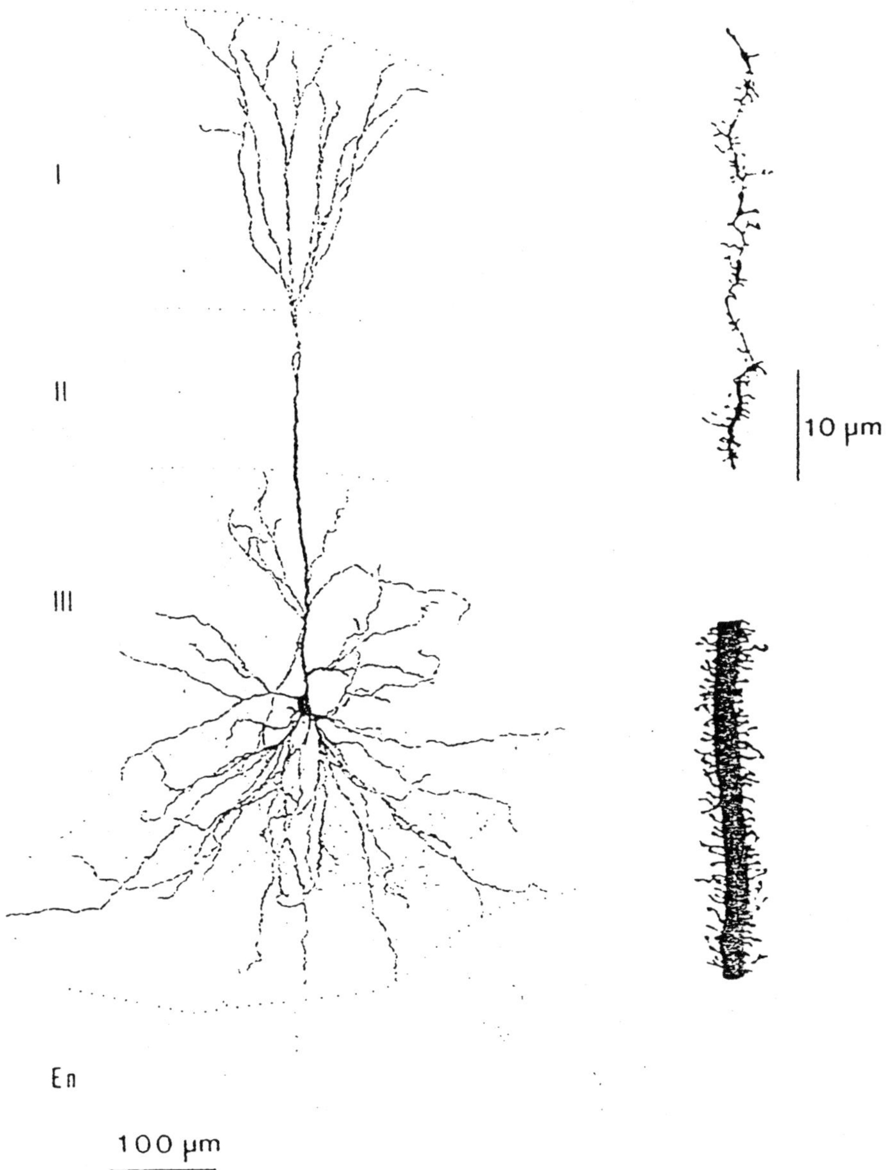

Figure 3.1: Deep pyramidal cell in layer 3 of the piriform cortex stained by intracellular injection of Lucifer Yellow with antiserum intensification. Fine processes are axon collaterals. Details of distal and proximal apical dendrites are at right (From [27]. Reproduced by permission of The American Physiological Society.)

update that neuron by the update rule:

$$u_i(t+1) = \theta\left[\sum_j w_{ij}u_j(t) - t_i\right] \tag{3.1}$$

where $\theta[x]$ denotes the theta-function introduced in Chapter 1, with value $\theta[x] = 1$ for $x > 0$, $= 0$ for $x < 0$. Moreover the quantities w_{ij} were introduced to allow the effect of the jth neuron on the ith one to be felt, since otherwise the state variables u_j cannot be passed around in the net without modification. The quantity t_i is termed a threshold for the ith neuron and allows there to be less (for large threshold) or greater (for smaller threshold) sensitivity of the neuronal response to its inputs.

Equation (3.1) is the basic equation of all neural network theory. It contains the fundamental parameters t_i, w_{ij} and the output response function, which together describe:

1. the sensitivity of the response of the neuron to its inputs (t_i)

2. the effectiveness of the input from neuron j to neuron i (w_{ij})

3. the sharp dependence of the response of the neuron on the total summed activity arriving from the cell surface (the theta function response). These are the central elements of neuronal response.

Further complexity is observable if one considers again the cell in Figure 3.1. It has a considerably elongated set of dendrites. This geometry is not the same for all cells, and that leads one to ask if the cell geometry is important as part of the information processing that the cell can perform. That can only be answered by more detailed analysis of a more complex model than that of equation (3.1), which only has point-like neurons as its building blocks.

Further complexity arises on more detailed consideration of the manner in which cells transfer information between one another. A magnified view of the transfer point indicates that there is a definite gap or cleft, termed the synaptic cleft, between the end of one axon and the dendritic spine of the one to which it is handing information. This is shown in the enlargement in Figure 3.2. The greatest enlargement in Figure 3.2 indicates the presence of small pellets or 'vesicles' of material in the presynaptic ending of the axon, the so-called presynaptic 'button'. These vesicles are composed of a chemical which is released into the synaptic cleft, flows across to the postsynaptic side and causes a change in the membrane potential there in a manner which is increasingly being understood. These vesicles are even released, at a much lower rate, when there is no flow of nerve impulse down the axon to the next nerve cell. Moreover these releases (spontaneous or induced by a nerve impulse) are not always in the same amount, although they have a constant mean rate; there is a degree of noise in the system at this point.

The complexity noted above is compounded when one turns to consider in more detail the manner in which the nerve impulse is itself generated at or

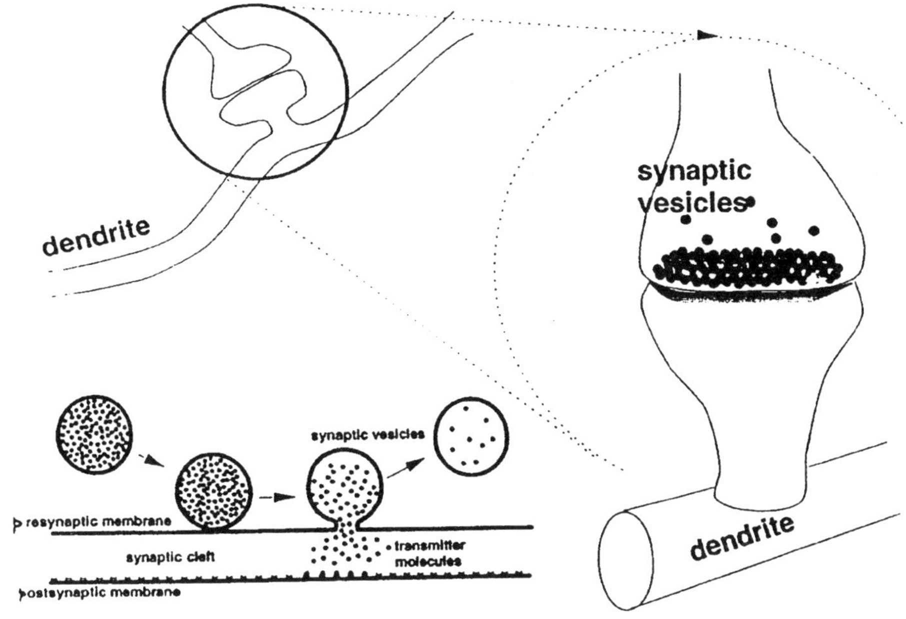

Figure 3.2: Details of the presynaptic ending of the axon of a neuron and the nature of the synaptic cleft.

near the cell body. A discontinuity like that of the theta function in equation (3.1) above is decidedly unlikely in nature, and a more careful analysis of nerve impulse generation shows that it is produced by a highly non-linear dependence of the ease of transmission of the electrical potential (the so-called conductance of the cell membrane) on the membrane potential itself. Inclusion of the details of this leads to very non-trivial dynamical equations.

The above features may be summarised as:

1. geometry,

2. synaptic noisy transmission,

3. non-linear nerve impulse generation.

Let us consider how some of these might be included into the mathematical analysis by extending equation (3.1) suitably. We cannot hope here to go into all of the details but at least hope to indicate how some aspects can be incorporated.

To begin with it is necessary to extend equation (3.1) to allow the neuron to have a history of its past membrane potential values instead of just updating from exterior signals from other neurons (or from itself if there is a self connection weight in the summation term in equation (3.1)). That can be done by adding to the term in the brackets on the right hand side of equation (3.1) a decaying sum of past activities:

$$\exp(-r/R)u_i(t-r) \tag{3.2}$$

where the summation over r in (3.2) is only over non-negative values. It is seen that the sum in (3.2) will be reduced by a factor of $1/e$ after about R terms, so corresponding to a finite lifetime for membrane activity. This extension (3.2) of (3.1) produces the more complicated equation:

$$u_i(t+1) = \theta\left[\exp(-r/R)\cdot u_i(t-r) + w_{ij}u_j(t) - t_i\right] \tag{3.3}$$

The form of neuron described by equation (3.3) is called a 'leaky integrator' neuron since it holds the activity on its surface over a lifetime R. Such a holding process allows neurons to keep a decaying past record of the activity they have received, so may help them in relating ongoing activity to that of the past. In other words it could help in learning temporal sequences of activity that the network processes, as has been used in [1] and by other workers.

The next feature is to take account more carefully of the way information is transferred between neurons, so as to take account of the synaptic noise. This can be done by making the synaptic weights 'noisy' or random variables. This can be done in a manner in which the right hand side of (3.3) is to be more properly regarded as the noisy variable stating whether or not the neuron will respond or not at that particular time for that particular random choice of the random connection weights w_{ij}. It is possible to obtain the probability that the neuron will respond, for a given set of inputs $\{u_i(t)\}$, by summing over all the random choices of the connection weights, given that set of inputs. This leads to a probability for the neuron responding (sending out a nerve impulse) given as a specific function of the incoming activities. These, for a net of N neurons, can be described by a vector of length N and composed of 0's or 1's- so a binary N-vector which we will denote by $\mathbf{u}$ (with components each of the activities of the separate neurons). Thus the effect of this is to obtain, for each neuron i, a number $\alpha^i_{\mathbf{u}}$ denoting the probability of the neuron i 'firing' (responding with a nerve inpulse) given it has had the set of inputs in the binary N-vector $\mathbf{u}$ at one time-step previously :

$$prob(u_i(t+1) = 1|\mathbf{u}(t) = \alpha^i_{\mathbf{u}}) \tag{3.4}$$

The expression (3.4) (read as the probability that $u_i(t+1)$ has the value 1 given the previous input $\mathbf{u}$) leads to a simple hardware implementation in RAM, in which at the binary N-vector address $\mathbf{u}$ in an N-RAM one inserts the real-valued quantity $\alpha^i_{\mathbf{u}}$ representing the probability of the ith neuron sending a 1 as output when the binary address $\mathbf{u}$ is accessed. This leads to the pRAM hardware neural chip (standing for 'probabilistic RAM' neuron) which has been implemented in VLSI form, including on- chip reinforcement learning, for 256 neurons [2].

The final effect of geometry can be accounted for by breaking up the total neuronal geometry into a set of cylinders connected to each other in a suitable (linear) manner. This leads to a detailed form of the extension of equation (3.3), where there is one equation for each of the compartments describing the manner in which the membrane potential develops in time. The very simplest one of these can be seen by defining the variable inside the theta function in equation (3.3) as the membrane 'potential' $V_i(t)$, which can be shown [3] to satisfy, in the limit of taking the time in the system to be continuous and not discrete as the updating of (3.1) and (3.3) supposes, to be the differential equation with respect to the time variable:

$$\tau dV_i/dt = -V_i + \sum_j w_{ij}u_j(t) \tag{3.5}$$

where the output of each neuron is related to its membrane potential by:

$$u_i(t) = f[V_i(t) - t_i] \tag{3.6}$$

with the non-linear function f including the non-linearities involved in the production of the nerve impulse, so having a considerably more complex form than contained in the sigmoidal function used earlier.

The detailed manner of this nerve impulse generation is from the presence of many channels which allow ions (sodium, potassium and others) to pass in and out of the membrane and so change its potential. The time courses of the openings and closings of these channels leads to complex cell reponse, such as a burst of nerve impulses, or on the contrary only one single nerve impulse for a given nerve impulse input. This difference adds enormously to the nature of the information transferred by each cell, making them more sensitive to other sources of modulation by other chemicals, such as dopamine (a particularily important one as far as schizophrenia and Parkinson's diseases are concerned).

The various extensions of the simplest point and deterministic binary neuron of equation (3.1) are summarised in Table 3.1; the effects they might achieve in processing are also noted in that table. It is clear that only the very surface of the subject of the complexity of the living cell has been viewed. The interested reader is advised to turn to the more complex discussion in the references for further reading [4].

3.3 Retinal Processing

Input arises from the retina as the ouptut of its only spiking cells, the retinal ganglion cells. These are of at least two forms in more advanced mammals, the alpha and beta ganglion cell types (together with others not considered here; see [5] for a more complete review, and [6] for a good overview by the acknowledged experts). The alpha-stream cells give a transient response to inputs, the beta cell ones a more durable one. Both sets of cells have a centre-surround response profile so that a central spot will enhance (reduce) the cell response compared to its spontaneous output and a more peripheral one reduce (enhance) the cell output. This allows the cells to signal edges of input stimuli rather than their more uniform centres. The difference between the two streams is that the alpha

Extension	Mathematical Structures	Capability Involved
1) leaky integration	Leaky integrator differential equations	Enhanced temporal sensitivity
2) noisy synaptic transmission	Probabilistic update equations	Noisy neuronal response for better generalisation
3) geometry	Compartmental description of neuron	Further temporal sensitivity
4) non-linear response	Separate chemical channels involved, with complex dynamics	Produces complex range of neuronal responses, such as 'bursting'

Table 3.1: Extensions of the Binary Neuron of equation (3.1).

cells have a considerably dendritic and larger receptive field than the beta cells, thought to be due to the larger degree of convergence of the input sensors (the cones in photopic vision) onto the former than the latter.

The alpha and beta cells also differ in their temporal responses as noted earlier: the alpha cells have a more transient response (phasic) than the beta cells (tonic). This can be explained in terms of the early processing in the outer plexiform layer of the retina, where the so-called horizontal cell layer averages the input over the retina and this is then subtracted from the local input to a given bipolar cell, acting as output of its relevant cone sensor (see Figure 3.3). The bipolar cell then gives a response which has redundancy of input removed from it. For the larger receptive fields (alpha cell) this corresponds to removal of all of the input (the averaged input being removed cancels the input and only allows an intial transient response before the average is calculated). On the other hand there is little spatial averaging for the beta cell, so its input will remain in (for a persistent input).

This processing has been analysed mathematically by setting up the coupled equations of the form of equation (3.5) for the so-called horizontal cells of the outer plexiform layer (OPL) [7], on neglect of any non-linearity in their response. It is possible to take the continuum limit of the cells of the OPL layer, where a new feature arises of great importance in understanding how redundancy can be removed from an input and only worthwhile information

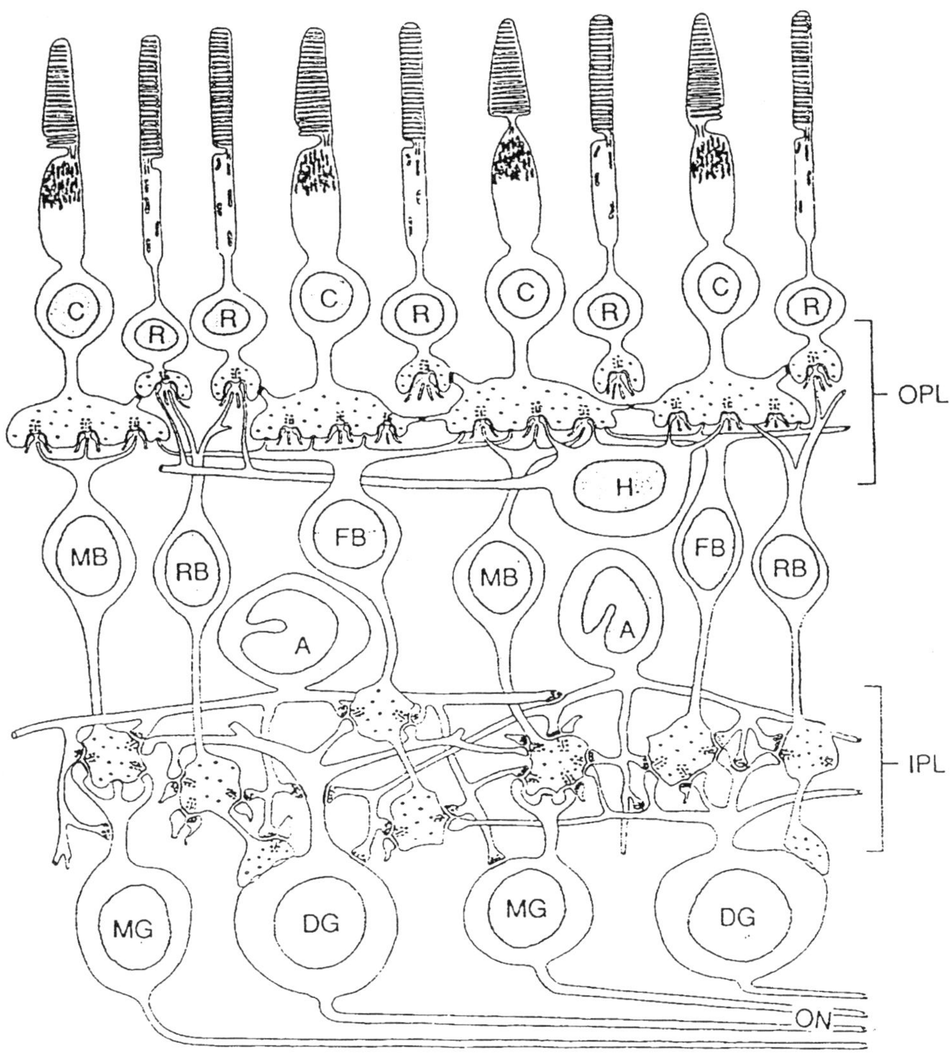

Figure 3.3: Diagram of the principle elements in the primate retina. R, Rod; C. Cone; MB, midget bipolar; ON, optic nerve fibers; OPL, outer plexiform layer; IPL, inner plexiform layer. (From [28]. Reproduced by permission of The Royal Society.)

passed on to later cells. The basis of this is very simple: it is termed lateral inhibition. Let us explore this in a little more detail.

Consider a line of cells, each responding linearly to their input, and passing on their output to an appropriate cell in a further line of cells parallel to the first, but also half that output, with inhibition, to the cells on either side of the given cell. Thus an input cell at the point x on the line will feed its ouptut, say $\lambda I(x)$ to the output cell on the next line at the same position x and the value $-(1/2)\lambda I(x)$ to the output cells at the next neighbourly positions, say at $x+a$ and at $x-a$, where a is assumed small. Then the total input to an output cell will be:

$$\lambda I(x) - (1/2)\,\lambda I(x+a) - (1/2)\,\lambda I(x-a) \tag{3.7}$$

For small values of a this may be written approximately by expanding $I(x+a)$ and $I(x-a)$ in powers of a about x as:

$$I(x+a) = I(x) + a\,\frac{dI(x)}{dx} + (1/2)\,a^2\,\frac{d^2I(x)}{dx^2}$$

so allowing formula (3.7) to be rewritten as:

$$\lambda a^2\,\frac{d^2I(x)}{dx^2} \tag{3.8}$$

Thus the response of the output cells, as given by their input (formula 3.8), is to the second spatial derivative of the input stimulus. If that were constant or linearly increasing then there would be no response; the redundancy in the input would have been eliminated by the use of the lateral inhibition feeding from the line of input cells to that of output cells.

This approach of taking a continuum limit results, for the retina, and using a reasonable approximation to the connectivity on the OPL, to the action of second derivatives in both directions (the OPL being a two-dimensional sheet of cells). The resultant response of bipolar cells, the cells at the next stage of processing in the retina after the cone/rod receptors, fits well with observed values in various classes of retinae. The principle of redundancy reduction by lateral inhibition can certainly be put to work to help guide the intricacies of the connectivity and functional structure in the retina (and elsewhere in the brain). It is also possible to consider the model, purely based on the main features of the observed retinal connectivity, as supporting the approach of redundancy reduction and allowing it to be more effectively observed at a quantitative level.

As a result of this earlier processing, the outputs of the alpha and beta cells can be approximately represented by a so-called 'convolution' of the input with a suitable 'difference of Gaussians' (DOG) kernel in space, together with a Gaussian window in time (to take account of adaptation):

$$Out(\text{ganglion at } x, t) = \int K(x-x', t-t')I(x', t')dx'dt' \tag{3.9}$$

cat	monkey	human	colour sensitivity
alpha	M	parasol	little
beta	P	midget	R-G, B-Y or Black-White opponent

Table 3.2: The comparable nomenclature across species, together with their colour sensitivity.

where I is the retinal input. The linearity assumed in (equation 3.9) may be an oversimplification for some cells but reasonably well describes the averaged output; it must be remembered that ganglion cells do output spikes (the only spiking cells in the retina, as noted above) so that the expression (3.9) should be interpreted as the probability of a given cell outputting a spike. In the lateral geniculate nucleus (a nucleus of the thalamus) there is segregation of the M & P streams, with the M in 2 layers and the P stream in 4 layers; each layer is a retinotopic map (see Figure 3.4). Thus the visual streams are already present at a retinal level; cortex is expected to adumbrate such a division further, as we will see. There is also further processing hinted at occurring in the LGN, as evidenced by the presence of 'lagged' cells, which cause a time delay to their output. Together with unlagged cells they can lead to temporal differencing, and so to velocity sensitive cell output. This may be in addition to the motion sensitivity of the M cell output.

3.4 The Cortical Streams

The development of a machine vision system as powerful as that of the human has not yet been possible. We can effortlessly recognise complex objects in our visual field and perform manipulations on them which no machine is anywhere near being able to achieve. Therefore human visual processing has been much studied over the years to attempt to understand how we manage such decidedly difficult tasks.

The results indicate that this is achieved by both breaking down the retinal image into component parts and recombination of those parts to activate previously stored templates so as to allow identification of objects and make appropriate action responses. Such processing has been found to have been achieved by splitting the visual input into a number of streams so as apparently to make the task simpler; how exactly that simplification is achieved is still unclear, but progress is being made in unraveling it. In the following sections the different streams will be considered, starting with streams identified

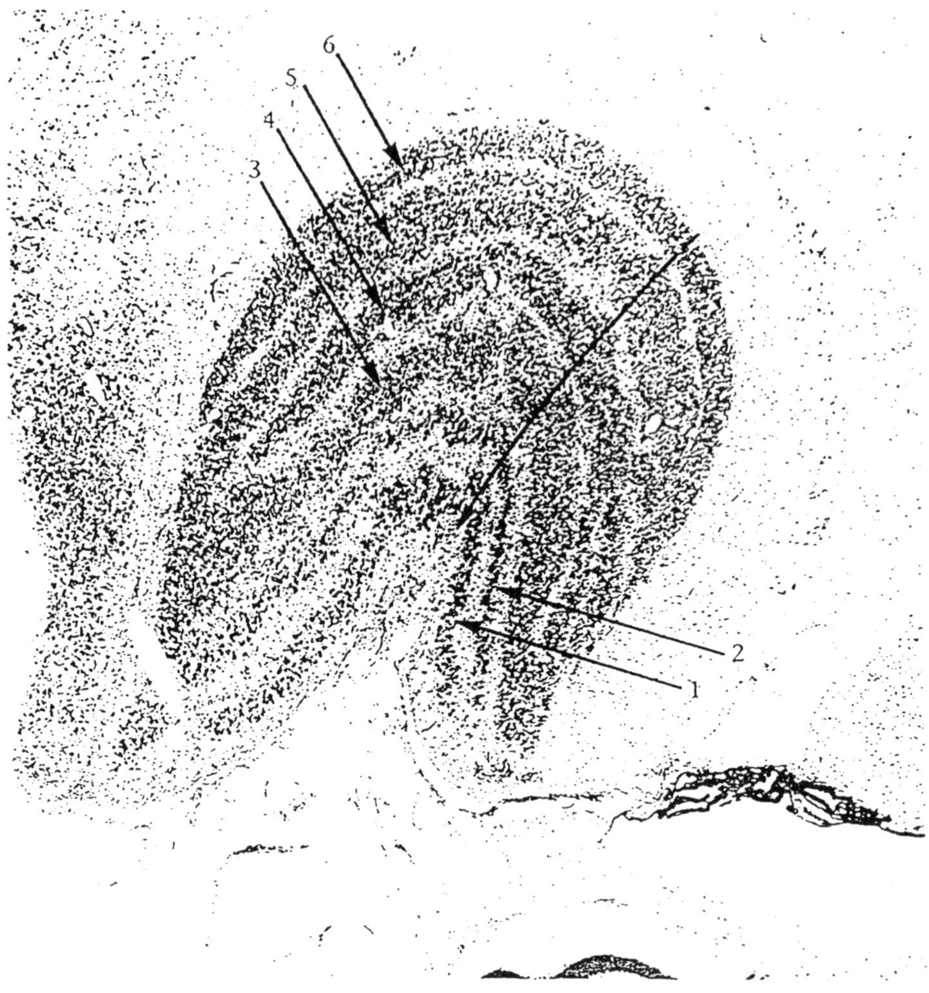

Figure 3.4: Lateral Geniculate Nucleus, or LGN, is located in a region of the brain thalamus. In primates the majority of retinal ganglion cell axons terminate in the thalamus. When stained with Golgi material the nucleus appears to have six separate layers. The four superficial layers have neurons with small cell bodies and are called the parvocellular layers. The two deep layers have neurons with large cell bodies and are termed the magnocellular layers. There are also cell bodies that fall between these layers and are called the intercalated zones. Each layer contains a map of the visual field, and the maps are in register. Hence the LGN neurons located near the radial arrow respond to stimuli as a common location in the visual field. (From [29]. Reproduced by permission of The Royal Society.)

at the sub-cortical level, since that is the most primitive and so is in some sense the bedrock of the total system. Cortical processing can be regarded in some sense as the 'icing on the cake'. We will then turn to the two main cortical streams, the 'what' for object processing and the 'where' for spatial and motion analysis. The manner in which these might be combined will then be described briefly; the story here is as yet very poorly understood. The problem of building a mathematical model of this processing will be considered after a description of some of the features involved in it.

Retinal output as P v M/ LGN compartments is now accepted as leading to an important separation of the further cortical visual processing into two streams, termed the dorsal (or M) stream and the ventral (or P) stream, fed from the corresponding streams from the retina and LGN mentioned earlier. These two streams are not completely separate in cortex - they have interchanges with one another, but in general seem to process separate aspects of the visual input in quite different ways. There is now understanding of this difference between the two streams from neuranatomical connections, from deficits in humans and primates, from single cell recordings and most recently by non-invasive techniques.

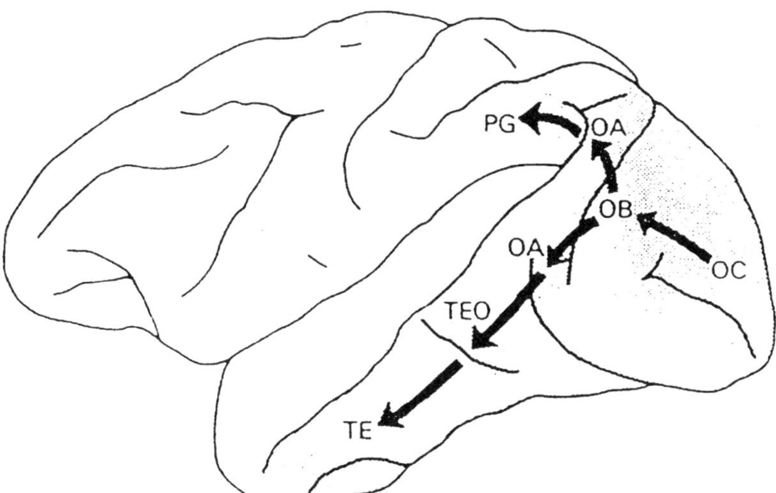

Figure 3.5: Lateral view of the left hemisphere of a rhesus monkey. The shaded area defines the cortical tissue in the occipital temporal and parietal lobes. Arrows schematize two cortical visual pathways, each beginning in primary cortex(areas OB and OA), and then coursing either ventrally into the inferior temporal cortex (areas TEO and TE) or dorsally into the inferior parietal cortex (area PG). Both cortical visual pathways are crucial for higher visual function, the ventral pathway for object vision and the dorsal pathway for spatial vision.

The differences between the two streams is shown in terms of the hierarchically arranged cortical areas through which they pass as shown in Figure 3.5 (from the fundamental paper [8]) and more diagrammatically from the information they carry in Figure 3.6 (from [9]). The first of these papers argues for such a separation from the deficits produced in monkeys in object discrimination (on removal of the temporal lobe) and in landmark discimination (on bilateral removal of posterior parietal cortex). The second of these references [9] then argues that object processing and recognition involves much interaction between these two pathways at several stages in the processing hiearchy. Let us consider each of the pathways in turn and then consider them in interaction.

3.5 The Ventral Stream

From Figure 3.7 it is clear that at the first stage of visual processing in cortex the streams are intermixed in V1, the striate cortex. Input from LGN consists of the two streams, termed the parvo (P) and the magno (M) possessing the properties they inherit from the P amd M streams of the retinal and LGN outputs described earlier.

As seen from Figure 3.7 the P stream, the so-called ventral stream (since it goes down ventrally in the brain to the inferotemporal lobe and the limbic system including hippocampus), travels to the so-called 'blobs' and the 'interblobs' in V1, as seen from their characteristic staining patterns observed when cytochrome oxidase has been applied to the cortex. As noted in Figure 3.7, the blobs contain cells mainly sensitive to colour, possessing wavelength sensitivity in the form of R-G or B-Y opponency (excitation to red shone centrally versus inhibition below the spontaneous level to green shone in the surround, for example). This strand sends outputs directly to the thin stripes (as determined by cytochrome oxidase staining) in the next area V2 and then on to the further inferotemporal area V4. This strand inputs to layer 4C of V1 and outputs through layers 2/3 (as is common in cortical processing from one area to the next one up in the processing hierarchy).

The P stream also sends input to orientation-sensitive cells in V1 (in the interblob areas in V1), some of which also have binocular response properties. These send their output to the interstripe areas (as observed by cytochrome oxidase staining) in V2 which preserve the properties of the V1 cell, at the same time giving more complex responses in some cases, such as through the 'end-stopped' cells, which are sensitive to the ends of lines and are thought to be involved in the construction of analysis of illusory conjunctions [10]. These cells also send their output mainly to the cells in V4 to combine with those of the blob/thin stripe cells of the colour-sensitive strand.

The assignation of V4 as 'the' colour area in the brain by Zeki [11] has been somewhat controversial but is now somewhat accepted (although it is very debatable that colour awareness actually arises from activity in V4). However the importance of V4 to colour vision is now well established both through deficits and non-invasive analysis. The story of the painter who had a very difficult time coming to grips with his loss of colour experience after suffering

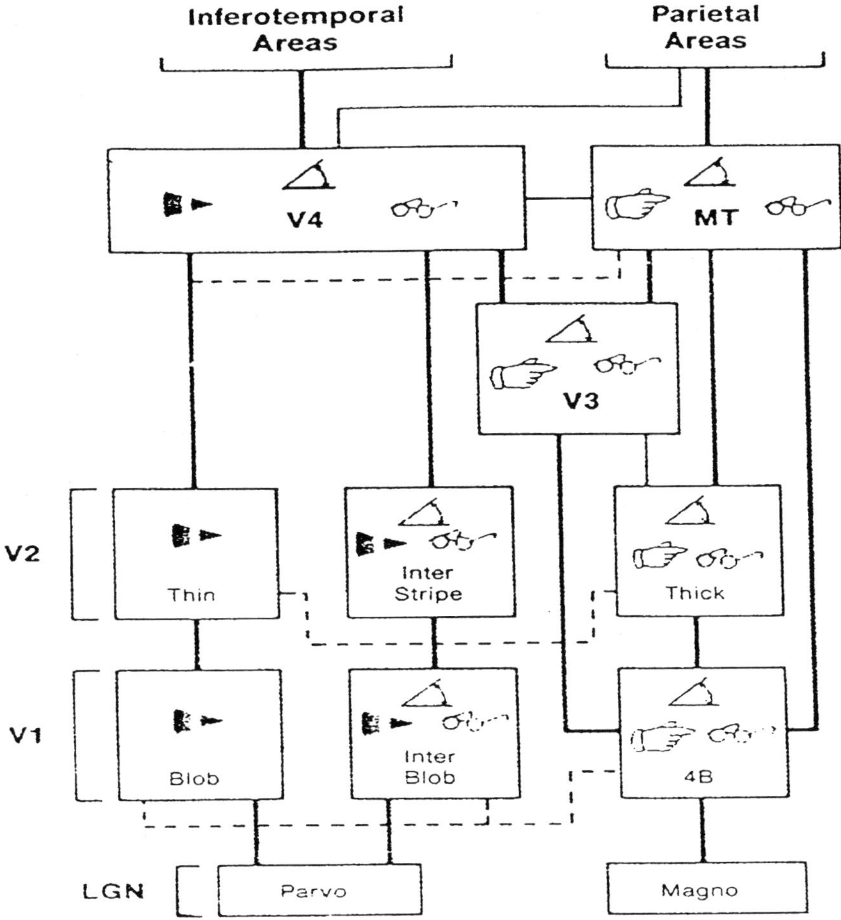

Figure 3.6: Schematic diagram of anatomical connections and neuronal selectivities of early visual areas in the macaque monkey. LGN+ Lateral geniculate nucleus (parvocellular and magnocellular divisions). Divisions of V1 and V2: blob = cytochrome oxidase blob regions; 4B = lamina 4B; thin = thin (narrow) cytochrome oxidase strips; interstripe = cytochrome oxidase-poor regions between the thin and thick strips; thick = thick (wide) cytochrome oxidase strips; V3 = visual area V3; V4 = visual area (s) V4; MT = middle temporal area. Areas V2, V3, V4, MT have connections to other areas not explicitly represented here. Area V3 may also receive projections from V2 interstripes and thin stripes. Heavy lines indicate robust primary connections, and thin lines indicate weaker more variable connections. Dotted lines represent observe connections that require additional verification. Icons: rainbow= tuned and/or opponent wavelength selectivity; angle symbol = orientation selectivity; spectacles = binocular disparity selectivity and/or strong binocular interactions; pointing hand = direction of motion selectivity.

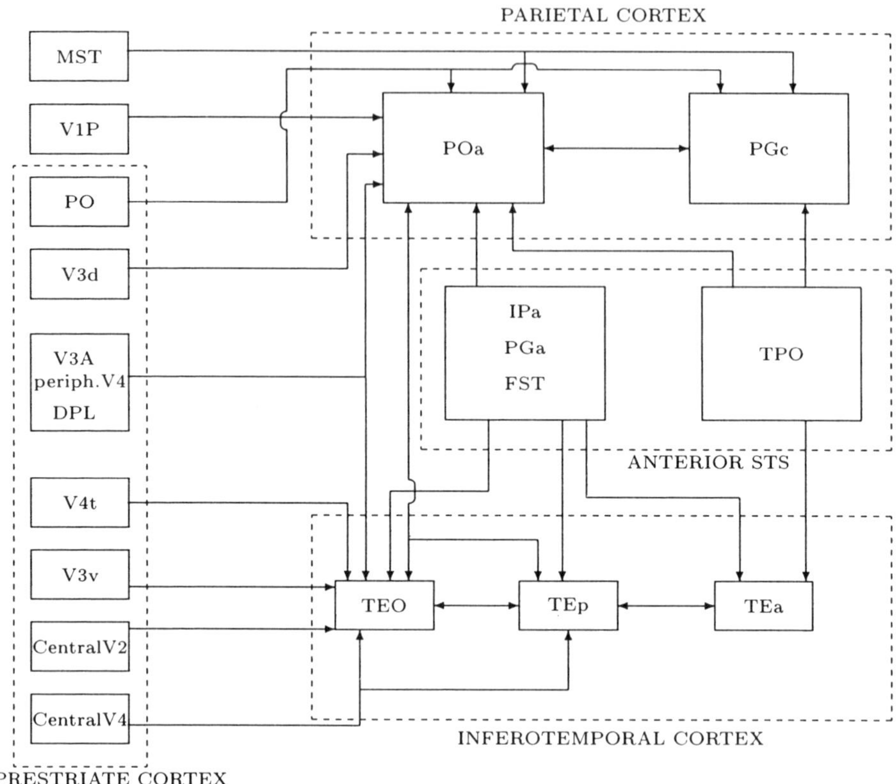

Figure 3.7: Summary diagram of the prestriate and anterior STS connections with the inferior parietal and inferotemporal cortex as observed in the present study. Central V2 and V4 refers to the regions of these areas coding the central visual field. Periph V4 refers to a part of V4 coding the periphery of the visual field.

a head injury in a car crash is well known [12], and just an extreme example indicating the importance of that region in colour sensitivity of cortex.

The end point of the ventral stream is thus posited as V4. This has cells which are variously colour sensitive, orientation sesnsitive and have binocular disparity selectivity or strong binocular interactions. In the process of input being transformed from one area to the next there is an increase of the size of receptive fields of the relevant cells from about 1 degree in V1 to 2-4 degrees in V2 to about 8 degrees in V4. This increase is thought to arise from simple serial fan-in as one proceeds up the processing hierarchy. The ventral stream arises from the foveal component of the retinal input (the inner 3-4 degrees of the visual field) with greatest acuity for analysis of the input. This is to be compared with the M stream which contains more peripheral input analysed with far less precision [5]. Let us turn to that now.

3.6 The Dorsal Stream

The motion-sensitive cells of the cortex start in layer 4B of the striate area V1, being fed by the retinal motion M stream from the LGN. This input is fed out directly from that same layer 4B in V1 (and is not processed in the higher layers 2/3 as the P stream input is) to the thick stripes in V2 and then on to the area MT (or V5) or directly to that area without interruption. Area MT is well known now to be crucial for the response to motion in various forms. Thus non-invasive experiments have shown very clearly the extreme form of sensitivity of the area to moving stimuli as compared to the earlier area V1 [13]. There is even a 60% response of MT cells to moving stimuli which only have a 1.5% contrast against background as compared to a few percent response of V1 cells to the same input.

The area MT is thought to be of importance in helping explain the response of patients with loss of V1, or parts of it, to moving images - the phenomenon of 'blindsight' [14]. In this phenomenon the subject, in spite of being blind in the sense of having no awareness of a moving spot in his/her blind field of view, will be able to indicate above chance level in which direction the spot has been moving. This may require some encouragement, since the patient is embarrased to 'guess' when they have no direct 'experience' on which to base their response. It is thought that there is direct input to MT which has been spared in whatever injury caused their loss of cortex, so that this region can allow knowledge to be obtained on which the 'guess' can be based. However there are some subjects who have direct awareness of the moving spot, in spite of that experience not allowing them to say what the spot looked like [15]. Thus MT is an important area in the processing hierarchy towards motion awareness.

The further areas in the dorsal stream help solve a number of tasks:

1. further analyse the moving input to achieve segregation into figure and ground (which in general will have different movement characteristics),

2. analyse the shapes of the various parts of the input,

3. analyse the positions of the parts of the input,
4. transform the retinal co-ordinates into body-centred ones for actions to be made,
5. give information to the frontal eye-fields (especially area 8) in order to direct eye movements to a new position in the visual field,
6. direct attention to any new and salient input which has recently peripherally entered the visual field,
7. co-ordinate or update visual input with actions about to be or in the process of being made. These (and others not mentioned) form an important but difficult set of tasks that need to be performed in order for actions based on the needs of the person and of their visual input to be made. That explains why there are numerous areas in the dorsal stream in extra-striate visual and parietal cortex. These are only now being explored fully; some of the latest results are contained in [16].

3.7 Object Construction

The manner in which objects are coded by the action of the visual streams described above is still unclear. The main set of representations are supposedly stored in later processing stages in the inferotemporal lobe, as determined by the loss suffered by patients with deficits in those areas. Thus loss of the dorsal stream will give rise to processing deficits involving action and motion [17] but not object recognition, whereas loss of temporal lobe will cause recognition deficits which tend to be severe, even though motion and action deficits may be low.

The work of Tanaka [18] has been important in elucidating the nature of some aspects of object representations. He has shown that in the monkey dorsal processing stream:

$$V1 \rightarrow V2 \rightarrow V4 \rightarrow TEO \rightarrow TE \tag{3.10}$$

there is a build up of 'elaborate' cells, which respond to comparatively complex stimuli, composed of circles with strips emanating from them or other non-trivial stimulus shapes. Moreover it ensues that there is roughly a hyper-column of cells, vertical to the cortical surface and with a diameter of about 400-600 microns, which codes for very similar 'elaborate' inputs. This is very reminiscent of the coding for orientation in V1, for which there are similar hypercolumns encoding orientation sensitivity to oriented bars of steadily changing orientation (but the same position in space) as one moves across the hypercolumn. This similarity in coding is to be expected, although the manner in which object representations are then constructed is not clear.

What is needed in this construction is:

1. segmentation of the object of interest from others in the input so that it can be processed at a higher level,

2. binding of the activities in the various codes (colour, motion, shape, texture, binocular disparity) in order to build a unified object percept,

3. activation of stored templates to compare with the so-constructed percept,

4. transfer of the resulting identified concept to frontal lobes to allow for suitable actions to be taken relevant to the goals of the subject.

Problem 1) is the famous 'binding' problem, for which numerous solutions have been proposed. Thus the use of coupled oscillatory neuronal activities (suggested as being at about 40 Hz) has been favoured for a time. This has now been reduced to the use of synchronous activations of the different codes for a given object stimulus [19]. Another alternative is to use motion as a basic segregator, since there are cells in MT which are sensitive to background motion relative to that of the background motion of the cell, and other cells there sensitive to total motion of the whole field. This would lead to relative inputs across these cells which would allow for very fast figure-ground segregation. Since there are good connections between the relevant dorsal and ventral pathways [20] then such information is very likely to be easily available to the ventral object stream, leading to activation of commonly moving parts of the visual field being sent from the dorsal to the ventral areas. These latter must then be trained (or genetically wired) so as to be sensitive to the correct set of common strong inputs; such a possibility is not difficult on use of Hebbian learning.

Moreover it is noted in [20] that 'both parietal and inferotemporal cortices receive common inputs from extensive regions in the anterior STS which may play a role in linking the processing occurring in these two cortical subdivisions of the visual system'. Such linking may allow common activations to achieve binding in a manner dependent on both the P and M stream common activations. This is seen in Figure 3.5 (for the monkey), where the anterior STS (regions denoted IPa, PGa, FST and TPO) connect together the dorsal and ventral streams. This result has been substantiated in [21]. It may even be appropriate to consider that there are three streams, the third one being this set of areas in the STS.

A final approach to the binding problem, with some overlap with the other techniques, is that of Triesman's 'Feature Integration Theory' [22], as shown in Figure 3.8. In this, attention applied to the master map of locations is used to bind together different activations on the feature map spaces, such as the colour and orientation maps shown in the figure, so that they are encoded in a temporary object representation file. This is then compared to a stored description of objects, held in permanent memory (in a task such as visual search for a given stimulus). Thus here binding is achieved by attention in an inexplicable manner; this leaves the final solution to the binding problem still open. However the notion of 'temporary object files' is very close to the associated one of a buffer store, supposedly occurring in posterior cortex as part of the working memory system [23].

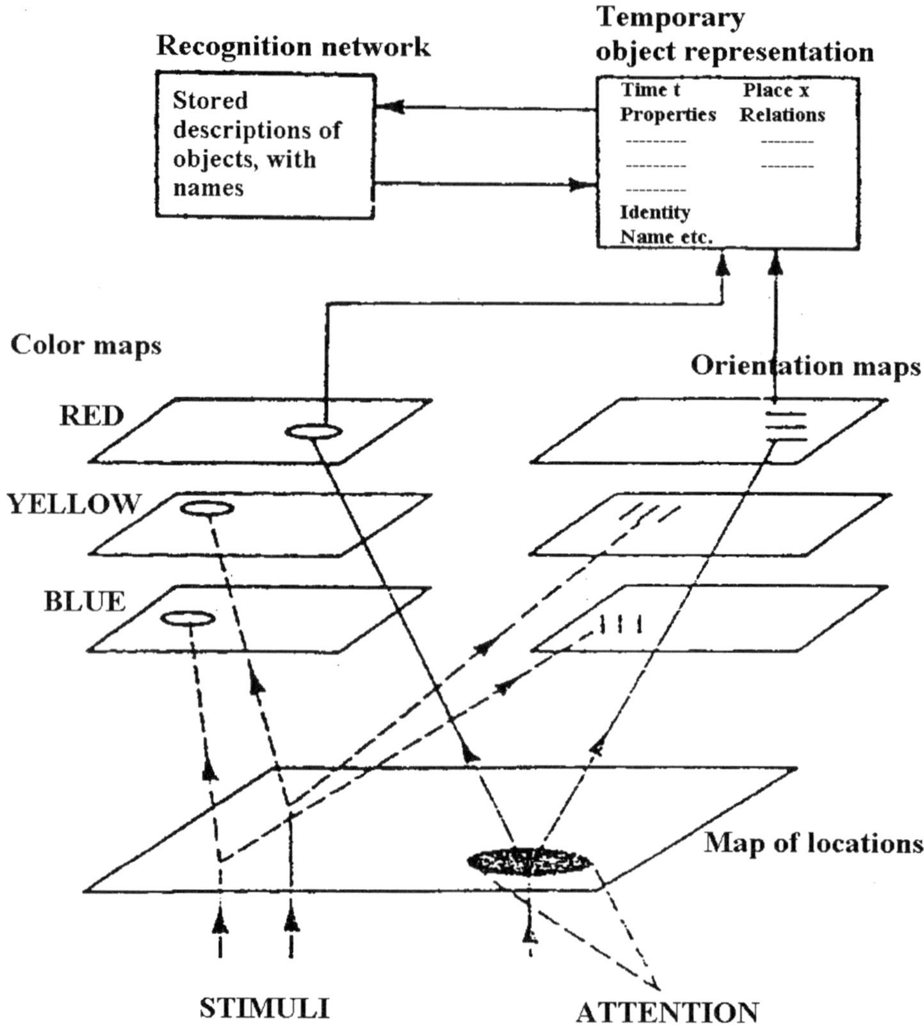

Figure 3.8: General framework relating hypothesised feature maps to a master map of locations through which focused attention serially conjoins the properties of different objects within the scene. (From [22]. Reprinted by permission of The Experimental Psychology Society.)

This still leaves open the sites of such object files, both for the buffer and for the permanent storage modes. The former have now been observed by non-invasive techniques in specialised posterior cortical sites, and the latter are thought to be placed in mesial temporal lobe systems, such as suggested in [22]. However it is emphasized there that such coding is not local and is more likely to involve representations which involve relationships with other objects rather than single object representations.

3.8 Modelling the Processing Streams

There have been many attempts to model the various stages of visual processing as recounted above. The problem is very hard, in spite of some partial successes. The main difficulty may be that we do not yet have a proper understanding of the nature of the object representation to be modelled, since activity associated with object or face recognition is distributed widely across a set of areas. However there appear to be two ways of tackling the problem:

1. insert all the possible connectivity structure between sets of modules of neurons, and attempt to determine, using simple learning laws (such as that associated with the Canadian psychologist Donald Hebb, that the synapse from one neuron to another is increased if they are both active together) and some simplifying assumptions, how the system will develop its transformations of suitably chosen inputs. This has been attempted, for example, by using the continuum limit which allowed us to read off its quantitative effect in equations (3.7) and (3.8) above on inputs.

2. A similar extension has been peformed for the manner in which a topographic representation of a set of inputs can build up in a continuum of cells in the case of one dimension [24] and more recently in two dimensions [25]. Such an approach needs more developed mathematics to analytically solve the problem of the nature of the distribution of orientation sensitivity in V1 and later areas, or of the manner in which motion or colour sensitivity will be distributed in MT and V4 respectively.

If these problems could be solved by the development of powerful enough analytic techniques then it should be possible to push on to the most difficult of all of these problems: that of object representation.

3.9 Conclusions

The theme throughout this chapter has been the attempt to build simple models of the processing systems of the brain based on suitable approximations. This program is somewhat different from the standard one in computational neuroscience (CNS) in which very large simulations are usually attempted to answer the questions being posed (as surveyed in [26] for the understanding of orientation sensitivity in V1). It is clearly appropriate to use both methods of attack, but dependence only on the simulations may hide the underlying principles.

3.10 Bibliography

[1] Reiss M and Taylor JG (1991) 'Storing temporal sequences', Neural Networks **4**, 773-787

[2] Clarkson TG, Gorse D, Taylor JG & Ng CK (1993) 'Learning pRAM nets using VLSI structures', IEEE Transactions on Computers, 41 (12), 1552-1561.

[3] Bressloff PC & Taylor JG (1994) 'Dynamics of Compartmental Model Neurons', Neural Networks 7, 1153-1165.

[4] Jack JJB, Noble D and Tsien RW (1975) Electric current flow in excitable cells., Oxford: Clarendon Press.

[5] Wandell BA (1995) Foundations of Vision, Sunderland, Mass: Sinauer Associates

[6] Livingstone M & Hubel D,(1988) 'Segregation of form, colour, movement and depth: Anatomy, physiology and perception', Science 240, 740-749.

[7] Taylor JG (1990) 'A Silicon Model of Vertebrate Retinal Processing', Neural Networks 3, 171-9.

[8] Mishkin M, Ungerleider LG & Macko KA (1983) , 'Object vision and spatial vision: Two cortical pathways', Trends in Neuroscience 6, 414-417.

[9] DeYoe EA & Van Essen DC (1988) 'Concurrent processing streams in monkey visual cortex', Trends in Neuroscience 11, 219-226.

[10] Peterhans E, von der Heydt R & Baumgartner G (1986) 'Neuronal responses to illusory contours stimuli reveal stages of visual cortical processing', in JD Pettigrew, KJ Sanderson & WR Levick (eds), Visual Neuroscience, pp 343-351, Cambridge: Cambridge University Press.

[11] Zeki S (1980) 'The representation of colours in the cerebral cortex' Nature 284, 412-8.

[12] Sachs O (1985) *The Man Who Mistook His Wife for a Hat*, (London, Picador)

[13] Tootell RBH et al (1995) 'Functional Analysis of Human MT and Related Visual Cortical Areas Using Magnetic Resonance Imaging', J Neuroscience 15, 3215-3230.

[14] Weiskrantz L (1986) Blindsight, Oxford: Oxford University Press.

[15] Weiskrantz,L, Babus JL and Sahraie A (1995) 'Parameters affecting conscious versus unconscious visual discrimination with damage to the visual cortex (V1)', Proc.Nat.Acad.Sci.**92**, 6122-6

[16] Workshop Proceedings: 'Parietal Lobe Contributions to Orientation in 3D-Space', ed P Thier & H-O Karnath, May 25-28, 1995, Univ of Tubingen.

[17] Goodale MA and Milner AD (1992) 'Separate visual pathways for perception and action', Trends in Neuroscience **15**, 20-25

[18] Tanaka K (1996) 'InferoTemporal Cortex and Object Vision', Ann Rev Neurosci 19, 109-39.

[19] Singer W (1996) Ann Rev Neuroscience; Fries P et.al. (1997), Soc. Neurosci. Abstr. **22**, 282

[20] Morel A and Bullier J (1990) 'Anatomical segregation of two cortical visual pathways in the macaque monkey' Visual Neuroscience 4, 555-578.

[21] Baizer JS, Ungerleider LG & Desimone R (1991) 'Organization of Visual Inputs to the Inferior Temporal Posterior Parietal Cortex in Macaques', J Neuroscience 11, 168-190.

[22] Treisman A 'Features and Objects: The Fourteenth Bartlett Memorial Lecture', Quart J of Exp Psychology, 40A (Psychology Press, 1988), pp 201-237.

[23] Baddeley A (1986) Working Memory, Oxford: Oxford University Press.

[24] Takeuchi A and Amari S-I (1979), 'Formation of Topographic Maps and Columnar Microstructures in Nerve Fields', Biol.Cybern. **35**, 63-72

[25] Taylor JG (1997), 'Neural Bubble Dynamics in Two Dimensions', King's College preprint

[26] Swindale NV, (1996), 'The development of topography in the visual cortex: a review of models', Network **7**, 161-248

[27] G.F. Tseng and L.B. Haberly, 'Deep Neurons in Priform Cortex I Morphology and Synaptically Evoked Responses Including Unique High Amplitude Paired Shock Facilitation', J. of Neurophysiology **62** (American Physiological Society, 1989),pp 369-385

[28] Dowling JE and Boycott BB, 'Organization of the Primate Retina: Electron Microscopy', Proc.Roy.Soc.Lond. **B166**, (The Royal Society 1966), pp 80-111

[29] Hubel D and Wiesel T, 'Functional Architecture of Macaque Visual Cortex' Proc.Roy.Soc.Lond. **B198**, (The Royal Society, 1977),pp 1-59

Chapter 4

Neural Network Control of a Simple Mobile Robot

4.1 Introduction

There is much interest in the development of intelligent machines which can learn from their environment. Various machines have been produced which have many sensors, sometimes including high resolution video, which require a great deal of computing power to process the information coming in to the machine, and still more processing is then required to determine suitable action in response to this information.

Researchers in the Department of Cybernetics at the University of Reading believe that it is best to start with simpler systems. Also, they believe that there is much to learn from nature and in particular the behaviour of simple organisms like insects.

Therefore, a number of simple mobile robotic 'insects' have been developed which are small, low power devices, which can operate rapidly in response to what they perceive in their environment.

Initially these devices, equipped with two simple ultrasonic sensors, were programmed to move around according to simple rules and information from the sensors. Then, a learning system was added to allow the robots to learn these rules [1]. Then an extra compound eye was added and a neural network used to allow the robots to learn different positions in the room. The next stage in the work is to use the eye and the sensors to allow the robot to navigate its environment and return to a recognised position. These different stages are all described below. Related work, not described here, considers communication between the various robots which allows the sharing of experiences and of teaching.

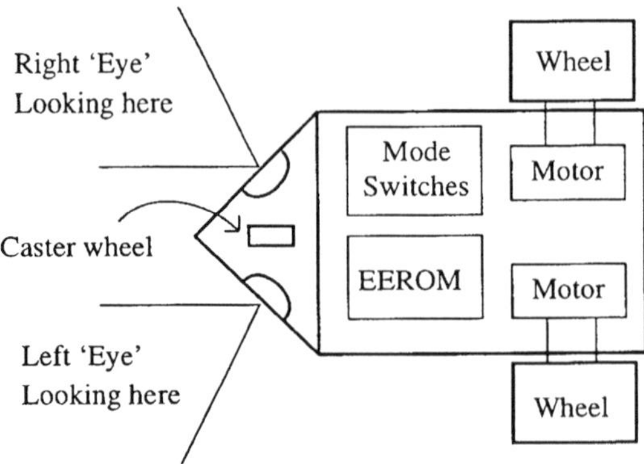

Figure 4.1: Block diagram of basic robot.

4.2 The Robot 'Insects'

A number of these robot insects have been built. The first device had motors which were so powerful that the robot travelled too quickly: the robot was set down in the laboratory, it set off at speed, saw a wall and turned to avoid it, by which time it saw another wall, which it tried to avoid, but it was going too fast. A table top robot was also built which could detect the edge of the table and move away from it. However, this was not suitable if the robot was to learn its behaviour as learning requires mistakes! As a result seven simple robots were built, smaller versions of the original robot, which have become known as the 'seven dwarves', and which contain no microprocessor. Subsequently, more advanced robots have been built, which are equipped with a low power microprocessor and a number of sensors.

The 'seven dwarves' robots have two ultrasonic sensors, which enable them to detect how far the nearest object is in front of a sensor, and two motors, each of which can be set to move forwards or backwards at a given speed. The actions of the robot are determined using a look-up table pre-programmed into an EEROM. A binary pattern, corresponding to the data from the ultrasonic sensors, is passed to the address lines of the EEROM, and the value at the addressed location specifies the speed and direction of each motor. Thus the contents of the EEROM define one set of simple behaviours. Different behaviours can be selected by using DIL 'mode' switches which provide extra address lines to the EEROM. A block diagram of the robot is shown in Figure 4.1.

The look-up table in the EEROM is generated using a special program which is used in the first year undergraduate laboratory. The student writes the rules determining what the robot should do given the information from the sensors, by encoding one procedure - the rest of the program generates

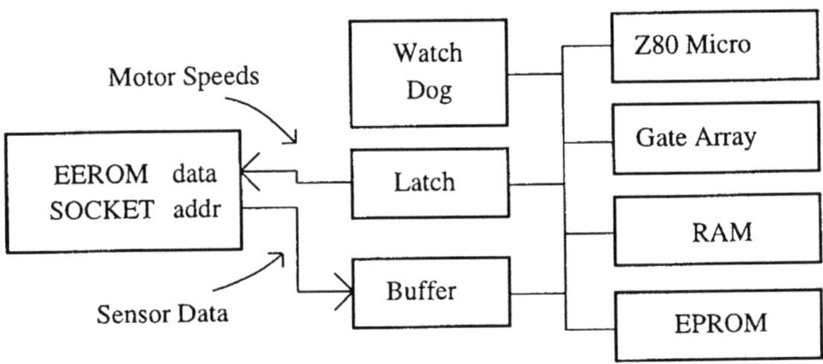

Figure 4.2: Z80 controller.

suitable calls to the procedure, the results of which are used to fill an array which is sent to the EEROM programmer. The EEROM is then loaded into the robot. A typical Modula-2 procedure for moving around avoiding obstacles is shown below: other behaviour can be programmed by writing the rules in other procedures.

```
PROCEDURE Mode (Range : INTEGER; LeftNotRight : BOOLEAN;
                  VAR LeftSpeed, RightSpeed : INTEGER);
(* Range is the distance of the nearest object (its value is 0..7);
   LeftNotRight indicates that the object is closest to the left eye, not the right eye;
   Given these suitable values are required for the speeds of the two motors:
   the speeds are in the range +4..-4: positive speeds mean go forward *)


BEGIN
  IF      Range = 7 THEN                          (* if no obstacle (or too far away) *)
          LeftSpeed := 4; RightSpeed := 4;        (* go forwards *)
  ELSIF   Range ¡ 4 THEN                          (* something very close *)
          LeftSpeed := -4; RightSpeed := -4;      (* go backwards *)
  ELSIF   LeftNotRight THEN                       (* nearest left eye *)
          LeftSpeed := 4; RightSpeed := -4;       (* turn right *)
  ELSE                                            (* nearest right eye *)
          LeftSpeed := -4; RightSpeed := 4;       (* turn left *)
  END
END Mode;
```

Programming the robots in this way has been very successful, providing an interesting exercise in problem solving and programming. However, the robots can only operate in a pre-programmed manner: they can not learn.

So, the robots were then enhanced with an extra Z80 based microprocessor circuit which replaces the EEROM, taking information from the sensors, processing that information and driving the motors. This means the robot can act as before, with the Z80 simulating the look-up table; or a suitable program on the Z80 can be used to allow the robot to 'learn' its actions. A block diagram of the Z80 is shown in Figure 4.2. The Z80 interfaces to the eyes of the insect via a tristate buffer and to the speeds of the insect via a latch: both the buffer and the latch are connected directly to the EEROM on the insect. The Z80 has 32K of EPROM and 32K of Static RAM, a watch dog circuit which allows the Z80 to enter a power-down state so as to consume little power, and a field programmable gate array, FPGA. The FPGA provides address decoding for the memory and peripherals (and allows for further expansion), a random number generator and a UART which allows for communication.

Subsequently, new versions of the insect have been built with the motors and ultrasonics connected in a more conventional manner, rather than via the EEROM socket. The more modular design of the new robots also allows the addition of further circuitry to provide extra sensors, such as the compound eye, and further communication systems.

4.3 Neural Network for Obstacle Avoidance

This section describes how a neural network is used to allow the insect to learn to move around its environment avoiding obstacles. The learning is achieved by a suitable program running on the Z80 emulating a neural network.

The technique used is, like the insects themselves, inspired by methods used in nature. When a baby is shown a toy it instinctively wants to grab the toy, but it does not know how. The baby is able to move its arms and legs but it is not aware which muscles control what part of the body. Initially, therefore, random movements occur. After a time, the baby stops moving its legs as it realises that the legs are not allowing the toy to be grabbed. A little later the baby learns to co-ordinate both hands, and finally is able to grab the toy. Thus the baby learns the appropriate action by trial and error: different actions are tried and the success or failure of the actions is used to determine the choice of later actions.

Therefore, the robots require various actions, a means of choosing the actions, criteria for assessing the actions and a strategy for reassessing the choosing of the actions. The technique used to implement these ideas came originally from the concept of fuzzy automata [2], but which can also be considered as a Hopfield type neural network [3] using modified Hebbian learning [4]. This is described in detail in [5], a brief summary is given below.

4.3.1 Learning Strategy

The basic idea is that the device has a number of possible actions and associated with each action is a probability of choosing that action. It is usual for the action with the highest probability to be chosen. Then the chosen action is performed, its success evaluated and this evaluation is used to adjust the probabilities: if the action was good then the probability of the action is increased and the probabilities of the other actions are decreased; if the action was poor, then its probability is decreased. Essentially therefore, good actions are rewarded, bad actions are penalised.

The insects are allowed nine possible actions - each motor can drive forwards, backwards or be turned off. Also, rather than always choosing the action with the highest probability, a 'weighted roulette wheel' technique is used - so the action with the highest probability is most likely to be chosen. One reason for using this technique is to try to stop the system getting trapped in local minima. Indeed, the probabilities are never allowed to be reduced to zero: there is therefore always a chance that each action is chosen at some stage.

Careful thought was required as regards determining whether the action is successful. The aim was to produce simple 'common sense' rules which did not directly 'tell' the insect how to behave. The rules chosen are as follows:

1. If the robot is in the open, then going forward fast is good.

2. If the robot is close to an obstacle, then moving away is good.

3. If the robot is quite close, then a combination of these rules are used.

These rules are encoded to give a 'goodness' factor, α, which can be positive (good) or negative (bad), and this factor is used to adjust the probabilities according to the following (where m is the number of actions, n is the number of the action chosen, p_n is the probability of the chosen action, and p_j is the probability of the jth action (where $j = 1 \ldots m$, and $j <> n$):

```
IF      α >= 0 THEN           (* action was successful *)
        p_n := p_n + α(1 - p_n)  (* increase probability of chosen action *)
        p_j := p_j(1 - α)        (* decrease probability of other actions *)
ELSE                          (* action was unsuccessful *)
        p_n := p_n - α           (* decrease probability of chosen action *)
        p_j := p_j + α/(m - 1)   (* increase probability of other actions *)
END
```

One problem with this technique is that the action which is best for moving around when there is no obstacle near the insect is likely to be different from the action required to move away from the object.

Therefore five sets of the nine actions are used, each with its own set of probabilities: no obstacle visible; obstacle quite close on the left; obstacle quite close on the right; obstacle very close on the left and obstacle very close on the right. At any instance the current information from the sensors determines the set of actions chosen.

This can be thought of as a Hopfield type network with modified Hebbian learning - as shown in Figure 4.3. The inputs to the network are boolean values determining the one set of actions chosen - in the figure only the fourth input is ON. All the elements in that column are the probabilities associated with that set of actions. The possible outputs are the state of each motor - which can be (B)ack, (F)orward or (O)ff. One action is chosen, based on the probabilities - in the figure this is BF - left motor back and right motor forward. That action is evaluated and if it is good the probability of the action in the chosen column is increased - the association of the chosen action with the given input is reinforced.

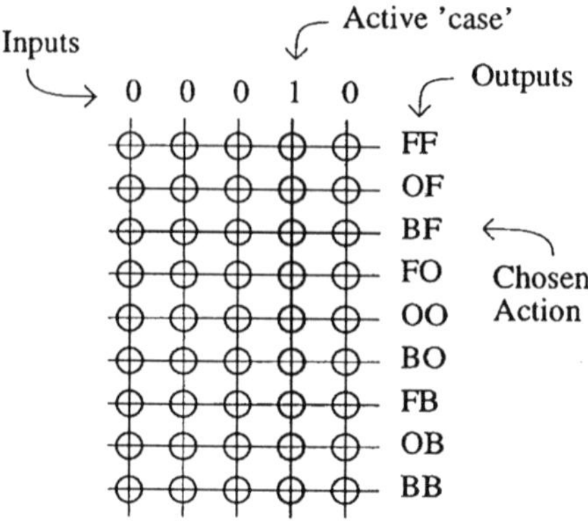

Figure 4.3: 'Hopfield' network.

4.3.2 Results

This technique was initially implemented in simulation, where it works very well, and then it was applied to the actual insects, where the technique is successful though not as clear cut as in the simulation. Partly this is because the simulation does not accurately model the dynamics of the insect. Also, the simulation assumes perfect response from the ultrasonic sensors. There is also a problem with the two ultrasonic sensors: when the robot approaches a wall on the left at an acute angle, turning the left motor forwards and the right motor backwards does, as one would expect, move the robot away from the wall. If, however, the robot approaches the wall at a larger angle, this same (correct) action initially causes the robot to move closer to the wall, although eventually the robot will move away from the wall. Thus two conflicting results occur for the same case. As a result, a third eye has been added between the two and this is used to generate the 'range' value in assessing the success of the chosen action.

Despite these problems, the robot learns rules similar to those outlined in the procedure given earlier in only a few minutes. In fact, the rules it learns are not identical each time (for instance there are three combinations of motor speeds which allow the robot to turn to the right: FB, FO or OB), and the rules which are selected are determined by the experience of the robot while it is learning.

It is worth commenting that this successful learning strategy is achieved with a 20 year old 8-bit microprocessor running a 5K byte program using about 100 bytes of static RAM.

4.4 A Compound Eye

The robot can do the simple task of exploring its environment avoiding obstacles despite, or perhaps because of, its very simple sensory system. It was decided however that a more advanced sensor was needed to improve the learning capabilities of the robot and allow it to learn more advanced tasks. The modular construction of the robot allows extra sensors to be added easily. The type of sensor chosen was again inspired by nature.

Given the complex behaviours prevalent in the insect world, an investigation was made into their visual organs with the intention of building an electronic analogue. It was noted that many insects possess a pair of compound eyes, bulging out of the head with a wide field of vision. The eyes cannot move, have a short visual range and are of fixed focus. Each compound eye may contain upward of 10000 simple photoreceptive cells, each with its own lens. A prototype eye was developed first, experiments on which demonstrated the feasibility of the technique, but also highlighted some problems, and so a more advanced eye was subsequently produced. The following sections describe the tests and techniques used for both eyes.

4.4.1 The First Compound Eye

To investigate how useful a simple vision system might be to the Cybernetic insects, a small compound eye was built. This eye consists of a three dimensional array of fifteen light dependent resistors (LDRs) mounted on the top fifteen faces of a truncated icosahedron. The circuit processing the output of the LDRs is mounted in a box under the eye. The eye is shown in Figure 4.4a.

Each LDR produces an output voltage dependent on the light falling on it. An analogue multiplexer selects which LDR is currently being measured and passes its associated voltage to an A/D converter which gives a digital output of this voltage: a block diagram of this circuitry is given in Figure 4.4b.

4.4.2 Estimating Robot Position

The output from the compound eye consists of an array of fifteen values digitised to the interval $[0 \ldots 255]$. These values need to be processed by the insect for a given application, and the chosen application was for the insect to be able to recognise its position within its environment.

This is a pattern recognition problem, so it was decided to apply a neural network to the data from the eye. Given that the insect was equipped with only a Z80 microprocessor, it was felt that a simple, non computationally intensive neural network was needed, one which learns rapidly and which could conceivably be implemented subsequently in hardware. The obvious choice, given the experience of researchers in the Department, was to use weightless or n-tuple networks.

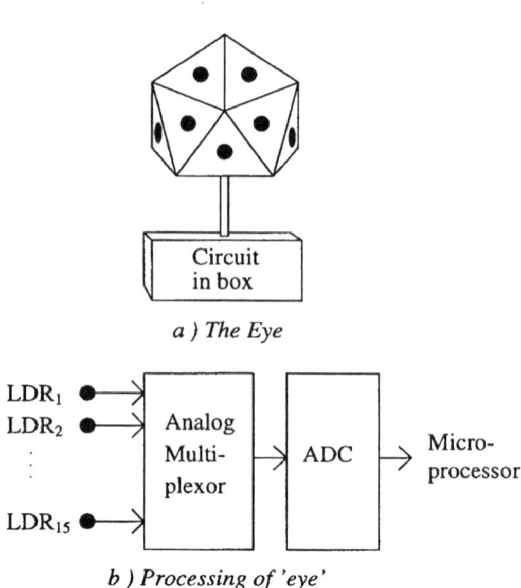

Figure 4.4: The first eye.

4.4.3 Weightless Networks

A weightless network [6] comprises many neurons where each neuron is a simple random access memory (RAM) which processes part of the input data. Initially all locations in each RAM contain '0'.

When the network is being taught an input, n bits are sampled randomly from the input data to form a tuple, and this tuple is used to address the first RAM, and a '1' is written into the addressed location. In this way, by one presentation of the data, the RAM neuron has learnt that tuple. The process is repeated, sampling the input to form the next tuple and then learning that tuples in the next RAM, until all the RAMs have been taught a tuple.

When the network is analysing its input data, tuples are formed as before, and a count is made of the number of RAMs which recognised their tuples, that is, how many RAMs had a '1' in their addressed location.

If the input to the network is the same as one already taught, then 100% of the RAMs will 'fire'. Such a neural network can therefore recognise inputs it had been taught. However, the system should be able to recognise similar inputs it has not been taught: it should be able to generalise.

This is achieved by teaching the network a number of similar inputs. Then, for instance, when an input is presented which it has not been taught, the first RAM might recognise the first tuple from the third input it was taught, the second RAM might recognise the second tuple from the seventh input it was taught.

4.4.4 Multi Discriminator Network

One set of RAMs is called a discriminator. Such a discriminator can be taught various sets of data and can report whether it recognises any of these data, but it will not be able to discriminate between them. However, if the network has many discriminators, then each data set is taught into its own discriminator only, and the system can report which data set, if any, from which a given input comes.

For the robots, the different data sets are different positions in the room: each position is taught into one discriminator. Thus the whole network should be able to report which part of the room the insect is in by seeing which discriminator outputs the highest response.

4.4.5 Processing Grey Level Data

The tuples sampled from the input consist of binary values. The inputs to the eye are 8-bit values; each value must be converted in some manner to a binary value. Three methods are commonly used, thresholding, thermometer coding [7] and Minchinton cells [8]; recent work has shown the advantages of Minchinton cells [9].

With thresholding, a value is taken from the input and compared with a constant value, the threshold, producing a logic '1' if the input value exceeds the threshold. With Minchinton cells, a given bit in the tuple is found by comparing two samples at random from the input data, returning a '1' if the first sample exceeds the second sample.

4.4.6 Weightless Networks and the First Eye

Attempts to use a standard weightless network with thresholding proved unsuccessful, so the data from the eye were sampled by an array of Minchinton Cells whose outputs were passed to a 150 RAM multi-discriminator network.

To quantise an environment to one of n positions, n discriminators were used, with each discriminator being taught to respond strongly to the light patterns characteristic of a particular position within a 4×2m section of the laboratory. First experiments used sixteen discriminators in a $[4 \times 4]$ array, spaced at 0.5m intervals along the x-axis and lm intervals along the y-axis. To achieve rotational invariance, each discriminator could be taught at five rotations of 72° around this point (see Figure 4.5). A tuple size of nine was selected to give a sharp discriminator response. Each discriminator was trained with eleven sets of data collected at random intervals throughout a week. No attempt was made to normalise laboratory lighting, window blind positions or other activity in the laboratory.

This process was then repeated using the same data, but this time training four discriminators in a $[2 \times 2]$ regular array, spaced at lm intervals along the x-axis and 2m intervals along the y-axis.

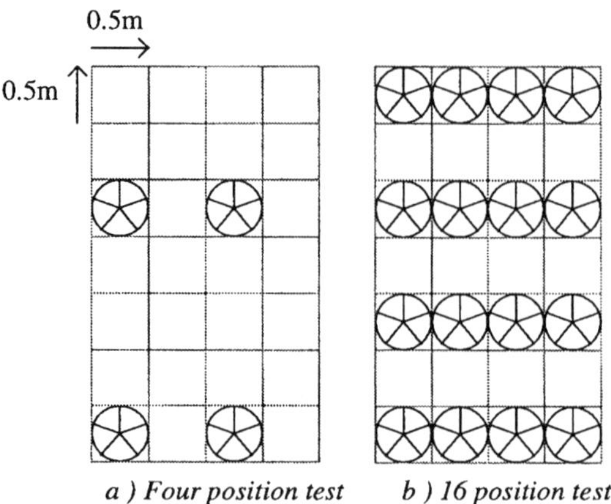

Figure 4.5: Discriminator positions.

4.4.7 Results

Figure 4.6 shows the output from a typical discriminator for both tests: on the left is the response from the $2 * 2$ test, on the right the response from the $4 * 4$ test. The figures show the response of the discriminator at various positions within the environment: the lighter the area the higher the response from the discriminator. For the $2 * 2$ test, the discriminator is that for the top left position; the response from the eye is at a maximum when the eye is in that position, and as the eye moves away from that position, the response decreases.

For the $4 * 4$ test, the discriminator is that for a position near the middle. Again, the response is highest when the eye is at that position and it falls off monotonically as the eye moves from the trained position.

Similar responses are found for the discriminators trained at the other positions.

4.4.8 The Second Eye

In the results described above, rotation of the insect was ignored. When, however, the insect was rotated, the performance of the system was worse than was expected. This was due to mismatches in the LDRs used, giving rise to large variations in output for a given incident light level. This meant that the RAM neurons had too many different tuples taught - the neurons saturated, which meant that the neurons 'recognised' too many input patterns, so there was insufficient discrimination between the positions in the room. It was also felt that 15 elements were insufficient.

Therefore a second generation eye has been produced. This has 32 phototransistors, mounted on the faces of half 'buckie balls', with the associated electronics mounted inside the ball. The phototransistors are well matched, so they should provide better performance when the insect is rotated, and have a

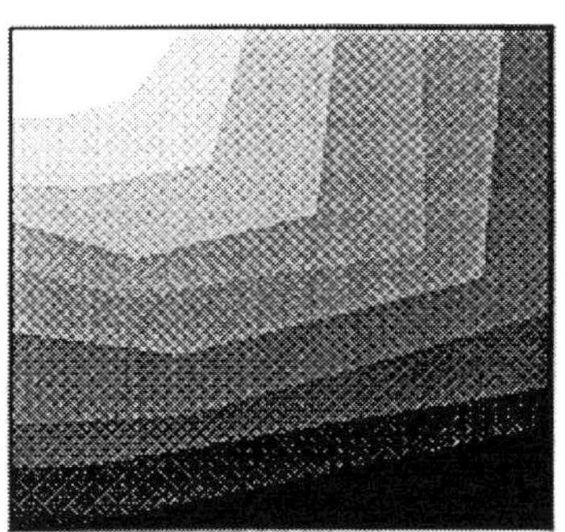
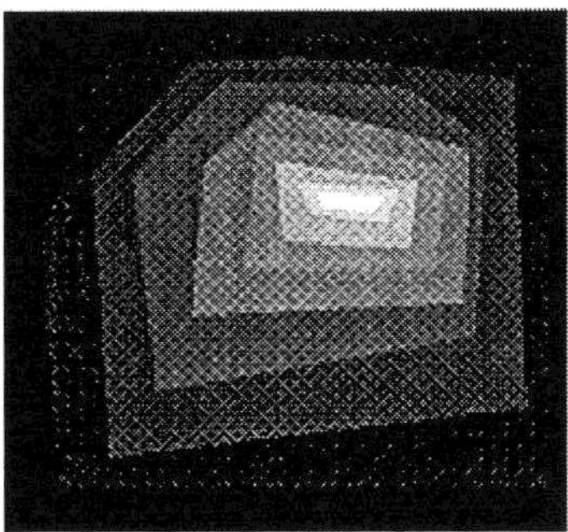

Figure 4.6: Output from one discriminator for each test.

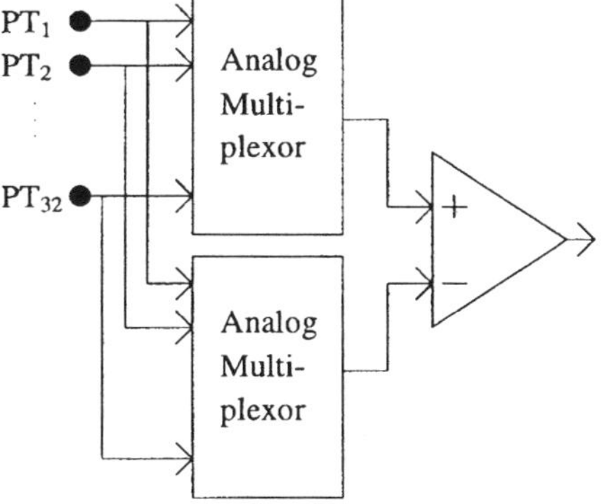

Figure 4.7: Circuitry of the second eye.

greater resolution.

The circuitry processing the signals from the phototransistors contains two sets of multiplexors and a comparator replacing the ADC, thereby incorporating the Minchinton cell processors. Using the comparators rather than an ADC eliminates quantisation noise. It also reduces the mains supply frequency noise which could otherwise cause fluctuations in the data with increased response time of the phototransistors.

A block diagram of an eye and associated circuitry is shown in Figure 4.7. The eyes were mounted on the robot and the Z80 on the insect was configured with a single discriminator weightless network. The insect was then taught one position in the room and then the response of the network found when the robot was in other positions. The results produced are consistent with the responses shown in Figure 4.6, but they also successfully cope with rotation of the insect.

4.5 Navigation

The output from the eye network gives both an indication of the approximate position of the room and an indication of the direction the robot must travel in order to reach a given position. In addition the ultrasonic sensors provide information which allows the robot to avoid obstacles. The next stage in the work to be described is how all this information is used to allow the robot to travel to a given position whilst avoiding any obstacle in its way. This is to be achieved using an algorithm inspired by the klinotaxic behaviour used by insects to navigate [10]. This behaviour inspired the Chemotaxis algorithm [11]: the basic idea of which is also used as an alternative to back propagation in multi-layer perceptron neural networks.

A key point in the results of the eye is that the output of the network decreases monotonically as the insect moves from the taught position.

The idea for navigation is as follows. The robot is set to move in a random direction, and it continues to move in that direction until it determines that it is moving away from its desired destination. This it can decide using the compound eye: if the response from the discriminator trained on the desired position decreases, the insect is going in the wrong direction. At such a time, the insect chooses another direction randomly, and repeats the process. Eventually the robot reaches the desired position.

However, an obstacle may be in the way. Therefore, a priority system is employed: if there are no obstacles visible to the robot, the 'chemotaxis' technique is used, but if there are obstacles, obstacle avoidance using the ultrasonic sensors is used to determine how the robot should move.

So far, this technique has been established in simulation only, though the learning rules and the compound eye have been successfully employed on the real robots.

4.6 Discussion

This chapter has discussed some of the work done in the Cybernetics Department on its simple mobile robots, in particular the research using neural networks. First a Hopfield-type network, which is used to allow the robot to 'learn' how to move around its environment avoiding obstacles. Next a compound eye is described with its neural network processing circuitry. Finally, a technique is outlined, based on an algorithm used in neural networks, for combining both networks to allow the robot to navigate back to a position it recognises within its environment.

4.7 Bibliography

[1] R.J.Mitchell, D.A.Keating and C.Kambhampati, Learning System for a Simple Robot Insect, Proc. Control '94, pp: 492-497, 1995

[2] K. Narendra and M.A.L. Thathachar, Learning Automata: an introduction, Prentice-Hall, 1989.

[3] J.J.Hopfield, Neural Networks and physical systems with emergent collective properties. Proc. Nat.Acad.Sci., USA 79. 2554-8, 1992

[4] D.E.Hebb, The Organisation of Behavour, Wiley, 1949.

[5] R.J.Mitchell, D.A.Keating and C.Kambhampati, Neural Network Controller for Mobile Robot Insect, Proc. EURISCON '94, pp: 78-85, 1994.

[6] I.Aleksander and T.J.Stonham, Guide to pattern recognition using random- access memories, IEE proceedings pt E, 2, 1, pp29-40, 1979.

[7] I.Aleksander and M.J.D Wilson, Adaptive windows for image processing, IEE proceedings pt E, 132, 5, pp233-245, 1985.

[8] J.M.Bishop, P.R.Minchinton and R.J.Mitchell, Real Time Invariant Grey Level Image Processing Using Digital Neural Networks, Proc IMechE Conf. EuroTech Direct '91, pp: 187-199, 1991.

[9] R.J.Mitchell, J.M.Bishop, S.K.Box and J.F.Hawker, Comparison of methods for processing 'grey' level data in weightless networks, Proc Weightless Neural Network Workshop 1995, pp76-81, Univ. of Kent, 1995.

[10] P.J.Gullan and P.S.Cranston, The Insects: An Ouline of Entomology, Chapman and Hall, 1994.

[11] H.J.Bremermann and R.W.Anderson, An alternative to Back-Propagation: a simple rule for synaptic modification for neural net training and memory, Internal Report, Dept of Maths, Uni of California, Berkeley, 1989.

Chapter 5

A Connectionist Approach to Spatial Memory and Planning

5.1 Overview

This chapter describes the design and testing of a biologically inspired vision-based model of spatial memory (SM).

Firstly, three theories of biological SM are discussed: the Stimulus-Reaction theory (e.g. Hull (1932) [29]), the Cognitive Map theory of Tolman (1948) [78] and the Topological Network-Map theory of Byrne (1979) [14]. The third one is the most promising theory of SM in humans and potentially allows for the most flexible planning at the lowest computational and memory cost in robot implementations.

Secondly, the Topological Network-Map theory is translated into general principles: i) successive perceptual states are linked, ii) pairs of states are associated with the action responsible for the transition between them, iii) links are exploited for building goal-directed sequences of states during planning and iv) backward planning is preferred to forward planning.

Two forms of connectionist implementation of these principles are discussed, using a sparse or a distributed representation of states. The sparse approach is the simplest and leads to a SM in the form of a view-graph (Schölkopf and Mallot, 1994 [70]). The sparse neural network model described here is characterized by: i) the use of a resistive-grid paradigm for backward planning, transforming the view-graph into a dynamic value-map; ii) the use of a separate network representing an inverse model of the robot - environment interaction; and iii) the use of state - object associations to drive planning by object-fetching tasks.

Thirdly, planning and map-learning experiments are performed with a robot. The perceptual states are defined as images from a narrow-angle on-board video camera. Objects are represented by letters drawn on cards placed in the environment. The instruction "fetch object A" is translated into "locate letter A".

These experiments reveal three problems with the implementation of the view-graph principle: i) local-minima due to interference between disjoint sequences of views built during exploration, ii) "representational dead-ends" in the form of views without a link to other views and iii) aliasing. The causes of these problems are discussed and solutions proposed. Solutions include a more complex vision system and a mechanism for internal integration of intermediate steps of plans.

5.2 Introduction

Spatial memory represents information on the location of objects in space in a form usable for spatial tasks. For instance, humans can be asked to indicate route (Ward et al., 1986 [84]), draw a map (Blades, 1990 [6]), estimate a distance (Sadalla and Magel, 1980 [66], Vann Bugmann, 1995 [81]) or manipulate shapes mentally (Huttenlocher and Newcombe, 1984 [32]). Mobile robots need spatial memory and planning capabilities in several situations: operation in environment without real- time remote control possibility (due to poor radio transmission of data or long light round trip delay in the case of planetary rovers) or no real time control possibility due to the inexperience of the user (household robotics). One of the most common uses of spatial memory is the planning of routes between start and goal places. We would like to give this latter ability to mobile robots to enhance their capability of executing high level commands such as "go to place X" or "bring me object Y".

Brain research is expected to provide clues on how to design an artificial spatial memory. There is a huge body of literature on psychological and neurophysiological aspects of spatial memory, the later notably centred on the hippocampal system (Burgess et al., 1995 [12]; Burgess and O'Keefe, 1996 [13]; Rolls, 1996 [65]) although recent imagery data indicate the involvement of wider systems (Mellet et al., 1995 [46], Maguire et al, 1997a [42], 1997b [43]). So far, there is no generally accepted theory of how spatial memory works in animals. Three schools of thought have emerged over the years (see section 5.3): those viewing spatial behaviour as sequences of reactions to stimuli (S-R), those believing that spatial behaviour results from cognitive processes based on internal maps, and those proposing that topological relations, not physical distances, are represented internally. In this chapter we report on experiments with a model implementing topological principles which promise more flexibility. We have not attempted to incorporate anatomical details but there may be advantages in doing so, as discussed in appendix 5.10.

An object fetching task requires vision for object recognition. We explore here the concept of also using vision for navigation. Potential long term advantages are: i) reduced number of sensors, ii) new knowledge from biology may more readily be transposed in the design, iii) users may give verbal route instructions to the robot using natural language, using visual landmarks in a way that communicates the "feeling" of taking the route (Ward et al., 1986 [84]).

Currently, there are four families of models of spatial navigation based on visual information[1]: i) homing navigators, ii) grid based planners, iii) view based navigators and iv) view-graph based planners.

Homing navigators use visual information to compute the direction of the goal or directly the motor command leading to the goal. Such systems are found in insects (Wehner et al., 1996 [87]) and in some models of navigation in rats (Burgess and O'Keefe, 1996 [13]) or for robots (Gaussier et al., 1996 [25], Franz et al., 1996 [22]). These systems are essentially "attracted" by the goal and can get caught in local minima for some obstacles configurations. On the positive side, they can also flexibly accommodate new starting positions and take shortcuts.

View based navigators are associating a particular action with a view. Thereby they can learn to follow a route based on views recognised along the route (McNaughton, 1989 [52]; Verschure et al, 1992 [82], Buehlmeier et al., 1996 [8]; Röfer, 1997 [64]) or based on signs placed along the route (Joulain et al., 1996 [34], Gaussier et al, 1996 [25]). These systems are "behaviour based" controllers (Brooks, 1986 [7]), which do not have the necessary flexibility for planning.

Grid based planners are based on the subdivision of the space into a fine grid, and vision is used for self localisation purpose. Each node in the grid is given a value, for instance the number of steps to the goal, that is then used for route planning. Various techniques can be used for computing the values (Barto et al., 1995 [5]). One of the fastest and most flexible is the resistive grid or Laplacian method (Conolly and Burns, 1993 [15], Siemiatkowska, 1994 [68], Bugmann et al. 1994 [9], 1995 [10], Althöfer, 1996 [1], Glasius et al., 1996 [27]). With grid based planners, a viable route is always found if it exists, between any two points on the grid. In closed environments of limited size, such techniques have been used successfully by mobile robots (see e.g. Kortenkamp et al., 1993 [36]). For navigation over large distances or planning in high dimensional spaces, these techniques are not practical, using too much memory and computation time.

View-graph based planners use temporal adjacency between views[2] from given places to encode the topology of the space (Schölkopf and Mallot, 1994 [70]). By associating the performed actions to each pair of successive views as well, goal-directed sequences of actions can be planned, if an uninterrupted chain of links from start-view to goal-view has been progressively assembled during past experiences. The model presented here is unique in: i) its scheme for encoding actions, ii) its use of a fast generalised resistive grid planning

[1]This is the most frequent biological case, and also the most interesting for autonomous robots applications. Other navigation models are based on ultrasound sensors (Recce and Harris, 1995 [62], Yamaguchi and Langley, 1996 [90], Thrun and Bücken, 1996 [77]) or optical range finders (Cox, 1991 [16]).

[2]A view is, in its simplest form, a picture of the environment taken by the robot.

method, iii) the possibility of planning towards objects found in views and iv) the use of narrow-angle images as views. This paper describes the first planning experiments with a view-graph in which views are images captured with a video camera on-board a real robot. These experiments reveal that view-graph based planners can suffer from a local minima problem (section 5.5). Solutions are suggested in section 5.6.

The design investigated in this paper is a view-graph based planner evolved from an earlier grid based planner described in (Bugmann et al., 1995 [10]). The proposed planning method is a generalisation of the resistive grid method aimed at extending the applicability of the resistive grid method into high dimensional problems, while preserving the good properties of grid based methods, namely the guaranty to find a solution, and reducing the computational load of the vision system. It should also avoid the self-localisation problem. In this generalised resistive grid method, nodes in the grid represent perceptual states, as defined by an image taken by a camera ("a view"), and do not require a Cartesian position to be known. The grid is constructed during exploration and has no predefined connectivity. Planning can be done by current spread along learned links, as in a standard resistive grid (Bugmann et al., 1995 [10]). Further, as the raw image defines a state, and implicitly the position, there is theoretically no need to extract features belonging to landmarks[3]. Obstacle detection is still needed, for avoidance purposes, but can be achieved by any low-level control mechanism, without the complexity of visual recognition and localisation. The long-term storage of the positions of obstacles is not needed, as only states/views corresponding to the free space are encoded and used for planning.

In traditional grid based techniques, where nodes represent a point in the problem space, the higher the dimension of the problem, the larger the number of nodes. This is the main limitation of these methods. Even 6 dimensional problems like a simple industrial robot arm require too many nodes for a precise control of the movement (Althöfer, 1996 [1]). Real life planning problems are of much higher dimension and are totally out of the reach of grid based methods. However, with the approach proposed here, nodes can code for points in very high-dimensional spaces but there are only as many nodes as new states discovered during the exploration process. Nevertheless, this number can become very large, as the experience of the robot increases. Some measures aimed at saving memory are discussed in section 5.6. These may involve off-line reorganisation of the spatial memory.

As for the organisation of the chapter, in section 5.3 the current knowledge of biological spatial memory is summarised, with emphasis on humans. Some of the difficulties with Stimulus-Response and cognitive map models are described. The principle of a network-map (topological) model is selected for implementation.

In section 5.4, an implementation of a network-map model is described. First, the principles of a spatial memory in form of association between states

[3]Although this may save memory, as discussed in section 5.6.

and pairs of states and actions are formalised. In particular, recall procedures are defined for planning. Then, the potential problems associated with connectionist implementations using distributed or sparse representations of states are discussed. A connectionist implementation of a sparse representation is preferred for feasibility reasons. Finally the neural network is described in details along with the learning and planning procedures.

In section 5.5, the application of the model to the control of a real robot with a video camera is described. As noted by Touretzky et al. (1994) [79], perception issues become important in real robot implementation. Accordingly, a simple visual set-up was used. Despite that, the visual image processing system, which is briefly described in appendices 5.8 and 5.9, is of greater complexity than the spatial memory network. The details of planning by the generalised resistive grid method are given. Finally, the learning and planning experiment are described. These reveal problems of local minima due to interferences between disjoint sequences of views built during exploration, representational dead-ends in the form of views without a link back to other views and aliasing.

In section 5.6 possible solutions to the problems encountered are discussed. These include remote place recognition and internal multi-step planning and integration mechanism. The conclusion follows in section 5.7.

5.3 Biological Spatial Memory

How is spatial information stored in the brain, and how is it used for navigation and route planning ? Three approaches are briefly outlined below:

5.3.1 Stimulus-Response Theory

The so called "behaviourist" school of thought saw spatial navigation as a sequence of response-movements caused by conjunctions of internal drives and external stimuli encountered along the path. Learning of the stimulus response pairs (S-R) is driven by reinforcements received during exploration or when reaching a goal (Spence, 1950 [72]). As the response R is produced by a conjunction of a stimulus S and a drive D, the approach should really be termed S,D-R, but the abbreviation S-R is commonly used. The S-R approach raised a number of questions. Some are due to a literal interpretation of the abbreviation "S-R": how can the same stimulus lead to different responses in case of different needs? Other are justified. For instance: how can two stimuli with similar appearances (e.g. different intersections in a maze) be associated with different actions? How does latent learning occur (learning of spatial relations without associating a need with each landmark)? How can a basis of S,D-R triplets be used to respond to new problems? Most of these questions were answered with complex theoretical constructs that were not general enough and difficult to confirm or infirm experimentally (see for example Hull, 1932 [29], 1934 [30], 1935 [31]).

O'Keefe and Nadel (1978) [55] argue that a route based navigation system is fragile. Routes are taken to be a list of landmarks with instructions on how to go from one to the next. If an instruction is lost or if the appearance of a landmark changes, unrecoverable navigation errors can occur. O'Keefe and Nadel (1978) [55] noted that experienced migratory birds have a robust navigation system. Also, monkeys can head straight to a point that they have reached once using a route with numerous detours.

McNaughton (1989) [52] and McNaughton and Nadel (1990) [53] have suggested that the Hippocampus might be part of the route based navigation system, by associating scenes with actions. Evidence of the respective involvement of the Basal Ganglia and the hippocampus in the learning of S-R pairs are reviewed by Curran (1995) [17].

In their present state of development, S-R concepts may account for stereotyped "habits" or "automatic" behaviours but they cannot explain the freedom of choice and creative problem solving that (sometimes) characterises human behaviour. Therefore, the S-R approach has not been accepted as a general model of biological spatial memory. However, it is still used to model insects (see for example Webb, 1995 [86]) and is investigated for the control of simple robots (see for example Brooks, 1986 [7]).

In recent S-R models for robot navigation, routes are stored in the form of sequences of S-R pairs such that the sensory input in one part of the route generates the action allowing to reach the next part of the path (Verschure et al., 1995 [83]; Buehlmeier et al., 1996 [8]; Rao and Fuentes, 1996 [61]). With a repertoire of such routes, planning can theoretically be performed (Hull, 1935 [31]). In models by Arbib and Lieblich (1977) [3] and Schmajuk et al. (1993) [67], drive-reduction values are associated with intermediate Stimuli in a route (as in Hull (1932) [29]) so that routes can be selected on the basis of the drive of the moment. Such systems do not have the flexibility to execute instructions such as those mentioned in section 5.2, because a Stimuli would need to to have a value attached for each possible goal.

5.3.2 Cognitive Maps

In reaction to the behaviourist thinking, Tolman (1948) [78] introduced the concept of cognitive map where an action is the result of a "cognitive" process performed on a corpus of internal knowledge ("cognitive field"). Perception is seen as an information gathering process which modifies the cognitive field rather than driving the behaviour directly. Using this internal knowledge was seen as analogue to reading a geographic map to plan a route. Tolman described cognitive maps as "relatively narrow and strip-like or relatively broad and comprehensive":

> "In a strip-map the given position of the animal is connected by only a relatively simple and single path to the position of the goal. In a comprehensive-map a wider arc of the environment is represented, so that, if the starting position of the animal be changed, or if variations in the specific routes be introduced, this wider map will allow the animal still to behave relatively correctly and choose the

appropriate route".

It has recently been noted that the rat experiments cited by Tolman (1948) [78] in support of his concept did not require the use of spatial maps. The tasks could also been realised using "cue guidance" strategies (O'Keefe, 1991 [57]) or "associative" principles (Wishaw, 1991 [88]). However, other experiments with monkey support the existence of cognitive maps. In these, locations of food caches are shown in a random order. The monkey usually retrieve the food in some economically ordered path (Menzel, 1973 [48]), which is consistent with the idea that an internal map is being used. Thompson (1980) [74] reports that blindfolded subjects can avoid previously seen obstacles, by updating their position on an internal map of the environment. This map seems to "fade away" after 8 seconds.

Support for a geographic-map-like spatial memory can also be sought in experiments showing that subjects can draw maps "as seen by air". However, this seems easier based on a verbal description of a scene (Ferguson and Hegarty, 1994 [21]) than based on personal experience navigating in an environment (Blades, 1990 [6]).

Other findings cannot be accounted for by a map model, such as the curious fact that the time taken to estimate the distance between two towns in different states is smaller than for two towns in the same state. This was taken as evidence that spatial memory is organised hierarchically (see a review by Tversky, 1992 [80]). Further, the number of intersections of the route linking two places (Sadalla and Magel, 1980 [66]), or the subjective "importance" of the places (Vann Bugmann, 1995 [81]), influences the estimated travel distance between these places.

On one hand these observations support the notion that large-scale spatial memory is not represented in a strictly "geographic map-like" way[4]. On the other hand, the data reported by Thompson (1980) [74] suggest that, at least for the immediate environment and temporarily, a map preserving the exact geometry may be used.

Presson et al (1989) [60] distinguish two forms of cognitive maps. The first, an episodic memory, contains information in a picture-like format and with a specific orientation determined at the time of perception. The second is a stable world model in which the subject can move and observe. The orientation of the model is in register with the subject's physical orientation, but can be rotated mentally in an effortful process (Rieser, 1989 [63]).

A world model is closer to the nature of our sensation when living in the 3-D world, representing the point of view of an actor immersed in the environment (Presson et al., 1989 [60]), while a cognitive map in its literal interpretation is a picture of a map (Levine et al.,1982 [39]). Acting in the real world may recruit brain mechanisms similar to those used for mental operations in the world model (Rieser, 1989 [63]; Maguire et al., 1997b [43], Mellet et al., 1995 [46], 1996 [47]).

[4]Formally, this conclusion is not very strong, because the mechanism for retrieving the information may also be affected by these factors.

In the hippocampus of the rat, cells are found which fire when the rat is in a given place (egocentric place representation) (O'Keefe and Speakman, 1987 [56], Speakman, 1987 [71]). In the postsubbiculum, near to the Hippocampus, cells are found which indicate the direction of the head relative to the cues in the environment (Burgess et al., 1995 [12]). Both place cells and head direction cells give the kind of information needed to align an internal world model and visual sensory perception (see for example Recce and Harris, 1996 [62]).

Recently, Rolls (1996) [65] found that cells in the Hippocampus of the monkey respond to views of specific places, providing an allocentric space representation. These cells may be used to encode the position of objects.

The map concept is attractive for theoretical reasons: a map has the property to be observer independent, can be rotated and read from various positions and can be used to solve route planning problems. In a map, relations between objects are encoded, not relations between observer and objects. O'Keefe and Nadel (1978) [55] proposed that a cognitive map is located in the Hippocampus.

In robot applications, a strict transposition of the map concept leads to encoding the positions of objects and the robot using Cartesian coordinates. For planning, the obstacle-free space must be calculated on a detailed map, this is then divided into cells, then a standard grid based planning technique can be used (Latombe, 1991 [37]). This approach is associated with the same self-localisation problem[5] as for grid based planners (section refs1). Planning over long distances, e.g. between two parts of a town, requires a very detailed map and huge computation time. Most autonomous mobile robots of today use Cartesian planners.

5.3.3 Topological Network-Maps

The Network-Map theory of spatial memory proposed by Byrne (1979) [14] states that the mental representation only represents topological connectedness, but not two-dimensional distance information.

Moar and Carleton (1982) [49] studied the acquisition of spatial knowledge as inferred from a sequence of pictures representing views along two routes with a common segment[6]. They found that after only one presentation of the two sequences, judgements of distances or directions were as accurate between two points belonging to the same route as those belonging to different routes. This was taken as evidence that the knowledge about the two routes was combined from the start in a single network of routes, in support of the network-map representation proposed by Byrne (1979) [14].

Such a representation can be used to plan routes (although only along known segments linking places) and may be compatible with the distortions of the perceived distances noted above (although a detailed explanation remains to be produced).

[5]The availability of the Global Positioning Satelite System (GPS) may solve this problem in some applications.

[6]The routes formed a H and their common segment formed the horizontal bar of the H.

However, Moar and Carleton (1982) [49] also found that the direction judgements[7] of the subjects improved greatly with experience. This was taken as evidence that although a network-map model may account for the initial state of knowledge during the learning phase, a more vector-like representation may emerge later, in which straight-line distance and relative directions of places are represented.

Evans et al. (1981) [20] found evidence that landmarks are used as initial anchor points and that only their relative positions[8] were stored. With increasing experience, an increasingly dense network of routes linked these landmarks[9] and provided more positional constraints, enable subjects to become more precise in estimation directions and Euclidean (straight-line) distances. Thorndyke and Hayes-Roth (1982) [75] found a similar increase in accuracy as a consequence of repeated navigation through the environment

In contrast, the accuracy of straight-line distance estimations by taxi drivers in Paris is similar to that of office workers who have lived for the same time in Paris (Peruch et al., 1989 [59]). This suggests that experience is not the factor that develops a (better) survey knowledge.

How is the length of a route encoded? Measurements show that subjects can estimates distances relatively precisely. When asked to estimate the length of a route, they sometimes spontaneously give the travelling time first (Vann Bugmann, 1995 [81]), as if time is more readily retrieved than distance. There is a linear relation between estimated time and distance, suggesting that subjects may infer the distance from the travel time. It is not known where and how time is stored in the brain[10] nor how it is retrieved by subjects to produce their verbal responses.

Muller et al. (1996) [50] have proposed that the Hippocampus might implement a "cognitive graph" in which the links between places-cells corresponding to neighbouring positions are strengthened.

In robotics, the use of topological network-maps has only recently started to be explored (Schölkopf and Mallot, 1994 [70]; Bachelder and Waxmann, 1995 [4]). It may be noted that, although the links between landmarks in network-maps may be implemented in the form of S-R-S chains, they represent task-independent spatial relations. This is to be contrasted with S-R systems in which some "goal-specific utility" is encoded in the links or nodes[11]. Thus

---

[7] "Assume that you are in the same position as the photographer who took this picture. Point in the direction of landmark x."

[8] Relative position refers to "A is closer than B" or "A is to the East of B", etc

[9] It commonly assumed that there is something distinctive about certain elements of a landscape which makes them good candidates to become landmarks. This aspect of the theory is rather vague. Formally there is a difference between a place and a landmark, a place being best defined by the configuration of objects or landmarks seen from that place. One can stay at a place but see a landmark. Thus route are formally links between places rather than landmarks.

[10] Animals can learn time intervals and conditioning experiments have been performed with durations of over 15 minutes (Lejeune and Wearden, 1991 [38]). A model of a neural mechanism for timing is found in (Bugmann, 1997 [11]).

[11] With resistive grid planning method described in section 5.4.5, the value of a Stimulus is assigned dynamically, in a task dependent way.

S-R systems can only be used for reaching predefined, built-in goals.

5.3.4 Planning

How do humans plan their movements? Human route planning is a multi-criteria optimisation process, involving considerations such as safety, scenery, speed, cost, etc. It is not just a problem of linking source position to goal position. Planning social behaviour is even more complex, due to uncertainty regarding current situation and also the outcome of actions.

In the early days of AI, methods were designed to mimic the way humans solve problems like mathematical puzzles, demonstrate equations or play chess (Newell and Simon, 1972 [54]). In all these activities, the main problem was to discover a set of operations and intermediate states which allow an initial state to be linked to a goal state. In the spatial domain, this is equivalent to exploration of an environment with unknown connectedness. In contrast, planning consists of finding a path in a situation where links and intermediate states are already established. There are no indications on how this process is performed by humans. For instance is forward or backward planning used (see section 5.4)? Gray (1995) [28] argues that planning is a subconscious process where only the result is available to consciousness.

Not much is known on the neural mechanisms of planning (Altman, 1995 [2]). Jeannerod (1994) [33] proposes that planning is the assembling of sequences of movement schema stored in posterior parietal cortex. Shallice (1988) [69] gives evidence that planning is crucially dependent on the prefrontal cortex.

The notion of planning by spread of activation has been mentioned several times in relation to network-map models (Byrne, 1979 [14]; McNamara et al., 1989 [51]) or in S-R models (Mataric, 1990 [45]; Maes, 1989 [41]).

5.3.5 Summary

In summary, there are more questions than answers regarding biological spatial memory. Topological maps seem to stand the best chance of forming the basis of a plausible model biological spatial memory. However, they need to be made consistent with the availability of map-like (vector-like) knowledge to experienced subjects and the memory experiments cited by Thompson (1980) [74].

One approach is to assume that the relatively simple routes used in the experiments of Moar and Carleton (1982) [49] may have enabled subjects to integrate the paths mentally and infer the bird's eye view of the local topography. Thus its is conceivable that a picture-like map may be constructed using mental inference, similar to a map built from verbal descriptions (Ferguson and Hegarty, 1994 [21]), possibly on a temporary basis (Presson et al, 1989 [60]). Another possibility, which is touched on in the discussion (section 5.6), is that the inference of straight-line distances may lead to the incorporation of straight-line links into a topological spatial memory which thereby may gain vector-like characteristics.

Speculatively, if the network of links then becomes dense enough, it may be possible that a topological map becomes functionally similar to a world model. This is in the sense that, to whatever object the mind's eye devotes attention, the map will produce an associated view and indication of direction. Conversely for attention devoted to a direction. It is unclear if the memory experiments cited by Thompson (1980) [74] can be explained along similar lines.

Topological network-maps could perhaps also be made consistent with the existence of habits. These are mainly revealed when something in the environment has been changed so that a usual route becomes inadequate. Taking the usual route in such a situation may be due to planning being done with obsolete information or, in other cases, with information provided by the memory rather than by visual perception. Thus the expression of habits may be due to planning with a "lazy" information management system rather than the use of a separate S-R system.

Regarding robot implementation, cognitive maps and S-R models have a number of known limitations. Thus we will explore further the concept of topological network-map in the remaining of the paper.

5.4 Connectionist Implementation

5.4.1 Principles of a Network-Map Based Artificial Spatial Memory

The following general principles are proposed to describe a network-map model:

i) The sensory experiences of a robot are called states (or "scenes" or "view"[12] when those experiences are predominantly visual): S_{t1}, S_{t2}, ... The indexes t1, t2 indicate different instants, separated by any time interval.

ii) The states are represented in a linked way, such that if S_{t2} is the next state experienced after S_{t1}, the fact that S_{t2} can be reached from St1 is encoded and exploitable for planning.

iii) The action $a(S_{t1}, S_{t2})$ performed to reach S_{t2} from S_{t1} is also memorised, and encoded in such a way that it can be recalled and re-executed if S_{t2} has to be reached again when the system is in state S_{t1}. An action can be a "ongoing task" rather than an action with a predefined duration. For instance, the statement of a goal S_{t2} and the instruction "move forward", can be taken to mean "move until S_{t2} is encountered".

iv) Backward planning, to reach a state S_{tn} is performed by recalling all states S_{tm} which have been immediate precursors to S_{tn}, then by recalling all states precursors to the states S_{tm}. The process is repeated iteratively until one the recalled states corresponds to the current state. The first action to be done in the plan is the one that is memorised as having caused the transition from the current state to the one following it in the recalled sequence.

It should be noted that backward planning differs from forward planning which operates by recalling all known successors to the current state, then

[12] The terms "view", "state" and "scene" will be used interchangeably.

the successors of these successors, and so on until the target state is found (Schölkopf and Mallot, 1994 [70]). The advantage of backward planning over forward planning is that, at the end of the search, the action to be done is immediately known because it is inferred from the last updated states. Forward planning only tells that a path (sequence of actions) exists.

An "ongoing task" could be implemented by a "Behaviour" in the sense of Brooks (1986) [7][13]. An example is to "walk on the pavement in this direction until...". Using "ongoing tasks" allows memory to be saved, as illustrated by the problem of planning a route that may include a train trip. In that case, there is no point in creating a memory of all places along the route or using this information for planning. By using an ongoing task like "staying in the train", a single link suffices between the starting station and the arrival station. Indeed, a control system is needed to determine when to encode or not to encode a scene. Alternatively, sequences of stored action / perception could be "folded" at a later time to reorganise the memory and clear useless sub-sequences. Therefore, the principle v) is proposed.

v) The content of the spatial memory can be examined after a series of experiences in such a way that a sequence of states $\{S_{ti}, S_{tj}, ..., S_{tk}, S_{tl}\}$ can be "folded" and a direct link between S_{ti} and S_{tl} can be created. Concurrently, a "procedure" p(S_{ti}, S_{tl}) is created which replaces the sequence of actions $\{a(S_{ti}, S_{tj}),\ ....,a(S_{tk}, S_{tl})\}$. The states $S_{tj}, .., S_{tk}$ and the associated actions can then be removed from memory.

In our model, only the principles i) to iv) will be implemented. The need for principle v) and constraints on its implementation are discussed further in section 5.6.

5.4.2 Sparse Versus Distributed Representation of Views

A view is a set a of measurements made on the environment which are characteristic of a given position and orientation of the robot. The view can be a set of marks on the floor (Mallot et al., 1995 [44]), a 360 degree image provided by a conic mirror (Franz et al., 1996 [22]) or a rotating photo sensor (Röfer, 1997 [64]), a 360 degree panoramic view log-transformed for maximum detail in front (Bachelder and Waxman, 1995 [4]), a set of objects associated with the viewing angle (Gaussier and Zrehen, 1995 [23]), a recognition grid based on ultrasound reflectometry (Yamauchi and Langley, 1996 [90]) or a set of ultrasound distance readings covering 360 degree (Recce and Harris, 1996 [62]). In the system used here, the view is a 30 degree narrow angle image produced by a video camera.

The spatial memory system to be designed is a memory of linked states (views), associated with action information. One may see it as a memory of sequences of $state_i \rightarrow action_{ij} \rightarrow state_j$ but there is more than sequence storage in this problem. The states and actions must be represented in such a

[13]In the sense of Brooks (1986) [7], a "Behaviour" is a low level S-R program dealing with tasks like "keeping balance", "avoiding obstacles", "follow a wall", etc. The combination of several "Behaviours" is supposed to result in the observable behaviour in the usual sense.

way that the system allows planning and extraction of action information. Two approaches to that, distinguished by a distributed or a sparse representation of states, are now discussed briefly.

Distributed representation:

In a distributed representation, a state is represented by the activity of a large number of nodes. The most distributed representation is the raw image, as all pixels contribute to the definition of the scene.

Figure 5.1a illustrates a distributed approach. A scene is encoded by the activity of a set of nodes N_1, representing features extracted from an image or being a copy of the image itself. A neural network "Linking" learns to reproduce the past scene S_{t1} from the current scene S_{t2} and projects it to N_2. Another neural network "Transition" learns that action $a(S_{t1}, S_{t2})$ has caused the transition between the two scenes. Replaying a sequence of scenes proceeds by setting the activity in N_1 to S_{t2}, and choosing an action a in the Transition network. The combination of S_{t2} and the action a allows then the "Linking" network to reconstruct in layer N_2 the scene S_{t1} having preceded S_{t2}. Although N_1 and N_2 are represented here as two separate layers, they could also be merged into a recurrent architecture of the type seen in (Schmajuk et al., 1993 [67]) and (Bachelder and Waxman, 1996 [4]), although these authors do not use distributed representations.

Backward planning is done by setting the activity in N_1 equal to that of the goal scene. Then all possible predecessors are replayed until the current state is found. As the actions distinguish the predecessors, the linking must be action-dependent. For forward planning, the system should be modified to reproduce S_{t2} from S_{t1}. Planning is done by replaying all scenes that are possible successors of the current one until the goal scene is found. Planning by sequence replay is a sequential process taking a time proportional to the depth and the number of branches of the tree to explore. There is also a need for some form of memory of scenes already replayed. So far, there is no working example of a planner based on a distributed representation.

Sparse representation:

In a sparse representation, the activities of a small number of nodes represent the scene. The most sparse representation is the one where a single node represents a scene.

Figure 5.1b illustrates a sparse approach. A scene is represented by one node at the output of a "decoder" network. Directional lateral connections encode the experienced transitions between scenes. A "transition" network encodes the action that was done to cause the transition.

For backward planning, these links allow S_{t1} to be activated from S_{t2}. Planning is done by activating the goal state and letting the activation propagate until the current state is activated. For forward planning, the links must activate S_{t2} from S_{t1} and planning is one by activation propagation from the current state to all the states until the goal state is activated. If the view graph is implemented as a resistive grid, current gradients or potential gradients can

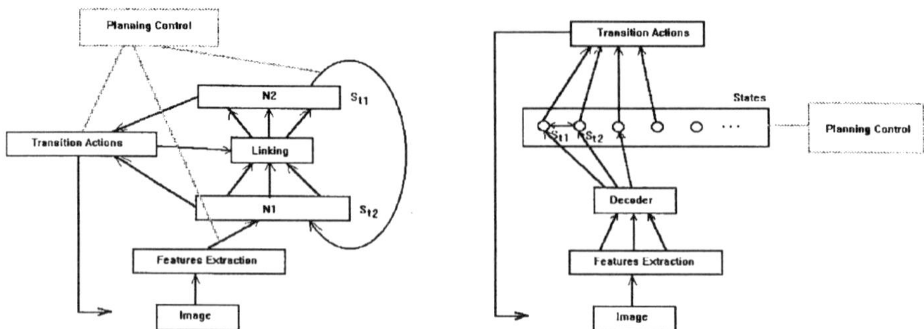

Figure 5.1: **a) Left:** Distributed spatial memory. **b) Right:** Sparse spatial memory. The operation principles of these models are described in the text.
i) In both cases the raw image is first pre processed to highlight relevant features, e.g. regions of high contrast. In the distributed case, these features are copied into the layer N_1 and represent, as a whole, a state. In the sparse case, the features are classified and the state is represented by one active node at the output of the decoder which activates its corresponding node in the "State" layer.
ii) During exploration, in the distributed case, the "Linking" module learns to reproduce the past state from the current state and from the action just done. In the sparse case, lateral connections are built in the state layer to encode transitions and the action having caused the transition is encoded in the "Transition Actions" layer.
iii) During planning, in the distributed case, the planning control module must manage the sequential replay of predecessors of a state S_{tn} (goal state). In the sparse case, the controller manages the spread of activation from the goal state nodes in the grid to all other connected nodes in the grid. In both cases, planning ends when the current state is reached (this is an example of backward planning).
iv) The state which was activated just before the current state becomes the next subgoal of the plan. The pair of states (current state; state which was activated just before) defines the action to be executed.

be used to determine the neighbour nearest to the goal. In that case, backward and forward planning are parallel search processes exploring all branches at the same time and taking a time proportional to the depth (length of sequence to find). The use of spiking neurons for planning is discussed in appendix 5.10.

A sparse approach was used in (Schölkopf and Mallot, 1994 [70]) for encoding the links. However, backward planning was used. For that purpose, the action nodes were connected to the relevant links, in order to modulate their strength. During planning, each action node was activated in turn to determine which neighbour of the current state node the action allowed to reach. Activity was then spread forward from this node to all connected nodes until the goal was reached. The neighbour that was nearest to the goal, in terms of spreading time, was then chosen as the next to be reached. This search process had to be repeated for each intermediate state and each possible action. With backward planning using a resistive grid method, a single current spreading process can suffice to set-up a potential landscape that is usable in each state and does not need to be recomputed until the goal is reached (Bugmann et al., 1995 [10]).

In (Bachelder and Waxman, 1995 [4]), conjunctions between the current state and the current action are used as input to a network that is trained to predict the next state. It is a representation of spatial knowledge in the form of S,R-S triplets. This form of spatial memory could support forward planning with the some of the complications described above for distributed representations.

A variant of sparse representations are topological maps[14] (see for instance Gaussier and Zrehen, 1994 [24], Buehlmeier et al., 1996 [8]; Zimmer, 1996 [91]). These maps are usually two or three-dimensional grids of nodes. States with similar sensory signatures are assigned to nodes with positions close to each other in terms of physical distance in the grid. This property may be used to generalise the properties discovered in one place to another place (Gaussier and Zrehen, 1994 [24]). For planning however, this generalisation may cause difficulties: similar sensory patterns do not necessarily correspond to similar positions in the world (Bachelder and Waxman, 1995 [4]; Rao and Fuentes, 1996 [61]). This is partly avoided in (Zimmer, 1996 [91]) where the estimated position is part of the "sensory" pattern[15]. Overall, the approach may deserve more investigation. One reason is that it makes sense to reduce the high dimensional feature space into the low dimensional problem space (position and orientation).

We may note that all existing models use a form of sparse representation. For practical implementation and testing with a robot, there are a number of reasons to prefer a sparse representation over a distributed one:

i) Planning time is shorter in the sparse model than in the distributed case because only single nodes need to be activated, taking at most one time step per pair of scenes. In the distributed case, the reproduction of scenes involves at least one extra-layer of nodes in the network "linking". Thus a pair of scenes

[14]These "topological maps" inspired by Kohonen (1988) [35] are not to be confused with topological network-maps or viewgraphs.

[15]This in turn requires to estimate ones position, one of the problems that were at the origin of the view-graph based approach...

takes at least 3 time steps, including the time to copy N_2 into N_1. In simulation of the distributed case on a serial computer, extra time is used to loop through all the neurons representing a scene.

ii) Planning in the distributed case requires a more complex control circuitry that in the sparse case, to manage the sequential search process. In the sparse case, a simple resistive grid algorithm can perform the search (Bugmann et al., 1995).

iii) The distributed case requires a powerful "Linking" module that can generate one image from another. The only architecture that can learn fast without catastrophic interferences is a layer of RBF nodes where one node is recruited per association to learn. This uses potentially more nodes than the sparse representation where there is one node per state (image) and each association is represented by a link[16].

iv) Every aspect of a sparse implementation is realisable with existing neural network tools and there seems to be no functional advantages to justify the design of the more complex distributed system. Details of the implementation of a sparse model are given in the next section.

5.4.3 Implementation of a Sparse Model.

Figure 5.2 summarises the model investigated here. The view nodes represent a low resolution version of the image of the environment seen by a video camera (see appendix 5.9). Each state node is trained to respond to only one view, as described below. The focus nodes represents the part of the image in which an object is found. In our tests, objects are characters. These are surrounded by a frame, which simplifies the focusing process, as described in appendix 5.8. Each "character" node learns to responds to a character when it is first met during exploration (see appendix 5.8). State nodes and Characters nodes use Gaussian Radial Basis Functions (RBF) (see equation 5.1) which can be trained from a single presentation.

The identities of the characters found in the scene are not explicitly used to define a state. Characters are already part of the visual features defining the scene and there is no need for additional high-level "semantic" concepts, as represented by the outputs of "character", to be used to define states. However, the high-level representation of letters needs to be associated with the states in which they are observed so that they can activate the corresponding state nodes as part of the planning process described further below (section 5.4.5).

Due to a peculiarity of RBF nodes, the association between characters and states needs to be done via a second layer of node: the output y_i of a RBF node (5.1) is significantly activated only when all its inputs x_j have an activity close enough to the learned activations w_{ij}, typically much closer than the half

[16]The sparse model ends up requiring one node per transtion too, in the action-encoding module (see section 5.4.4).

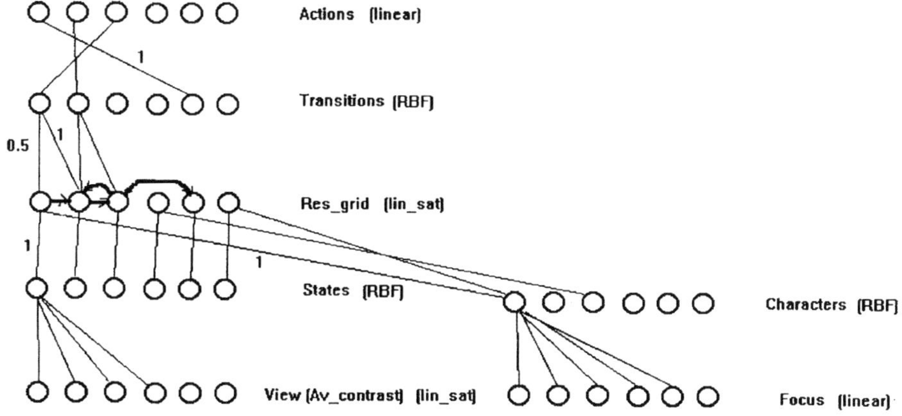

Figure 5.2: The type of neurons used in each layer are indicated and the weights of the connections. This network is built during exploration and increases in size as the spatial knowledge increases. See details in the text.

width σ (see appendix 5.9):

$$y_i = \exp(-\frac{\sum_j (x_j - w_{ij})^2}{2\sigma_i^2}) \tag{5.1}$$

Therefore an active "character" node would not alone be able to activate a state node and a view would not activate a state node if the associated character was not activated beforehand. For this technical reason, a second layer of nodes is used which is a copy of the "states" layer (see Figure 5.2). This layer is the "resistive_grid". It has nodes with linear-saturating transfer functions:

$$\begin{aligned} y_i = & \quad \textstyle\sum_j x_j w_{ij} \quad if\ 0 < \sum_j x_j w_{ij} < 1 \\ y_i = & \quad 0 \quad if \sum_j x_j w_{ij} < 0 \\ y_i = & \quad 1 \quad if \sum_j x_j w_{ij} > 1 \end{aligned} \tag{5.2}$$

The control of this system in Figure 5.2 is not implemented using connectionist techniques. A relatively complex procedure written in a Basic-type language determines when given layers of the neural network are updated. Decisions like switching to an exploratory behaviour are based on tests done on the activity of nodes at given times. More details on the robot-user interaction scenario built into the procedure are given in section 5.5.2.

A purely connectionist implementation of the procedure would represent an extremely complex neural network. The case of the planning procedure is discussed in the appendix 5.10 where it is suggested that the use of spiking neurons may simplify greatly the control circuit. In contrast, the present connectionist implementation involves neurons with graded responses or binary gates.

5.4.4 A Constructive Learning Procedure for States and Actions

How are views and actions selected for building a view graph ? This is the problem of exploration and novelty detection which has not yet been solved in a conceptually satisfactory way. In the model of Yamauchi and Langley (1996) [90], new state nodes are created when the robot has moved a given distance from the last known position. In (Bachelder and Waxman, 1995 [4]), and (Recce and Harris, 1996 [62]), the space is divided into regular squares and it is attempted to create only one place node per square. In (Franz et al., 1996 [22]) novelty is determined by the response level of nodes. None of these methods is optimal. It has been suggested that predictive mechanisms, possibly based on egomotion, could be used to detect new (defined as "unpredicted") situations and drive the learning of new states (Denham and McCabe, 1995 [18], 1996 [19]). We have experimented with an approach based on the principle of maximum information gain (see a discussion in Franz et al., 1996 [22]), in which the preferred next action was one that had never been executed in the current state. This was too time-consuming to be practical (section 5.5.2). Consequently, we simply used repeated displacements of small amplitude. New views also appeared during execution of a task (section 5.5.2).

A number of the connections in the network in Figure 5.2 are constructed during exploration or during the task, to encode new knowledge regarding {state - letters} associations and {state pairs - action} associations. Thereby, the size of the neural network increases with experience. For most of the existing connections, the weights are set during the task. Only the connection weights between the states and the resistive grid are preset to 1.

When the robot meets its first scene, the first node in "state" is trained to recognise that scene (with the RBF nodes used here, this corresponds simply to set the input weights equal to the input activity pattern). If a letter is also found in the scene, a connection is created between the corresponding character node and the node in the resistive grid corresponding to the current state, with weight 1. When an action results in a new scene to be found, the next state node is recruited and trained for this new scene. A scene is considered to be new if no state node responds with an amplitude above a certain threshold (see appendix 5.9).

A connection from the first node to the second node in the resistive grid is made, with weight 0.1. These two nodes are also connected to a transition node with weights 0.5 from past state-node and 1 from current-state node (transition nodes are of RBF type with $\sigma = 0.3$). Finally, the transition node is connected to the node coding for the action just performed, with weight 1.

The connections from resistive grid to action nodes are designed in such a way that, after the successor of the current state is identified by the planning process (see below), setting the activations of the res_grid node corresponding to the current state to 0.5 and that of its successor to 1 leads to the activation of the action known to have caused the transition between the two states. The connections in the resistive grid may be bi-directional, if there are corresponding actions. With lateral weights of 0.1, repetitive updating of the grid does usually not lead to a saturation of the activities.

At early stages of the design of the system, a N_s x N_s matrix was used to encode the actions linking any two pairs of the N_s possible states[17]. While experimenting with the robot, it turned out that several actions could cause the same transition. For instance, a "forward" command could result in a leftward rotation if one of the wheels was blocked by some small obstacle on the table. This would erase the previous information stored in the matrix, because the "left" motion previously encoded as linking the current view to the view to its left would be replaced by the "forward" motion. It was therefore necessary to enable the encoding of several possible actions as linking two states. The constructive encoding procedure proposed here enables several transition nodes to be assigned to the same pair of states.

A future refinement of the encoding should reflect the probability that an action causes a given transition between two states, requiring a neural version of Q-learning (Watkins and Dayan, 1992 [85]) to be developed for planning.

5.4.5 Planning and Search Procedure

In the task assumed here, letters are objects that the robot may be asked to retrieve. As part of the planning procedure, the node in "character" corresponding to the target letter is set to an activity of 1. This activates the nodes in the resistive grid corresponding to the states where the letter has been observed. In addition, the node in the resistive grid corresponding to the current state is inhibited, with an activity set to zero.

For planning, the resistive grid is then updated several times, until a stable distribution of activations is obtained (Bugmann et al., 1995 [10]). If the current-state node has a neighbour with non-zero activity, this indicates that there is a path leading to one of the states where the target letter has been observed. Otherwise, the program assumes that no path is known and enters into an exploratory mode which, at present, consists of asking the user for a rotation direction.

If during this exploration, a state is encountered that is part of a path linking to the goal, exploration is abandoned and plan-driven motion is resumed.

Any new state encountered is encoded in a new state node and new links are created to transition nodes. Any new letter encountered is learned and a link is formed between the corresponding character nodes and the current state node in the resistive grid.

[17] Despite the constructive principle, the size of the array had to be declared in advance. We used $N_s = 100$.

Figure 5.3: Example of the experimental set-up. A robot made of LEGO components carries a VVL video camera with a viewing angle of 30 degrees and faces a number of "objects" represented by letters surrounded by a frame. These are drawn on cards of size A6.

5.5 An Experiment with a Robot

5.5.1 Task and Experimental Set-up

The simple task designed for this robot is analogous to the task in which a household robot has to fetch a named object. As our robot has only a video camera and no gripper, it is assumed that "looking at" the object, i.e. localising it in the image, is equivalent to gripping it. To simplify the vision aspect of the problem, "objects" are letters drawn cards of size A6 and placed on a table (Figure 5.3).

The initial tests reported here use only rotation movements. Spatial memory is tested by asking the robot to "look at" a given letter by rotating towards it from any initial orientation. A maze-like set-up is planned for future experiments. Therefore, the artificial spatial memory system is designed to be compatible with translation movements.

The Basic-type program that controls the operation of the network in Figure 5.2 was written to realise following scenario:

> A robot is asked to retrieve a given object (character).
>
> If the character is unknown, ask the operator to place a character in front of the camera, and the learn the character.
>
> If the current view is unknown, create a new view-node and learn the view. If there is no path to the character, ask the operator for a direction in which to start searching.
>
> If the character is known and there is a path, perform planning, execute the first step of the plan, recognise the new view and perform planning again, until the goal character is in sight.

Such a procedure results in an interesting interactive demonstration (see below) where the robot starts with knowledge of neither the shape of characters nor the appearance of the experimental set-up, and ends up being (hopefully) able to execute the command of looking at a character from any starting position. As shown below, most experiments actually failed at the final planning stage, but that helped us to better understand view-based spatial memory models and propose improvements (section 5.6).

5.5.2 Learning and Planning Test

Exploration:

The experimental set-up was similar to the one in Figure 5.3, except that letters were distributed over approximately 120 degrees. Exploration was done by successive steps of leftwards or rightwards rotation. Initially an "information maximising" procedure was designed, by which the next chosen action was the one for which no outcome was known (no corresponding link). However, this led to time consuming random-looking moves. Therefore, for these initial tests, simple rightwards and leftwards sweeps were used. Each time an exploration procedure was required, the robot asked the operator for a direction of rotation.

To build the spatial memory, the robot was first pointed towards the most rightwards letter, a "C" and was asked, by calling the relevant subroutine in the neural network simulation environment CORTEX-PRO, to "look at" "A". The robot first noted that "A" was not known and asked for an example to be shown. It then noted that there was no path connecting "A" to any of the nodes in the resistive grid and asked for a direction in which to rotate to "explore" the environment. It then created a view node coding for the current scene and found a letter in the scene that it did not know. It asked the name of this letter which it then linked to the first state node. It then rotated leftwards by approximately 15 degrees, found that the view was unknown and recruited the second state node to code for it. A link between the two corresponding nodes in the resistive grid was created and a transition neuron recruited to encode the "left" action linking the states. The letter "C" was also visible in that scene and was linked to the state 2. Figure 5.4 illustrates the history of this exploration. A chain of states 1 to 10 was generated until the goal letter "A" was observed.

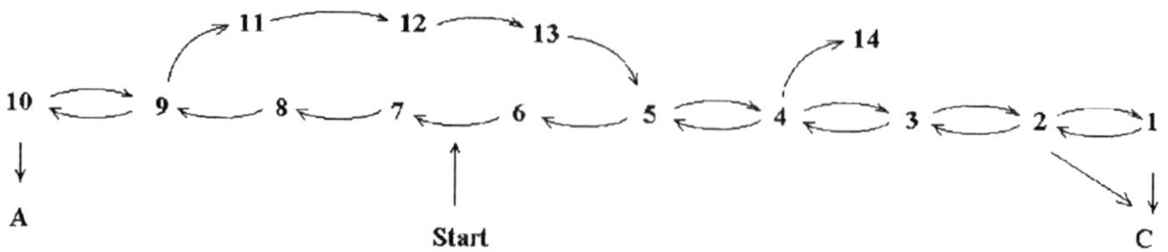

Figure 5.4: Sequence of states constructed during leftward exploration (1-10), rightward exploration (11-13) and during planning (14). Each state is represented by a node in "states" in Figure 5.2. The arrows indicate the direction of rotation and the opposite direction of the links built into the network (for backward planning, the activity must spread from the goal to its predecessors).

The robot was then asked to "look at" C. He knew what "C" looked like, and found that it was linked to the resistive grid. It started the planning routine but could not activate any neighbouring node of state 10 (remember that the connections are in the direction opposite from the one of the arrows drawn in Figure 5.4). Consequently the robot asked again for an exploration direction, which was given as "left". The first view encountered corresponded to view 9 and a "left" link was created between node 10 and 9. At the next step, for some reason, the node 8 was not activated strongly enough and the robot recruited a new node for a new view 11. A succession of 3 new state nodes were created in that manner, until the state node 5 recognised its view. The robot recognised then all expected views and created "right" links between views 5, 4, 3, 2 and 1. Figure 5.5 shows the computer screen at the end of an exploration process. The character "C" is in sight and the number of connections reflects the acquired spatial knowledge.

Planning:

The robot was then rotated by hand to face a direction indicated by "Start" in the Figure 5.4 and was "told" to "look at" "A". The recognised view was view 12. The planning procedure indicated the state 13 as the next one to be reached and a rightwards movement was executed. The next view was recognised as view 6. The planning procedure indicated state 7 as the next state and caused a leftwards movement. Then view 12 was recognised and a rightwards movement occurred. This repeated itself several times: The robot was stuck in a local minima! Such a problem has not been mentioned before.

Another problem that was found is the creation of state 14. When the start position faced view 5 and the goal was the letter "C", there was no possibility of a local minima in the structure of the spatial memory. However, after two rightward steps, the state 3 was not recognised and a new state 14 was created. As there was no known path from this state to the states 1 or 2, where "C" is found (or to any other states), the robot decided to initiate an exploration procedure.

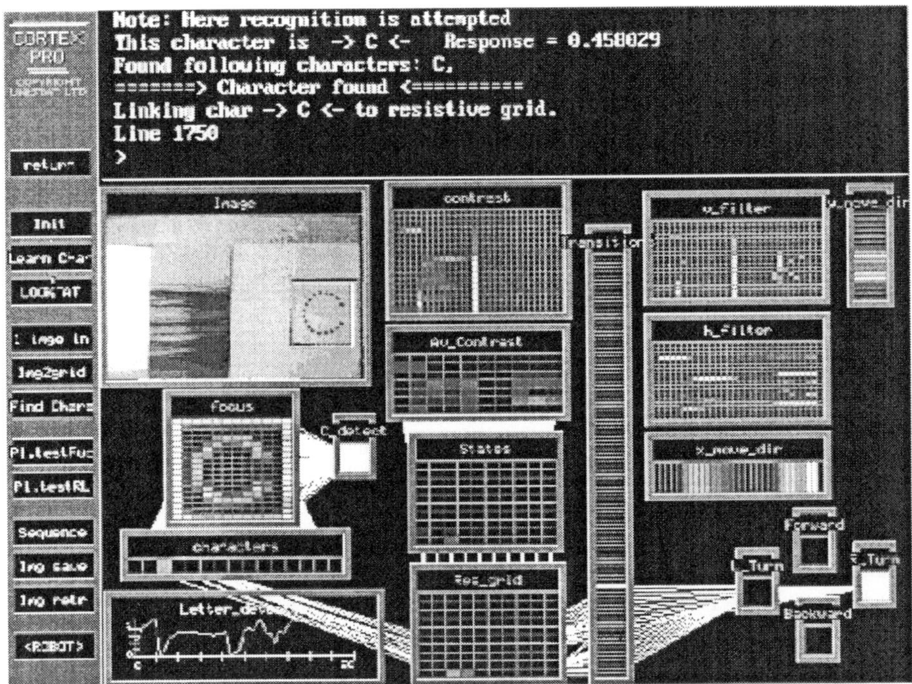

Figure 5.5: Screen showing the vision and spatial memory system (A pile of papers is behind the "C"). The grids have following X x Y sizes: Av_Contrast: 10 x 7, Contrast: 30 x 18; States and Res_grid: 10 x 10, Transitions 1 x 100, Characters: 15 x 1, Focus: 12 x 18; v_filter and h_filter: 30 x 18; x_move_dir: 30 x 1; y_move_dir: 1 x 30. The simulations are done using the package CORTEX-PRO. The operations of the systems vision and state recognition systems are described in detail in appendices 5.8 and 5.9.

Finally, the aliasing problem occurred frequently. It occurred that different views appeared sufficiently similar to be assigned the same node during exploration, which resulted in a view-graph with erroneous links. During planning also, a view could be recognised by the wrong node in Figure 5.2 which led to an inadequate action plan.

5.6 Discussion

In the experiment described above, a form of local minima occurs, resulting in repeated left and right rotations. A second problem is the occurrence of a representational dead-end when a new state is created. A third problem is aliasing. The experiments used only rotations "on the spot" possibly similar to eye saccades, but the observed problems are general and can also occur during translations.

Local minima, disjoint paths and internal path integration:
The causes of the local minima problem may lie in the existence of disjoint paths in Figure 5.5 and/or in the planning process.

Disjoint paths can arise when the state space is not densely covered during exploration. In our case, as exploration is not done with very small rotation steps, "holes" in the graph do occur. These allow a second disjoint path to be built during a later exploration sweep. Disjoint paths are probably a characteristic property of view-based graphs, and can be avoided only when a very comprehensive exploration process is practically possible (and affordable in terms of memory costs).

Considering planning, if a planned path is followed step by step, interferences with other paths built during exploration may occur. That is what causes the local minima problem. It also leads to a peculiar behaviour of reproducing the exploratory movements to reach a goal. Normally, humans or monkeys tend to use direct trajectories.

This behaviour and the local minima problem can be avoided if the successive steps of the plan are integrated internally before being executed. In the case shown in Figure 5.4, a single rotation to the target would result, whether state 6 or state 12 is recognised at the start. Exploration would then not need to be done in regular sweeps or in small steps. Thus neighbourhood in the real world would not necessarily need to correspond to neighbourhood in the view graph, although this might be corrected during off-line reorganisation (see below).

If a plan contains combinations of translations and rotations, as for mazes, successions of rotations "on the spot" and successions of collinear translations need to be integrated and executed separately. If these integrated movements were also stored in the view graph, the planning process would progressively contain less internal steps (or "detours"), at least for frequently used routes.

The different practical constraints existing on rotations and translations may affect the final organisation of a view-graph. One view can be linked to many other views by simple rotations on the spot, while a translation can link a view to only one other destination view. As a result of the integration process, view graphs may thus become enriched in links encoding rotations from given points but have a more limited number of links encoding translations between views attached to these points (see also Rieser, 1989 [63]).

Some thought should also be given to the practical meaning of "translation". Robots can not keep moving in straight line for a long time. It may be more practical for them to head towards some landmark, which would then provide visual feedback on the quality of the trajectory. Thus "translations" would become "heading procedures" of a certain complexity, very different in nature of simple ballistic rotations on the spot. Thus there might be a good case for using two different action-encoding schemes.

Representational dead-ends and visual recognition of distant places:
When the new state 14 is created, the encoded link and action only carry information on how to reach state 14. There is no information on how to reach

the rest of the graph from state 14.

It is tempting to solve this problem by using bi-directional links. However, not all actions are reversible. There are one-way roads and exit doors.

Another approach is to use vision in a more natural way, for recognising distant places and inspect possible routes to these places. This may involve the extraction of the 2-D layout of the space ahead (see for example Onillon et al, 1995 [58]), then a "virtual displacement" on this "local map" to a previously visited place and the establishment of new links between the current view and a view seen from a position reached by this virtual displacement. Thus a temporary map of limited size may be needed.

A less geometric approach would require the visual system to perform size-invariant view recognition. This may enable the robot to recognise a distant view, and encode a link possibly associated with a heading-based translation procedure (as introduced above), using a feature of the distance view as a goal. In any case, more work on the visual component of the model is needed.

Aliasing problem and priming:

Another problem found during the experiments was the aliasing problem (Whitehead and Ballard, 1991 [89]). Typically, in maze situations, identical local configurations occur in different positions. Thus a purely view based system does not provide unequivocally the position and orientation information needed for route planning. This is the classical reconstructibility problem in control theory. The solution usually proposed is to use the history of the measurements to define the current state (see for example Rao and Fuentes, 1996 [61]). Using this approach in our experiments, a view would be recognised as an image <u>and</u> a sequence of previously recognised views. However, if a surrounding is explored with random visual saccades, it is unlikely that any useful sequence may be stored.

Alternatively, if the robot has a purposeful behaviour, then any view is the goal of the previous movement, and it could be primed as such (see also Denham and McCabe, 1995 [18], 1996 [19]). Even random motion may predict views if the connectivity in the view graph is sufficiently dense. In practice, priming would be done via back-projections from the action-encoding part of the network to the state layer in Figure 5.2. Further work along these lines is needed.

Memory requirements, landmarks, and off-line graph reorganisation:

The upper part of the network in Figure 5.2 is a partial inverse model of the effect of actions on views, constructed progressively during experience. As one node represents each transition, the inverse model has a potential for using a large memory space.

A related potential problem with our approach is the increasing number of state-nodes when the experience of the robot increases. For instance, assuming that a robot sees a new unknown scene every second, it will build 2.6 million view-nodes in a month (to these must be added the corresponding transition nodes). If the views are encoded at a low level, by the intensity of each pixel of a video image, a month-long experience would represent 600 GBytes of memory[18]. Most robots would probably not experience such a fast changing and unpredictable environment. Nevertheless, some care is needed to maintain the memory size within manageable limits.

One way to a reduce the memory load is to reduce the number of encoded perceptual states and keep only those that are really useful. One approach might involve attentional mechanisms or novelty detection (Denham and McCabe, 1995 [18], 1996 [19]). Alternatively, visually salient features or landmarks (Thrun, 1996 [76]) could be used as focus points around which views are memorised, similarly to (Gaussier et al., 1997 [26]).

In addition, the problem of deciding if a landmark is "salient" or "memorable enough" may be solved automatically during an off-line "cleanup" process. For instance, if a simple ongoing action or procedure can be discovered to link distant views, then intermediate views may become redundant, and eliminated according to principle v) in section 5.4.1. Eventually, the only surviving nodes may be those shared by several route and those where low-level procedures fail to produce the adequate action.

Other models:

The model presented here is functionally similar to that used in mobile robot experiments by Mallot et al., (1995) [44]. However, the experiments reported here were done under different conditions, which revealed problems that Mallot et al, (1995) [44] could not have observed. For instance, Mallot et al., 1995 [44] did not use vision to recognise views, but bar-code signs on the floor which were read during translation towards a junction. The maze was hexagonal so that there were only 3 possible actions: "rotate left 60 degree then advance until junction", "rotate right 60 degree then advance until junction" or "reverse until junction". During translation, the robot was following the corridor linking two junctions. Neither aliasing, nor disjoint paths, nor representational dead ends could occur.

In Franz et al., (1996) [22], the coverage of the space was dense enough to enable the robot to "sense" neighbouring views and test, by moving towards the position of the neighbouring view, if a link could be added to the memory. Aliasing was the only possible problem.

A model of biological spatial memory?:

How does the model compare to biological spatial memory? The model conforms to the idea that spatial memory may take the form of a network of routes. During exploration, views and the action linking them are encoded, which defines a route in the usual sense. If views are common between two route, a network of route forms automatically, as observed by Moar and Carleton (1982) [49].

A feature that needs to be added is the encoding of travel distance or time. In conjunction with the proposed internal integration, the model may then be used to estimate directions and distances. If all actions of a plan were integrated

[18]Assuming an image with 320x240 pixels, the color of each pixel being coded in 24 bits.

using a form of internal dead reckoning, this would result in an estimation of the straight line distance to the goal and its direction. This information could also be added to the view graph, but the associated action should then be a straight-line flight! Mataric (1990) [45] proposed a similar method for finding shortcuts. Thus, the evolution of the capabilities of the model with increasing experience may parallel a similar evolution in human subjects, from sequence specific knowledge to vector-like knowledge.

Finally, it is noted that:

- Due to the structure of the model, comprising views and displacements, it may be possible to add nodes and links to this spatial memory by using verbal route instructions. This would be facilitated by the introduction of "heading" translation procedures and size independent landmark / feature recognition.
- One problem that needs to be considered is planning with constraints. To include constraints on the type of places that are part of a route, it may be beneficial to use a biologically plausible implementation (appendix 5.10). However, to include constraints on the type of action (e.g. vehicle type, or to avoid the "flight-links" mentioned above), some restructuring of the model may be needed, possibly using feedback for priming view nodes.
- Real world experiments submit the model to constraints such as noisy sensory information, unreliable actuators and timing constraints. Such constraints affect the design of a model. They have for instance led to a modification of the action encoding scheme described in section 5.4.4.

5.7 Conclusion

The original motivation of the model presented was its potential for avoiding Cartesian maps, avoiding the dimensionality curse and avoiding complex visual information processing.

However, robot experiments revealed a number of problems that were neither readily predictable from the principles formulated in section 5.4.1, nor from the connectionist structure described in section 5.4.3, nor from the procedures in sections 5.4.4 and 5.4.5, nor from the literature in the field. Consequently, a number of modifications to the model are proposed.

To eliminate representational dead-ends, it is proposed to use vision for remote recognition of places and assessment of the route to that place, so that links can be established between the current view and a view from the remote place. Technically, this may require to replace the concept of "photographic" view by that of group of visual features linked by geometrical relations. Concurrently, a procedure for "heading towards a visual feature" needs to be added to the repertoire of translation actions.

To avoid local minima during planned movements, it is proposed to use internal integration to plan rotation movements. Such a mechanism may also enable off-line restructuring of the view-graph so that an "inactive" robot may progressively refine the content of its spatial memory.

To reduce the aliasing problem, it is proposed to investigate mechanisms for priming view nodes by actions, in a predictive way.

The proposed solutions still avoid Cartesian maps and the dimensionality curse, but call for an increase in the complexity of the the visual system, the planning system and the action representation.

5.8 Appendix: Vision system for letter detection and recognition

This part of the vision system is described briefly. Its function is to precisely localise letters in the image to enable their recognition with a simple neural network (a single layer perceptron). The letters displayed in the environment are surrounded by a frame of 5 x 5 cm drawn in a full line. The letters are drawn with a dotted line. The local contrast of the image is calculated and copied into a grid of size 30x18 named "contrast" (Figure 5.5). This coarse image is filtered with vertical and horizontal centre-ON local operators and copied to two grids "v_filter" and "h_filter". Average intensities are computed along respectively horizontal or vertical axis and copied into the layers "y_move_dir" and "x_move_dir".

Using the size of the regions of high activity in these layers, the position of the frame surrounding the letter is determined approximately. Two frames are created successively for each image. If neither of them contains a letter, it is assumed that the image contains no letter. After a first frame has been found, the corresponding region in the grid-"contrast" is reduced in intensity. This allows other regions to attract the "attention" of the localisation process for the second frame.

The content of a frame is copied to the grid "focus". As the initial video frame does not fit exactly the frame surrounding the letter, a routine shifts the video frame slightly until the frame in the image fall exactly on the most external nodes of the grid "focus". Therefore the letter to be recognised always has the same position in the grid, independent of the distance and angle of view. However, a positional jitter of the size of one pixel in the grid focus is still possible. To reduce the effect of fluctuation of intensity and of the jitter, the image is blurred artificially. It is then recognised by RBF nodes (see appendix 5.9) in the layer "characters". These nodes are only connected to the central region of "focus", where the relevant information lies. A two-node wide belt of nodes not used for recognition. This belt is connected to a node "C_detect" which responds maximally when a frame has been placed correctly. Its activity is plotted in the graph "letter_detector" in Figure 5.5.

Recognition is accepted if a node in the layer "character" wins with an activity more than twice the activity of its nearest competitor, and with an absolute activity higher than 0.18. These settings are found by trial and error.

The number of letters that the systems knows is predefined, and each node in "Characters" is assigned to a letter. However, the appearance of a letter is not pre-encoded. Thus a detected character may be unknown to the system.

This situation is characterized by a high activity of the node "C_detect" and an insufficient response in the layer "Characters". This triggers a learning procedure in which the program asks the user to identify the letter, then the weights of the corresponding node in "characters" are set so that the current visual pattern elicits a maximal response.

5.9 Appendix: Vision system for state recognition

The vision system is shown in Figure 5.5. The part assigned to letter detection and recognition is described in the appendix 5.8. This appendix focuses on the recognition of states based on the current view. To reduce the number of connections to the state nodes, the activity of the "contrast" grid is copied into a coarser grid named "Av_Contrast". Each node of that grid represents the average activity over a receptive field of relative area 0.12 x 0.12 in the grid "contrast" (of size 1 x 1). Thus blurring occurs in the transformation, in addition to resolution reduction. Blurring makes the recognition process more tolerant to translation but also reduces the discrimination between views. An more robust approach would be to make the recognition of a state explicitly shift invariant, e.g. by use of shift maps (Bugmann et al., 1994 [9]), which would allow a more selective view recognition method. State neurons are of RBF type, with a width of $\sigma = 0.75$, and a response y_i given by (5.3):

$$y_i = \exp(-\frac{\sum_j (x_j - w_{ij})^2}{2\sigma_i^2}) \tag{5.3}$$

where x_j is the activity of a node in the layer "Av_Contrast". A state is assumed to be recognised when the activity of one of the RBF-node in the grid "states" exceeds 0.25, for a maximum of 1. If a view is not recognised, a new node is recruited and trained to recognise the new state. This is done by copying the outputs x_j of the nodes in "Av_Contrast" into the "weights" w_{ij}. Note that in RBF nodes, weights indicate the centre of the receptive field and do not have their usual multiplicative role.

5.10 Appendix: Notes on a biologically plausible implementation

i) Parts of the system are not implemented using connectionist techniques. For instance, after the resistive grid has been updated, the next state is selected by scanning through the nodes connected to the current-state node, not all the nodes. Its activity is then set to 1 and that of the current state to 0.5. Using a purely connectionist design, a quite complicated circuit needs to be designed to realise the same function. An added complication comes from the fact that current-state nodes are changing all the time. Thus, although the basic principles of this model are simple, a fully connectionist implementation involves a complex control circuitry. On the other hand an implementation with networks of spiking neurons (Bugmann, 1997 [11]) may again be simpler, as discussed below.

ii) The incomplete coverage of the state space by state nodes and the use of on-going actions calls for some caution in the planning process using current spread: in a graph where places are at unequal distance, a resistive grid with equal resistances between all nodes will only minimise the number of intermediate nodes, not indicate the shortest path. A solution to that is to set the resistances in proportion to the distances between places. In this case, the path of least resistance will be indicated by the link carrying the highest input current (Althöfer, 1996 [1]).

If Leaky Integrate-and-Firer (LIF) Neurons were used in place of the artificial neurons (5.3), they would fire earlier with a large input current than with a small one. The activation of the goal node would lead to a spread of spiking activity through the net. Using only the wavefront, the first node that activates the current-state node would be the one that has to be reached by the next action. Using transition nodes with sequence detection capability (Bugmann, 1997 [11]) would automatically lead to the correct action to be executed (or integrated). Thus, with biological neurons, the circuit in Figure 5.2 would be almost all that is needed. The only notable addition is probably a feedback inhibition scheme to control the persistence of activity in the grid and preserve only the wave front.

iii) An additional potential advantage of an implementation with spiking neurons is the capability for multicriteria planning. For instance, "finding a route between A and B which maximises the number of views of the sea" is a task that can be achieved by increasing the readiness of firing of all view-nodes associated with a view of the sea. This would increase the speed of the wave front along routes including these nodes. With a connectionist implementation of the resistive grid as used here, it is a very delicate matter to modify the potential distribution so that certain nodes are more likely to be part of the path.

5.11 Bibliography

[1] Althöfer K. (1996) "Neuro-Fuzzy Path Planning for Robotic Manipulators", PhD Thesis, Dept. of Electronic and Electrical Engineering, King's College London, Strand, London WC2R 2LS, UK

[2] Altman J. (1995) "Deciding what to do next", Trends in Neuroscience, 18, 117-118.

[3] Arbib M.A. and Lieblich I. (1977) "Motivational Learning of Spatial Behaviour", in Metzler J., (ed) "Systems Neuroscience", Academic Press, New York, pp. 221-239. (a more accessible reference may be Lieblich and Arbib, 1982)

[4] Bachelder I.A. and Waxman A.M. (1995) "A View-Based Neurocomputational System for Relational Map-Making and Navigation in Visual Environments", Robotics and Autonomous Systems, 16, pp. 267-289.

[5] Barto A.G., Bradtke S.J. and Singh S.P. (1995) "Learning to Act using Real-Time Dynamic Programming", Artificial Intelligence, 72, 81-138.

[6] Blades M. (1990) "The Reliability of Data Collected from Sketch Maps", J. of Environmental Psychology, 10, pp. 327339.

[7] Brooks R.A. (1986) "A Robust Layered Control System for a Mobile Robot", IEEE Journal of Robotics an Automation, RA2, pp. 14-23.

[8] Buehlmeier A., Duerer H., Monnerjahn J., Noelte M. and Nehmzow U. (1996) "Learning by Tuition, Experiments with the Manchester 'FortyTwo'", University of Manchester, Computer Science Technical Report Series, UMCS-96-1-2, Can be downloaded from: ftp://ftp.cs.man.ac.uk/pub/TR/UMCS-96-1- 2.ps.Z

[9] Bugmann G., Taylor J.G. and Denham M.J. (1994) "Sensory and Memory Based Path Planning in the Egocentric Reference Frame of an Autonomous Mobile Robot" Research Report NRG-94-01, School of Computing, University of Plymouth, Plymouth PL4 8AA, UK. http://www.tech.plym.ac.uk/soc/research/neural/neur20/pub licat/papers/nrg9401.zip

[10] Bugmann G., Taylor J.G. and Denham M.J. (1995) "Route finding using Neural Nets", in Taylor J.G (ed) "Neural Networks", Alfred Waller Ltd, Henley-on-Thames, pp. 217-230.

[11] Bugmann G. (1997) "Biologically Plausible Neurocomputation", Biosystems, 40, 11-19.

[12] Burgess N. Recce M. and O'Keefe J. (1995) "Hippocampus: Spatial Models", in Arbib M.A. (ed) "The handbook of Brain Theory and Neural Networks", Bradford Books/MIT Press, pp. 468-472

[13] Burgess N. and O'Keefe J. (1996) "Neural Computation Underlying the Firing of Place Cells and their Role in Navigation", Hippocampus, 6, pp. 749-762.

[14] Byrne R. (1979) "Memory for urban geography", Quarterly Journal of Experimental Psychology, 31, 147-154.

[15] Connoly C.I. and Burns J.B. (1993) "A Model of the Functioning of the Striatum", Biological Cybernetics, 68, pp. 535-544.

[16] Cox I.J. (1991) "Blanche - An experiment in Guidance and Navigation of an Autonomous Robot Vehicule", IEEE Transactions on Robotics and Automation, 7, pp 193-204.

[17] Curran T. (1995) "On the neural Mechanisms of Sequence Learning", Psyche, Vol. 2 (on-line journal: http://psyche.cs.monash.edu.au/volume2-1/psyche-95-2-12 sequence-1-curran.html).

[18] Denham M.J. and McCabe S.L. (1995) "Robot control using temporal sequence learning", Proc World Congress on Neural Networks (WCNN'95), Washington DC, USA, July 1995, INNS Press/Lawrence Erlbaum Associates, vol 2, pp 346-349.

[19] Denham M.J. and McCabe S.L. (1996) "Biological basis for a neural model of learning and recall of goal-directed sensory-motor behaviours", Proc World Congress on Neural Networks (WCNN'96), San Diego, USA, September 1996, INNS Press/Lawrence Erlbaum Associates, pp. 1283-1286.

[20] Evans G.W., Marrero D.G. and Butler P.A. (1981) "Environmental Learning and Cognitive Mapping", Environment and Behaviour, 13, pp. 83-104.

[21] Ferguson E.L. and Hegarty M. (1994) "Properties of Cognitive Maps Constructed from Texts", Memory and Cognition, 22, pp. 455-473.

[22] Franz M.O., Schölkopf B., Georg P., Mallot H.A., and Bülthoff H.H. (1996) "Learning View Graphs for Robot Navigation", Technical Report no 33, Max-Planck Institute für Biologische Kybernetik, 72076 Tübingen, Germany. Can be downloaded from the website: http://www.mpik tueb.mpg.de/projects/techr/list3.html

[23] Gaussier P. and Zrehen S. (1995) "PerAc: A neural Architecture to Control Artificial Animals", Robotics and Autonomous Systems, 16, pp. 291-320.

[24] Gaussier P. and Zrehen S. (1994) "A topological Neural Map for On-Line Learning: Emergence of Obstacle Avoidance in a Mobile Robot", Proc SAB'94 (Brighton, UK), pp. 282-290.

[25] Gaussier P., Joulain C., Revel A. and Banquet J.-P. (1996) "Are Shaping Techniques the Correct Answer for the Control of Visually Guided Robots ?", Proc. Control'96 (Exeter, UK), IEE Publ. London, pp. 1248-1252.

[26] Gaussier P., Joulain C., S. Zrehen, Banquet J.-P. and Revel A. (1997) "Visual Navigation in an Open Environment without Map", Submitted to IROS'97, Grenoble.

[27] Glasius R., Komoda A. and Gielen S. (1996) "A biologically Inspired Neural Net for Trajectory Formation and Obstacle Avoidance", Biological Cybernetics, 84, pp. 511-520.

[28] Gray J.A. (1995) "The contents of consciousness: A neurophysiological Conjecture", Behavioural and Brain Science, 18, pp. 659-676

[29] Hull C.L. (1932) "The goal Gradient Hypothesis and Maze Learning", Psychological Review, 39, pp. 25-43.

[30] Hull C.L. (1934) "The Concept of Habit-Family Hierarchy and Maze Learning", Psychological Review, 41, pp. 33-52.

[31] Hull C.L. (1935) "The Mechanism of the Assembly of Behavior Segments in Novel Combinations suitable for Problem Solution", Psychological Review, 42, pp. 219-245.

[32] Huttenlocher J. and Newcombe N. (1984) "The Child's Representation of Information about Location", in Sophian C. (ed) "Origin of Cognitive Skills", Lawrence Erlbaum Associates, NJ, pp. 81-111.

[33] Jeannerod M. (1994) "The Representing Brain: Neural Correlates of Motor Intention and Imagery", BBS, 17, pp. 187- 202.

[34] Joulain C., Gaussier P. and Revel A. (1996) "Apprentissage de Catégories Sensory Motrices par un Robot Autonome" Research Report ENSEA/ETIS, Paris. Can be downloaded from: http://bunny.ensea.fr/Pages_Perso/Arnaud_Revel/articles/nsi2.ps

[35] Kohonen T. (1988) "Self-Organisation and Associative Memory", Springer Verlag, Berlin

[36] Kortenkamp D., Huber M., Cohen C., Raschke U., Bidlack C., Bates Congdon C., Koss F. and Weymouth T. (1993) "Integrated Mobile-Robot Design (Winning the AAAI'92 Robot Competition)", IEEE Expert, 8, pp. 61-73.

[37] Latombe J.C. (1991) "Robot motion Planning", Kluwer Academic Publ.

[38] Lejeune H. and Wearden J.H. (1991) "The Comparative Psychology of Fixed-Interval Responding: Some Quantitative Analyses", Learning and Motivation, 22, pp. 84-111.

[39] Levine M., Jankovic I.N., Palij M. (1982) "Principle of Spatial Problem Solving", Journal of Experimental Psychology: General, 111, pp. 157-175.

[40] Lieblich I. and Arbib M.A. (1982) "Multiple Representations of Space", The Behavioral and Brain Sciences, 5, 627-659.

[41] Maes P. (189) " How to do the right Thing", Connection Science, 1, pp. 291-323.

[42] Maguire E.A., Frackowiak, R.S.J. and Frith C.D. (1997a) "Learning to find your way - a role for the human hippocampal region", submitted.

[43] Maguire E.A., Burgess N., Donnett J.G., O'Keefe J. and Frith C.D. (1997b) "Knowing where things are: Parahippocampal involvement in encoding object locations in virtual large-scale space", Submitted.

[44] Mallot H.A., Bülthoff H.H., Georg P., Schölkopf B. and Yasuhara K. (1995) "View-Based Cognitive Map Learning by an Autonomous Robot", Proc. ICANN'95 (Paris), Vol. 2, pp. 381- 386. A number of papers related to this work can be retrieved at the following website: http://www.mpik tueb.mpg.de/projects/navigation/navigation.html

[45] Mataric M.J. (1990) "Navigating with a Rat Brain", Proc. of Simulation of Adaptive Behavior: From Animals to Animats (SAB'90), MIT Press Bradford Books, pp. 169-175.

[46] Mellet E., Tzourio N., Denis M., and Mazoyer B. (1995) "A positron Emission Tomography Study of Visual and Mental Spatial Exploration", J. of Cognitive Neuroscience, 7, pp. 433-445.

[47] Mellet E., Tzourio N., Crivello F., Joliot M., Denis M., and Mazoyer B. (1996) "Functional Imagery of Spatial Mental Images Generated from verbal Instructions", J. of Neuroscience, 16, pp. 6504-6512.

[48] Menzel E.W. (1973) "Chimpanzee Spatial Memory Organisation", Science, 182, pp. 943-945.

[49] Moar I. and Carleton L.R. (1982) "Memory for routes", Quarterly Journal of Experimental Psychology, 35, pp. 381-394.

[50] Muller R.U., Stead M. and Pach J. (1996) "The Hippocampus as a Cognitive Graph", J. Gen. Physiol., 107, pp. 663-694.

[51] McNamara T.P., Hardy J.K. and Hirtle S.C. (1989) "Subjective Hierarchies in Spatial Memory", J. of Experimental Psychology: Learning, Memory and Cognition, 15, pp. 211-227.

[52] McNaughton (1989) "Neural Mechanisms for Spatial Computation and Information Storage", in: Nadel L., Cooper L.A., Culicover P. and Harnish R.M. (eds) "Neural Connections, Mental Computation", MIT Press, Cambridge MA, USA, pp. 285- 350.

[53] McNaughton B.L. and Nadel L. (1990) "Hebb-Marr Networks and the Neurobiological Representation of Action in Space", In Gluck .M.A. and Rumelhart D.E. (eds) "Neuroscience and Connectionist Theory", Lawrence Erlbaum Associates, Hillsdale NJ, USA, pp. 1-63.

[54] Newell A. and Simon H.A. (1972) "Human problem Solving", Prentice Hall, Inc, Englewood Cliffs, NJ

[55] O'Keefe J. and Nadel L. (1978) "The Hippocampus as a Cognitive Map", Clarendon Press, Oxford.

[56] O'Keefe J. and Speakman A. (1987) "Single Unit Activity in the Rat Hippocampus during a Spatial Memory Task", Experimental Brain Research, 68, pp. 1-27.

[57] O'Keefe J, (1991) "An Allocentric Spatial Model for the Hippocampal Cognitive Map", Hippocampus, 1, pp. 230-235.

[58] Onillon V., Bugmann G., Simpson A. and Nurse P. (1995) "Artificial vision for micromouse" Research Report NRG-95-05, School of Computing, University of Plymouth, Plymouth PL4 8AA, UK, can be downloaded from http://www.tech.plym.ac.uk/soc/research/neural/

[59] Peruch P., Giraudo M-D. and Gäling T. (1989) "Distance Cognition by Taxi Drivers and the General Public", J. of Environmental Psychology, 9, pp. 233-239.

[60] Presson C., DeLange N. and Hazelrigg M. (1989) "Orientation Specificity in Spatial Memory: What Makes a Path Different From a Map of the Path ?", Journal of Experimental Psychology: Learning, Memory, and Cognition., 15, pp. 887- 897.

[61] Rao R.P.N. and Fuentes O. (1996) "Learning Navigational Behavior using a Predictive Sparse Distributed Memory", To appear in "Proc. from Animals to Animats'96: The Fourth International Conference on Simulation of Adaptive Behavior", MIT Press

[62] Recce M. and Harris K.D. (1996) "Memory for places; A navigation model in support of Marr's Theory of hippocampal function", Hippocampus, 6, pp. 735-748.

[63] Rieser J.J. (1989) "Access to Knowledge of Spatial Structure at Novel Point of Observation", Journal of Experimental Psychology: Learning, Memory, and Cognition, 15, pp. 1157- 1165.

[64] Röfer T. (1997) "Controlling a wheelchair with image based homing", in N. Sharkey and U. Nehmzow (eds.) "Spatial Reasoning in Mobile Robots and Animals", University of Manchester Technical Report UMCS-97-4-1

[65] Rolls E.T. (1996) "The representation of space in the primate hippocampus, and its role in memory", in Ishikawa K., McGaugh J.L. and Sakata H. (eds) "Brain Processes and Memory", Elsevier: Amsterdam, pp. 203-227.

[66] Sadalla E.K. and Magel S.G. (1980) "The perception of travelled distance", Environment and Behaviour, 12, pp. 65- 79.

[67] Schmajuk N.A., Thieme A.D. and Blair H.T. (1993) "Maps, Routes and the Hippocampus: A neural Network Approach", Hippocampus, 3, pp. 387-400.

[68] Siemiatkowska B. (1994) "Cellular Neural Networks for Mobile Robot Navigation", Proc. 3rd Int. Conf. on Cellular Neural Networks and their Applications (CNNA94), Rome, Italy, pp. 285-290.

[69] Shallice T. (1988) "From Neuropsychology to mental Structure", Cambridge University Press, Cambridge, UK. (see section 14.6 on planning and frontal lobe lesions).

[70] Schölkopf B. and Mallot H.A. (1994) "View-Based Cognitive Mapping and Path Planning", Technical Report no 7, Max Planck Institute für Biologische Kybernetik, 72076 Tübingen, Germany. Can be downloaded from: http:// www.mpik-tueb.mpg.de/ projects/ techr/ list1.html Also availbable as: Schölkopf, B. and H.A. Mallot (1995) "View-based cognitive mapping and path planning.", Adaptive Behavior 3, pp. 311-348.

[71] Speakman A. (1987) "Place cells in the brain: Evidence for a cognitive map", Sci. prog., Oxf., 71, pp. 511-530.

[72] Spence K.W. (1950) "Cognitive Versus Stimulus-Response Theories of Learning", Psychological Review, 57, pp. 159-172.

[73] Tesauro G.J. (1995) "Temporal difference learning and TD Gammon", Communication of the ACM, 38, pp. 58-68.

[74] Thompson J.A. (1980) "How doe we use visual information to control locomotion", Trends in Neuroscience, October 1980, pp. 247-250.

[75] Thorndyke P.W. and Hayes-Roth B. (1982) "Differences in Spatial Knowledge Acquired from Maps and Navigation", Cognitive Psychology, 14, pp. 560-589.

[76] Thrun S. (1996) "A Bayesian Approach to Landmark Discovery and Active Perception in Mobile Robot Navigation", Technical Report CMU-CS-96-122, Carnegie Mellon University, Pittsburgh, PA.

[77] Thrun S. and Bücken A. (1996) "Learning Maps for Indoor Mobile Robot Navigation", Technical Report CMU-CS-96-121, Carnegie Mellon University, Pittsburgh, PA.

[78] Tolman E.C. (1948) "Cognitive maps in Rats and Men", Psychological Review, 55, pp. 189-208.

[79] Touretzky D.S., Wan H.S. and Redish A.D. (1994) "Neural Representation of Space in Rats and Robots", in Zurada J.M., Marks R.J. and Robinson C.J. (eds) "Computational Intelligence: Imitating Life", IEEE, pp. 57-68.

[80] Tversky B. (1992) "Distortions in cognitive maps", Geoforum, 23, pp. 131-138.

[81] Vann Bugmann D. (1995) "Human Spatial Memory: A Study of Cognitive Distance", Research report, Department of Psychology, University of Plymouth, UK.

[82] Verschure P.F.M.J., Kröse B.J.A. and Pfeifer R. (1992) "Distributed Adaptive Control: The Self-organization of Structured Behaviour", Robotics and Autonomous Systems, 9, pp. 181-196.

[83] Verschure P.F.M.J., Wray J. Sporns O., Tononi G. and Edelman G.M. (1995) "Multilevel Analysis of Classical Conditioning in a Behaving Real World Artefact", Robotics and Autonomous Systems, 16, pp. 247-265.

[84] Ward A.K., Newcombe N. and Overton W.F. (1986) "Turn Left at The Church, or Three Miles North: A Study of Direction Giving and Sex Differences", Environment and Behaviour, 18, pp. 192-213.

[85] Watkins C.J.C.H. and Dayan P. "Q-Learning", Machine Learning, 8, pp. 279-292.

[86] Webb B. (1995) "Using Robots to Model Animals: a Cricket Test", Robotics and Autonomous Systems, 16, pp. 117-134.

[87] Wehner R., Michel B. and Antonsen P. (1996) "Visual Navigation in Insects: Coupling of Egocentric and Geocentric Information", Journal of Experimental Biology, 199, pp. 129- 140.

[88] Whishaw I.Q. (1991) "Latent Learning in a Swimming Pool Place Task by Rats: Evidence of the Use of Associative and Not Cognitive Mapping Processes", The Quarterly J. of Experimental Psychology, 43B, pp. 83-103.

[89] Whitehead S.D. and Ballard D.H. (1991) "Learning to Perceive and Act by Trial and Error", Machine Learning, 7, pp. 45-83.

[90] Yamauchi B. and Langley P. (1997) "Place Learning in Dynamic Real-World Environments", Intl. Workshop on Learning in Autonomous Robots (RobotLearn'96), May 19-20, Keywest, Florida, pp. 123-129. (http://www.cs.buffalo.edu/~hexmoor/robolearn96/workshop.html). Possibly more accessible: Yamauchi B. and Langley P. (1997) "Place recognition in dynamic environments" J. of Robotics System, 14, pp. 107-120.

[91] Zimmer U.R (1996) "Robust World-Modelling and Navigation in a Real World", Neurocomputing, 13, pp. 247-260.

Chapter 6

Turing's Philosophical Error?

6.1 Turing's Machine

In 1936 the brilliant young English mathematician Alan Turing described a machine which he called the 'universal machine'; he argued that it could compute exactly the same things as any human 'computer'. The equally brilliant Austrian mathematician Kurt Gödel later praised Turing for this work. He said that it leads us to 'a precise and unquestionably adequate definition of the general concept of formal system' by analysing the concept of 'mechanical procedure'.

Turing's machine is known today as the *universal Turing machine.* Apart from the fact that it uses up arbitrarily large amounts of memory, it's quite small. A standard exercise for computer science students is to write programs that emulate it.

Many people find it shocking that Turing's little machine can calculate anything that any human being could ever calculate. Every calculation done by Newton, Gauss or Einstein can be copied and checked on this machine, and the same would be true even if Newton, Gauss and Einstein lived for ever and kept on calculating through the centuries. It doesn't say much for the power of human thought, does it?

Gödel spent much of his career looking for gaps in Turing's argument, places where the human spirit could slip out and prove itself cleverer than the machine. In about 1972, six years before he died, Gödel wrote a short note which was never published in his lifetime, consisting of three Remarks. The third Remark was headed 'A philosophical error in Turing's work'; I shall refer to it as Remark 3. In it Gödel suggested that human beings might be able to compute 'systematically', by 'mental procedures', some things which can't be computed by 'mechanical procedures'. This chapter considers what Gödel might have meant by this note.

6.2 Digital Problems

What did Turing mean by 'computing'? Remember that he was writing in 1936, before electronic computers were invented. In fact Turing's work was one of the breakthroughs that led to the construction of the first electronic digital computers in the late 1940s. When Turing wrote, computing was still a thing that mathematicians did when they were calculating a number by means of a formula.

For example we know how to compute the two solutions of the equation:

$$x^2 - x - 6 = 0$$

by the usual formula for solving quadratics:

$$x = \frac{1 \pm \sqrt{1 + 4 \times 6}}{2} = \frac{1 \pm 5}{2} = 3 \text{ or } -2.$$

Another example is the formula for computing π:

$$\pi = 4\left(1 - \frac{1}{3} + \frac{1}{5} - \frac{1}{7} + \ldots\right).$$

Turing had in mind examples like both of these, and particularly examples like the second, where we compute the decimal expansion of a real number. This example has a feature that was very important for Turing: *the calculation never finishes*. We can go on for ever, calculating π to more and more decimal places. In fact one can show that if we want to get within $1/2n$ of the true value of π, it's enough to add up the numbers as far as:

$$\pm\frac{1}{2n - 1}.$$

But π is an irrational number so that, however many decimal places we calculate, the answer that we get will not be the exact value of π.

Turing's notion of computing also included many things not at all like these two examples. For a start, he assumed that we always compute in one fixed finite alphabet; this alphabet could include the numerals:

0, 1, 2, 3, 4, 5, 6, 7, 8, 9

which would allow him to write down all the natural numbers:

0, 1, 2, 3, 4, 5, 6, 7, 8, 9, 10, 11, ...

in arabic notation. For definiteness I'll assume that the alphabet for all our calculations is ascii[1], which has the following symbols (listed in 'alphabetical order'):

!,",#,$,%,&,',(,),*,+,′,-,.,/,0,1,2,3,4,5,6,7,8,9,:,;,<,=,>,?, verb+@+,
A,B,C,D,E,F,G,H,I,J,K,L,M,N,O,P,Q,R,S,T,U,V,W,X,Y,Z, [,\,],ˆ,_,‘,
a,b,c,d,e,f,g,h,i,j,k,l,m,n,o,p,q,r,s,t,u,v,w,x,y ,z,{,|,},˜.

[1] *American Standard Code for Information Interchange.*

There are two reasons why this is a safe assumption. First, anything that can be said in English can be said in ascii. (There is just one problem: I haven't included the space character, because it doesn't show up as a symbol. We_can_use_the_underline_character_as_a_space_instead,_like_this. But for the sake of familiarity I shall write ordinary spaces rather than underline characters.)

And second, any alphabet whatever can be coded up using just the symbols 0 and 1. For example we can write:

$$\begin{array}{lll} 01 & \text{for} & \alpha, \\ 001 & \text{for} & \beta, \\ 0001 & \text{for} & \gamma, \\ 00001 & \text{for} & \delta, \end{array}$$

and so on. The coding (putting 01 for α etc.) and the decoding (putting α for 01 etc.) are both computations which are obviously mechanical; in fact one could design a simple machine to do them.

So from now on, all our calculations are in ascii. By a *string* we mean a finite string consisting of one or more ascii symbols. Since the ascii alphabet contains all the numerals 0 to 9, we can write each positive integer as a string.

We shall use the fact that all strings can be listed mechanically, by listing first the strings consisting of a single symbol, then the strings of length two, then the strings of length three, and so on. All the strings of a fixed length are listed in alphabetical order; I gave the alphabetical order for ascii a moment ago. One can make a machine which translates each positive integer n into the n-th ascii string, and another machine which does the opposite.

By a *digital problem* I shall mean a pair of strings, the first called the *question* and the second called the *answer*. Here are four examples of digital problems:

1. **Question:** 2 times (3 plus 4).
   Answer: 14.

2. **Question:** What is π correct to ten decimal places?
   Answer: 3.1415926536.

3. **Question:** $x^2 - x - 6 = 0$.
   Answer: $3, -2$.

4. **Question:** $x^2 - x - 6 = 0$.
   Answer: wildebeest.

Notice that we haven't said anything about the answer being 'correct'; the digital problem is just the pair of question and answer.

6.3 Turing's Thesis

Turing wasn't interested in problems one at a time. If you are given any digital problem whatever, you can construct a machine which gives the answer whenever the question is fed into it. This is completely trivial and tells us nothing at all about the nature of computing. What interested Turing was *infinite families of problems which call for a uniform method of solution.* For example the formula for solving quadratic equations gives the right answer for any quadratic equation, and there are infinitely many different quadratic equations. The method for calculating the decimal expansion of π works uniformly to calculate any number of decimal places.

So we need - Turing needed - a way to capture the notion of an infinite family of problems. The following idea works. A *digital family* is a set of digital problems, such that (1) the question of each digital problem in the set is a positive integer, and (2) each positive integer is the question of exactly one digital problem in the set. (The correct name for digital families is 'character-string-valued total functions of positive integers'.)

For example the problem of calculating the decimal expansion of π is a digital family: the answer to question n is the n-th digit after the decimal place. We can make the solution of quadratic equations (with integer coefficients) into a digital family too, but this is a little more subtle. Suppose n is a positive integer and S is the n-th string in the listing of all strings. If S consists of three integers a, b, c with semicolons separating them:

$$a; b; c$$

then the answer to question n is the solution to the quadratic equation:

$$ax^2 + bx + c = 0.$$

If S doesn't have this form, then the answer to question n is the word NO. Obviously we can check mechanically whether or not S does have this form.

Now we can say more precisely what Turing claimed to have proved. He described a machine called the *universal machine.* You start this machine by feeding a string into it. Then it calculates; either it will go on calculating for ever, or else it will eventually halt and output a string. He claimed that his machine has the following property:

> **Turing's thesis.** If F is a digital family, and there is a method of computing which will tell us, for every question n, what answer F gives, then there is a positive integer e (nowadays called a *Gödel number* of F) such that if n is any positive integer and the universal machine is started with input string $e; n$ then it will eventually halt and output the answer to question n.

This statement is painfully precise. But it has to be. If we make slight changes, Turing's claim can become either trivially true or provably wrong.

Suppose for example we wrote:

> **Turing's thesis, incorrectly stated.** If F is a digital family, and there is a method of computing which will tell us, for every question n, what answer F gives, then if n is any positive integer, there is a positive integer e such that if the universal machine is started with input string $e;n$ then it will eventually halt and output the answer to question n.

This happens to be true but uninteresting. If the answer to the n-th question is the string S, then all that's needed is a number e which tells the machine to print out S. The point is that in this incorrect version, the number e can depend on n, whereas in Turing's thesis the number e depends only on F and not on n.

Or again, suppose we wrote:

> **Turing's thesis, still incorrectly stated.** If F is a digital family, and for every question n there is a method of computing which will tell us what answer F gives, then there is a positive integer e such that if n is any positive integer and the universal machine is started with input string $e;n$ then it will eventually halt and output the answer to question n.

This is wrong because if F is any digital family whatever, then for every question n there is a method of computing which will tell us the answer to n; namely, write down the answer to n. (As we noted earlier, there is always a method of computing that gives the answer to a single digital problem.) But one can show that there aren't enough positive integers e to go around all the possible digital families.

One more try:

> **Turing's thesis, incorrectly stated yet again.** If F is a digital family, and there is a method of computing which will tell us, for some questions n, what answer F gives, then there is a positive integer e such that if n is any positive integer for which the method of computing does give the answer, and the universal machine is started with input string $e;n$ then it will eventually halt and output the answer to question n.

It takes some quite refined mathematics to show why this is wrong. But anyway it is wrong.

We can avoid these mistakes by introducing some new definitions. We define a *computable function* to be a digital family F such that there is a method of computing which will tell us, for every question n, what answer F gives to n. We also write $F(n)$ for 'the answer given by F to n', and we call $F(n)$ the *value of F at n*. We define a *Turing-computable function* to be a digital family F for which there is a positive integer e such that if n is any positive integer and the universal machine is started with input string:

$$e;n$$

then it will eventually halt and output $F(n)$. Then a correct statement, shorter than before, is:

Turing's thesis. Every computable function is Turing-computable.

It will be helpful to introduce one more notion. We say that a nonempty set X of strings is *Turing-enumerable* if X consists of all the strings $F(n)$ for some Turing-computable function F. The set of all natural numbers (in arabic notation) is Turing-enumerable. Another example of a Turing-enumerable set is the set of all strings

$$n; k_n$$

where n is a positive integer and k_n is the n-th digit in the decimal expansion of π.

6.4 How Turing Built his Universal Machine

The idea behind Turing's universal machine is that the number e is a coded description of a mechanical method of calculation. Given an input $e; n$, the machine first decodes e, and then uses the result as a set of instructions for operating on n. So e is in some sense a program for the universal machine. This idea of Turing is said to have put into John von Neumann's mind the notion that programs could be input into a computer just like data; in other words, programs can be software rather than hardware. Von Neumann helped to guide the design of early electronic computers, and so through him Turing's idea led to the creation of the software industry.

There are three parts to Turing's argument. The **first** part is to show that all methods for calculating computable functions can be put together out of the same small number of pieces - a kind of abstract Lego set of calculating steps. The **second** part is to show that we can describe any method for calculating a computable function F by listing the steps which it uses. If this list is the e-th string, then we can take e to be a Gödel number of F. The **third** part is to describe a machine which will decode any Gödel number and follow the instructions in the resulting list.

I shall only discuss the first part. The second follows more or less automatically from the first. The third part is the most complicated, but it needs patience rather than cleverness. In any case Turing himself didn't get it quite right (but nobody worries - he had all the right ideas).

According to Turing, (human) computing is done by writing symbols from some finite alphabet, normally on a two-dimensional sheet of paper. But if we only had a long thin strip of paper marked off as a line of squares, and we could only write one symbol per square, we could still get by; for example we could put a marker symbol into every 50th square, and count the squares between two of these markers as one line of the page. The strip of paper might have to be very long indeed. Turing was happy to allow it to be infinite, or at least extendible whenever we run out of squares. The (human) computer can look at finitely many squares at any one time; without loss of generality we can suppose that he (Turing's computer was male) can look at just one square at a time. The computer can recognise at once which symbol is in the square, and he can change it to any other symbol in the alphabet. He can also move to look at the next square on the left, or at the next square on the right.

So far, we have allowed our computer to do just a finite number of things to change the world, though he might do each of them many times over. He can write a given symbol in the square that he is looking at. He can move one square to the left. He can move one square to the right. Turing calls these the *simple operations.*

Up to this point, nothing is controversial. People are happy to accept that all human computation can be done with Turing's simple operations. The controversial question is the next one: at any stage in the calculation, what determines which simple operation the computer performs next? According to Turing, we can always arrange things so that the choice of the next simple operation depends on two things:

1. The symbol in the square that the computer is looking at, and

2. the *state of mind* of the computer.

You might throw up your hands in horror at this stage. 'State of mind' - what has this got to do with *computation*? If we bring human minds into the matter, anything goes. But wait: Turing demands just one property of states of mind. All he demands is that there are just finitely many of them. Given that, he can build his universal machine.

Here are Turing's exact words (reproduced by permission of the London Mathematical Society):

> The behaviour of the computer at any moment is determined by the [symbol] which he is observing, and his "state of mind" at that moment. ... We will also suppose that the number of states of mind which need be taken into account is finite. ... If we admitted an infinity of states of mind, some of them will be "arbitrarily close" and will be confused. ... the restriction is not one which seriously affects computation, since the use of more complicated states of mind can be avoided by writing more symbols on the tape.

I give warning that this is the passage which Gödel pointed to when he spoke in Remark 3 of Turing's 'philosophical error'.

Turing's last sentence is puzzling at first sight. Turing seems to be saying that the number of states of mind could be infinite without affecting the notion of computation; but this is certainly wrong. Nor is it true that we could compensate for losing infinitely many states of mind just by writing more symbols. Probably Turing means only that the exact finite number of states of mind is not important, because we can lessen the strain on our mind by writing more things down. (As we get older and our memories get weaker, we need to write longer and more precise shopping lists.) So there is some trade-off between the number of states of mind and the length of the calculation. In principle, brainy people can do calculations faster.

6.5 Gödel's Attempt to Break Turing's Thesis

Contrary to what you may read, it is not absolutely clear that Gödel ever accepted Turing's thesis. When he referred to Turing's work from 1947 on, he consistently praised Turing for his analysis of *mechanical* computation, or *finite* computation, or (what he took as equivalent) computation within a *formal system*. Gödel made one earlier reference to Turing's work, in 1946, and in 1965 he added a note to his 1946 text to make it clear that he was considering only computation within a formal system.

Turing intended his thesis to be broader than this. His arguments were meant to capture all functions 'which would naturally be regarded as computable' (that is, by a human being). Gödel never endorsed the idea that all human computing is mechanical. In fact he reacted against this idea with increasing vigour as he grew older.

In 1961 Gödel was elected to membership of the American Philosophical Society. The executive secretary of the Society wrote to Gödel inviting him to give a lecture to the Society. Gödel wrote the lecture but - as far as we know - never answered the invitation. (He travelled very little, claiming that his health was poor.) I shall refer to this lecture as the APS lecture. It is very revealing, and we shall see that his comment on Turing's thesis follows on directly from his thoughts in the lecture.

Gödel's first published work early 1930s was deeply influenced by some earlier ideas of the great German mathematician David Hilbert. In the APS lecture Gödel described how Hilbert's ideas seemed to him now, thirty years later. He detected two main influences on Hilbert. The first was the 'spirit of the time', which was fundamentally pessimistic and said that the only things we can know in mathematics are those which we can calculate. Hilbert expressed this spirit by asking for all mathematical problems to be reduced to the manipulation of symbols according to certain rules. (This aspect of Hilbert's theories, which I think Gödel has overstated a little for the sake of simplicity, is often called his 'formalism'. I avoid this word because Hilbert never used it himself; I believe it was planted on Hilbert by his enemy Brouwer as a deliberate insult, and somehow it stuck. There is a philosophical theory called formalism, which says that mathematics consists of nothing more than manipulations of symbols according to rules, and elementary statements about these manipulations. Hilbert's view was subtler than this.)

The other main influence was mathematics itself, which (said Gödel) is fundamentally optimistic. Hilbert responded to 'the mathematician's instinct' by clinging to the belief that:

> every precisely formulated yes-or-no question in mathematics must have a clear-cut answer.

Here Gödel was certainly thinking of Hilbert's declaration in 1926: 'In der Mathematik gibt es kein *Ignorabimus!*' (In English: For mathematicians there is no such thing as 'We shall never know'.) Hilbert seems to have meant not only that properly posed mathematical questions have definite answers; he meant also that mathematicians should assume that they can always find these answers if they try for long enough.

Now, Gödel said in his APS lecture, we know (from Gödel's own theorem in 1931—we shall come to this in a moment) that Hilbert can't have it both ways. There is no mechanical method of calculation which will solve all problems of arithmetic, let alone of all mathematics. So there are two possible reactions. The one usually adopted, since it is the only one that suits the spirit of the time, is to give up any hope of being able to solve all the problems of mathematics. But a person with a better historical perspective would react differently, and say that we should look for non-formalistic ways of solving problems. At this point Gödel says where he believes we should look for these non-formalistic ways. But before we consider this, we must look back at what Gödel proved in 1931, and see how it connects with Turing's thesis.

6.6 Yes-or-No Questions

You may have noticed that Gödel spoke as if all mathematical results are the answers to 'yes-or-no' questions. This looks wrong at first. The answer to the question 'What is $2 + 2$?' isn't either Yes or No. However, we get the same information from either of the two following statements:

> $2 + 2$ is equal to 4.
>
> The answer to the question 'Is $2 + 2$ equal to 4?' is Yes.

So there is really no loss of generality if we restrict ourselves henceforth to digital families where all the answers are either Yes or No. I call these *yes-or-no* digital families.

One way to describe a yes-or-no digital family F is to give a rule which allows you to calculate, for each positive integer n, a string $S(n)$ which can be read as a statement. Then $F(n)$ is Yes if $S(n)$ is true, and No if $S(n)$ is not true.

Example 1. For each n let $S(n)$ be the statement 'The n-th digit in the decimal expansion of π is 7'.

This describes a yes-or-no digital family which happens to be a computable function. We can compute the answer by calculating the decimal expansion of π until we get to the n-th place, and then looking to see if the number in that place is 7. As we saw earlier, there are mechanical formulas for working out the decimal expansion of π. In fact if n is not too large (say not more than a few hundred million), you can look up the answer on the Internet by going to the ftp site:

> `ftp : //www.cc.u - tokyo.ac.jp/`

You can download buckets of pages of the decimal expansion, starting with:

3.14159265358979323846264338327950288419716939937510582
0974944592307816406286208998628034825342117067982148086
5132823066470938446095505822317253594081284811174502841
0270193852110555964462294895493038196442881097566593344
6128475648233786783165271201909145648566923460348610454
3266482133936072602491412737245870066063155881748815209
2096282925409171536436789259036001133053054882046652138
4146951941511609433057270365759591953092186117381932611
7931051185480744623799627495673518857527248912279381830
1194912983367336244065664308602139494639522473719070217
9860943702770539217176293176752384674818467669405132000
5681271452635608 . . .

Example 2. This time we take $S(n)$ to be the statement 'The decimal expansion of π contains n copies of n in a row'.

Nobody knows whether this is a computable function. For all we know, it may be that the statement is true for *every* value of n, and in that case the answer is always Yes. But nobody has proved that, and maybe nobody ever will. For a particular value of n, we can run through the decimal expansion looking for n copies of n in a row. If they exist then eventually we shall find them, maybe after a billion and three years, and we can write down the answer Yes. If they don't exist, so that the answer is No, then we and our offspring may never know that. Maybe the human race will carry on till the end of time looking for 18 consecutive copies of 18:

181818181818181818181818181818181818

in the decimal expansion of π and never finding them, but never finding a proof that they can't be found.

For the first few million digits of the decimal expansion of π, you can try this out on the Internet too, by using Jeremy Gilbert's 'Search Pi' program at:

`http://gryphon.ccs.brandeis.edu/~grath/attractions/gpi/`

It found me seven consecutive 7's in the section between 3 and 4 million digits, but it failed to find twelve consecutive 12's in the first 9999907 digits. Maybe you can get better results.

6.7 Material and Formal Proofs

Before we turn to the third and most important example of a yes-or-no digital family, we need to spend a moment thinking about proofs.

A *proof* of a statement is something which rightly convinces people that the statement is true.

For example I can prove that I was born in Reading by producing my birth certificate; the certificate is a proof. I can prove that:

$$\frac{dx^3}{dx} = 3x^2$$

by writing down the definition of d/dx and doing the appropriate calculations; the definition and calculations are the proof. I can prove that if m is a positive integer, then out of any $m+1$ distinct integers strictly between 0 and $2m$ there are two with no common prime factors, by pointing out that two of the integers must be consecutive; my argument is a proof.

During the late 19th and early 20th centuries there was a strong move to reduce all mathematical arguments to some standard forms in a standard language. In fact several mathematicians argued that all mathematics should be carried out in a precisely defined formal language with precisely defined rules, just to be sure that there were no false moves. Broadly speaking, this was achieved in the period between the two world wars, when a number of mathematicians showed how virtually all of mathematics could be written in a formal language known as the first-order language of set theory, if one cared to do that. But it would be unreadable in that form, so we don't, but we are glad to know that it's possible.

Hilbert emphasised that an important part of this venture was to write down all the facts that were taken for granted and all the rules of deduction that were used. He and members of his school achieved this too. They wrote down rules of deduction which were shown to capture exactly the notion of 'proof in a first-order language'; you will learn these rules if you take an undergraduate course in logic. They also wrote down axioms of set theory, the so-called Zermelo-Fraenkel axioms, and showed that virtually all mathematical theorems can be deduced from the Zermelo-Fraenkel axioms using just the allowed rules of argument.

As a result, there was a new notion of 'proof', namely a deduction of a conclusion from some assumptions, in a first-order language, using just the first-order rules of deduction. An object of this kind was called a *formal proof*. One curious fact about first-order languages, which Hilbert often emphasised, was that the symbols in them don't have to be given any meaning. They can just be operated as a 'game with symbols'. So a formal proof can be a meaningless array of symbols; it need not be a genuine proof in the sense we defined earlier. Following Hilbert, let us say that a proof is *contentful* if it is a genuine meaningful proof which really does prove that some statement is true.

There are several connections between formal proofs and contentful proofs. One is that if we have a formal proof and give meanings to the symbols, then it becomes a meaningful argument and so it expresses a contentful proof that its conclusion follows from its assumptions. So one formal proof can give indefinitely many contentful proofs according to how it's interpreted.

Another very important connection between contentful proofs and formal proofs is this. If you have two strings of symbols in a formal language, then either you can prove (contentfully) that the first string is a formal proof of the second, or you can prove (contentfully) that it isn't. If it is, then you can work out mechanically what are the statements assumed in the proof; in the case of set theory, you can work out mechanically whether or not the assumptions are axioms of Zermelo-Fraenkel set theory. In short, if the two strings are in the first-order language of set theory, then either you can prove that the first is a formal proof of the second from the Zermelo-Fraenkel axioms, or you can prove that it isn't.

Now suppose ϕ is a sentence in the first-order language of set theory. If ϕ follows from the Zermelo-Fraenkel axioms, then we can (contentfully) prove ϕ by finding a formal proof of ϕ from the axioms; one can show that if there is such a formal proof, then a systematic search will eventually find it, though perhaps very slowly. On the other hand if ϕ doesn't follow from the Zermelo-Fraenkel axioms, then we may never know either way. We have no general way of showing that something *doesn't* follow from the axioms, unless we can show that its contradiction does follow.

Hilbert saw this as a fundamental problem in the foundations of mathematics: we can show when something does follow from given assumptions, but we can't show when it doesn't. In fact he proposed as a test problem for twentieth century mathematicians to work on:

> **Hilbert's Second Problem.** Show that it is impossible to deduce '$0 = 1$' from the axioms which we assume about the natural numbers.

He said that the proof should be *finitary*, i.e. it should just use properties of the symbols on the page. This was because he wanted the proof to be in some sense a justification of our axioms for arithmetic, and so the proof itself should use assumptions which are much weaker and safer than those of arithmetic. There is nothing special about the sentence '$0 = 1$'; one can show that this will be deducible if and only if some contradiction can be deduced from the axioms.

Now we can give our third example of a yes-or-no digital problem.

Example 3. This example is more complicated, and it brings us directly to Gödel's theorem. Just as there is a first-order language of set theory, so there is a first-order language of arithmetic; I call it L. The following axioms can all be written in L; they are known as the *first-order Peano axioms for arithmetic*, PA for short. They state or imply all the 'obvious' facts about natural numbers that we can write down in the language L.

1. There is no solution of $x + 1 = 0$.

2. If $x + 1 = y + 1$ then $x = y$.

3. $x + 0 = x$.

4. $x + (y + 1) = (x + y) + 1$.

5. $x \times 0 = 0$.

6. $x \times (y + 1) = (x \times y) + x$.

7. (For any 'first-order property' P:)
   Suppose 0 has property P, and whenever a number x has property P, $x + 1$ has it too. Then every number has property P.

For example the first axiom says that there is no natural number x such that $x + 1 = 0$. In axiom 7 (the *induction* axiom) a 'first-order property' means one that can be expressed in L. In fact axiom 7 is really infinitely many axioms, one for each first-order property. But the entire set of axioms of PA, though it's infinite, is clearly Turing-enumerable; we could fix up a machine to write down these axioms and only these axioms.

Now let T be any set of sentences of L. We can define a yes-or-no digital family F_T by taking $S(n)$ to be the statement: 'The n-th string is a sentence of L which can be deduced from assumptions in the set T'.

We can state Gödel's theorem of 1931 a little better if we use hindsight and imagine that Turing had already described his universal machine. What Gödel showed was (among other things):

> **Gödel's first incompleteness theorem, 1931.** Suppose T is a Turing-enumerable set of sentences of L, such that (1) every sentence in T makes a true statement about the natural numbers, and (2) T includes all the axioms of PA. Then F_T is not Turing-computable.

Why is this called 'incompleteness theorem'? Because it implies that the PA axioms are incomplete, in the sense that they aren't strong enough to provide a yes-or-no answer to every question about the natural numbers that we can ask in the language L. Suppose they were strong enough. Then we could compute F_{PA} (taking T to be PA) as follows:

> Let n be any natural number. First check whether the n-th string is a sentence of the language L. If not, give the answer No. If yes, let S be the sentence. For each natural number k in turn, check whether the k-th string is or is not a proof of either S or Not-S from the PA axioms. By assumption, since the PA axioms settle the truth of S, we will eventually reach either a proof S (in which case we give the answer Yes) or a proof of Not-S (in which case we answer No).

However, the first interesting fact for us is that the digital family F_{PA} is not computable. The second interesting fact is the devilishly clever proof which Gödel gave for his theorem.

What Gödel did was to show that statements about L can be coded up as statements about numbers. Then Gödel showed that we can write down formulas of arithmetic, in the language L, which express:

1. n is the code number of a sentence of L.

2. m is the code number of a formal proof of the expression with code number n, from assumptions in T.

3. k is the code number of the sentence got by writing the number m in place of the variable 'x' in the expression with code number n.

(For 2 we only need to know that T is Turing-enumerable.) Using these, Gödel wrote down a formula $E(x, y)$ of L which expresses:

> There is no formal proof, from assumptions in T, of the sentence which is got by putting the number x in place of the variable 'x' in the formula with code number y.

Since $E(x, x)$ is a formula of L, it has a code number e. Then $E(e, e)$ is the sentence which is got by putting the number e in place of the variable 'x' in the formula with code number e. So what $E(e, e)$ expresses is that $E(e, e)$ itself is not provable from assumptions in T.

So if $E(e, e)$ was not true, there would be a proof of $E(e, e)$ from assumptions in T, and we would have a proof of something false from assumptions which we have supposed are true. This proves that $E(e, e)$ must be true. But it also proves that there is no proof of $E(e, e)$ from assumptions in T.

In particular, taking T to be PA:

> We have found a sentence about the natural numbers, which is true but not provable from the Peano axioms.

6.8 The *Dialectica* Lecture

Gödel proved two incompleteness theorems in 1931. He sketched a proof of the second one at the end of his published paper, and promised a full proof in a sequel which was never written. Recall from the previous section that if T is a Turing-enumerable set of sentences of L, then Gödel showed how to write down a sentence ϕ of L which expresses 'It is impossible to deduce '$0 = 1$' from T.'

> **Gödel's second incompleteness theorem.** Suppose T is a Turing-enumerable set of sentences of L, such that (1) every sentence in T makes a true statement about the natural numbers, and (2) T includes all the axioms of PA. Then the sentence ϕ is not provable from assumptions in T.

Hilbert had asked for a 'finitary' proof of the consistency of the axioms of arithmetic, and the axioms in PA are certainly some of those axioms. What Gödel's theorem tells us is that there is no formal proof from PA of a certain sentence ϕ expressing that '$0 = 1$' is not deducible from PA. Does Gödel's theorem show that Hilbert's Second Problem can't be solved?

Certainly the sentence ϕ is true if and only if PA is consistent. But there are two other questions we need to ask.

First, could there be another sentence ψ of L which expresses (in some different way) that PA is consistent, but is deducible from PA? (Of course we wouldn't be able to prove from PA that ϕ and ψ said the same thing.) This is not a question that allows a straightforward answer, because it relies on our notion of what is a (contentful) proof, and that's something that nobody has yet managed to formalise. There is some slight evidence that Gödel himself changed his mind about it; possibly his uncertainty was one reason why he never completed Part II of his 1931 paper. I think the general view of the experts, who have tried out many possibilities, is that it's morally certain that there is no such sentence ψ, and that that's probably the best one can say about the matter.

In any case Gödel put his energies into the second question: can every 'finitary' proof be expressed by a formal deduction from PA? Gödel considered this question in a German paper published in 1958 in the journal *Dialectica*—I shall call it the *Dialectica* paper. Arrangements were made for an English translation to appear, but Gödel sat on the translation and never gave it the go-ahead. In the meanwhile Bruce Watson and Wilfrid Hodges published a translation soon after Gödel's death. Gödel's literary executors found the reason why he never gave permission for the translation; he wasn't satisfied with what he had said in the paper, and he had made several attempts to rework it in English. The notes which contained Remark 3 were found attached to the latest version of the *Dialectica* paper, together with a note referring to them as 'the subsequent paper'.

Near the beginning of the *Dialectica* paper, Gödel distinguishes between finitary and constructive proofs. A finitary proof can only consider:

> finite space-time configurations of elements whose nature is irrelevant except for equality or difference.

A constructive proof can consider these things too, but it can also consider *abstract concepts*, so long as these are 'exhibited, or obtained by construction or proof'. Gödel comments:

> By abstract concepts, in this context, are meant concepts which ... do not have as their content properties or relations of *concrete objects* (such as combinations of symbols), but rather of *thought structures* or *thought contents* (e.g., proofs, meaningful propositions, and so on), where in the proofs of propositions about these mental objects insights are needed which are not derived from a reflection upon the combinatorial (space-time) properties of the symbols representing them, but rather from a reflection upon the *meanings* involved. [Gödel's italics.]

Gödel then gives a constructive proof of the consistency of PA. Where does he find his 'abstract concepts which are obtained by construction or proof'? Here are some examples. We can build up the natural numbers by starting from 0 and adding 1 any number of times:

$$0, 0+1, 0+1+1, 0+1+1+1, \ldots.$$

Now we can define the operations + and $\times$ as in the Peano axioms:

$$x + 0 = x; \quad x + (y + 1) = (x + y) + 1.$$

$$x \times 0 = 0; \quad x \times (y + 1) = (x \times y) + x.$$

You may have noticed that we can go from $\times$ to exponentiation just as we went from + to $\times$:

$$x^0 = 1; \quad x^{y+1} = (x^y) \times x.$$

That being said, why not invent the next function in line? If $f_0(x, y)$ is $x + y$, $f_1(x, y)$ is $x \times y$ and $f_2(x, y)$ is x^y, then $f_3(x, y)$ should be defined by:

$$f_3(x, 0) = 2; \quad f_3(x, y + 1) = f_2(f_3(x, y), y).$$

And so on. The general formula should be:

$$f_0(x, y) = x + y; \quad f_{n+1}(x, 0) = n; \quad f_{n+1}(x, y + 1) = f_n(f_{n+1}(x, y), y).$$

This defines a method for getting new functions f_n out of a starting function +. We can generalise this to get second-order functions F_n; let g be any starting function, and write:

$$\begin{aligned} F_0(g, x, y) &= g(x, y) \\ F_{n+1}(g, x, 0) &= n \\ F_{n+1}(g, x, y + 1) &= F_n(g, F_{n+1}(g, x, y), y) \end{aligned}$$

And so on. These examples should be enough to give the idea.

Gödel defines a collection of numbers, functions and higher-order functions which are known as the *primitive recursive functions of higher type.* His proof of the consistency of PA uses these objects. But it also uses properties of these objects, and one property in particular: every value of any one of his primitive recursive functions of higher type can be calculated. This property is not provable using just the axioms of PA; but we can see that it holds by seeing how the primitive recursive functions of higher type are built up. We can see it by *thinking*, not by calculating.

What kind of thinking? Let me dwell on this for a moment, because I believe it is vital for understanding Gödel's Remark 3. Gödel used the word *Anschauung* for what happens when we can 'see' that a mathematical fact is true. Bruce Watson and I got some stick for refusing to use the usual translation of this word, which is *intuition*; we were wary of the various uses of this word in ordinary English (for example 'woman's intuition'). Instead we said *inspection* and gave an explanatory footnote. It was reassuring to discover that Gödel in his own translation also didn't translate *Anschauung* as *intuition* either. Instead he used the phrase *concrete intuition* and added a longer footnote than we did.

Gödel's footnote says that in his discussion of finitary mathematics, he takes the word *Anschauung* from Hilbert, and he understands Hilbert to mean substantially what the philosopher Immanuel Kant (German, late 18th century) meant by space-time *Anschauung*, 'confined, however, to configurations of a finite number of discrete objects'. At the risk of bringing down on my head the wrath of Kantian philosophers, let me try to say what this means. Kant was talking about the form of human consciousness. His point was that when we are conscious, we are conscious of being in a world of three dimensions and at a certain point of time. The various sensations that come to us all fit into this framework; we are aware of them as being in a certain place at a certain time. (Not so with the other senses; we don't sense everything as having a certain smell or taste, for example.)

On the view that Gödel is attributing to Hilbert and Kant, there are some facts about the world that we can just see, from the fact that they belong in space and time. For example we can just see that if we add 'c' to the right of the string 'ab', we get the same string as if we added 'a' to the left of 'bc'. We don't have to be looking at a particular set of symbols in order to see this; but we can see that it is so by a kind of introspection. Mental pictures will help, but they aren't essential. This way of knowing mathematical facts about configurations of symbols is what Gödel calls *concrete intuition*.

So far as I know (I am not a Kant expert), Kant never used any phrase like 'konkrete Anschauung'. But the phrase does occur in the writings of an early twentieth century philosopher, Edmund Husserl. According to Dagfinn Føllesdal's commentary on Gödel's APS lecture in Gödel's Collected Works:

> Gödel did not start to study Husserl until 1959 ..., but he soon became quite absorbed by the reading of Husserl's writings. He owned all of Husserl's main works, and his underlining and comments in the margins indicate that he studied them carefully.

According to Husserl, besides our 'concrete Anschauung' of physical objects, we also know abstract concepts through a 'categorical Anschauung'. Gödel must have recognised this as a kindred idea to what he had already suggested in the 1958 version of the *Dialectica* paper: that we can simply *see* that certain abstract objects have certain properties, by considering the meanings of their constructions. Already in 1958 he had the idea that we can jump outside the restricted world of finitary mathematics by using facts about concepts.

6.9 A Calculus of Concepts?

Now turn back with me to Gödel's APS lecture. Remember that Gödel was going to offer us some ground for optimism about solving problems that we couldn't solve by sheer calculation. Here is what he says:

> Obviously, this means that the certainty of mathematics is to be secured ... by cultivating (deepening) knowledge of the abstract concepts themselves ... In what manner, however, is it possible to extend our knowledge of these abstract concepts, i.e., to make these concepts themselves precise and to gain comprehensive and secure

> insight into the fundamental relations that subsist among them? ...Now in fact, there exists today the beginning of a science which claims to possess a systematic method for such a clarification of meaning, and that is the phenomenology founded by Husserl. Here clarification of meaning consists in focusing more sharply on the concepts concerned by directing our attention in a certain way, namely, onto our own acts in the use of these concepts, onto our powers in carrying out our acts, etc. But one must keep clearly in mind that this phenomenology is not a science in the same sense as the other sciences. Rather it is [or in any case should be] a procedure or technique that should produce in us a new state of consciousness in which we describe in detail the basic concepts we use in our thought, or grasp other basic concepts hitherto unknown to us. I believe there is no reason at all to reject such a procedure at the outset as hopeless. ...
>
> In fact, one has examples where, even without the application of a systematic and conscious procedure, but entirely by itself, a considerable further development takes place in the second direction, one that transcends "common sense". Namely, it turns out that in the systematic establishment of the axioms of mathematics, new axioms, which do not follow by formal logic from those previously established, again and again become evident. It is not at all excluded by the negative results mentioned earlier that nevertheless every clearly posed mathematical yes-or-no question is solvable in this way. For it is just this becoming evident of more and more new axioms on the basis of the meaning of the primitive notions that a machine cannot imitate.
>
>

Gödel seems to have had three kinds of example in mind for these 'new axioms'.

First, there are the basic axioms which underlie the whole of mathematics. We can take them to be the Zermelo-Fraenkel axioms of set theory. The Peano axioms for arithmetic can be deduced from them.

Second, there are the sentences which we can prove by the method of Gödel's incompleteness theorems. For example we accept that the Zermelo-Fraenkel axioms are true, because they state facts that in some sense we can 'see' about sets. Since this set of axioms is true, it is consistent. But it is also Turing-enumerable, and so by Gödel's first theorem we can prove the truth of a sentence of set theory which doesn't follow from the Zermelo-Fraenkel axioms. By the second theorem, this can be a sentence expressing that the axioms are consistent.

Third, and more controversially, there are the sentences of set theory which we get by assuming that 'there are infinite cardinal numbers so large that ...'. These sentences are known as *large cardinal axioms*; Gödel himself called them *axioms of infinity*. Insofar as we can speak of evidence, the evidence for the truth of large cardinal axioms is of a very different kind from that for the

Zermelo-Fraenkel axioms. As Gödel himself put it in his paper 'What is Cantor's continuum problem?', these axioms might be justified by the fruitfulness of their consequences. In the extreme case:

> There might exist axioms so abundant in their verifiable consequences, shedding so much light upon a whole discipline, and furnishing such powerful methods for solving given problems ... that quite irrespective of their intrinsic necessity they would have to be assumed at least in the same sense as any well-established physical theory.

There are a fair number of large cardinal axioms in circulation, but I doubt if many set theorists feel that they have quite met the test which Gödel proposes here. In any case, while many mathematicians claim that they can 'see' that the axioms of the first and second kinds must be true (in a sense that Gödel would probably have counted as Anschauung), it would be hard to make a convincing case that we can 'see' the truth of any large cardinal axioms.

6.10 Turing's Error

Now at last we can turn to Gödel's comment on Turing's argument. Here is what he says:

> *A philosophical error in Turing's work.* Turing in his [paper on the universal machine] gives an argument which is supposed to show that mental procedures cannot go beyond mechanical procedures. However, this argument is inconclusive. What Turing disregards completely is the fact that *mind, in its use, is not static, but constantly developing*, i.e., that we understand abstract terms more and more precisely as we go on using them, and that more and more abstract terms enter the sphere of our understanding. There may exist systematic methods of actualizing this development, which could form part of the procedure. Therefore, although at each stage the number and precision of the abstract terms at our disposal may be *finite*, both (and, therefore, also Turing's number of *distinguishable states of mind*) may *converge toward infinity* in the course of the application of the procedure. Note that something like this indeed seems to happen in the process of forming stronger and stronger axioms of infinity in set theory. This process, however, today is far from being sufficiently understood to form a well-defined procedure. It must be admitted that the construction of a well-defined procedure which could actually be carried out (and would yield a non-recursive number-theoretic function) would require a substantial advance in our understanding of the basic concepts of mathematics. Another example illustrating the situation is the process of systematically constructing, by their distinguished sequences $\alpha_n \to \alpha$, all recursive ordinals α of the second number-class.

I hope I've set things up so that most of this makes sense at once. (I'll ignore the technical example in Gödel's last sentence.)

Remember that Turing's paradigm example of a computer is a person calculating the decimal expansion of a real number, for example π. We generalised this to a person calculating one by one the values of the answers $F(0)$, $F(1)$, etc. to a digital family F. So Turing is describing a process which goes on for ever. In practice this means that the calculation won't all be done by one computer. When he dies, maybe his daughter will take over, and then her daughter, and so on for infinitely many generations. That is, unless the human race discovers the secret of eternal life. We have to suppose that the human race will evolve over this course of time, intellectually and genetically.

Now Gödel considers Turing's argument to show that the computer can have only finitely many states of mind. The argument was, if you recall, that otherwise some of the states of mind will be "arbitrarily close" and will be confused. But of course this only shows that the computer can have just finitely many states of mind at any one time. He/she might evolve to have more and more states of mind as time goes by, and then it is no longer clear that the same computation can be done by a machine with some fixed finite number of 'states of mind'. This clearly is a gap in Turing's argument.

How serious is it? Gödel claims that there could be 'systematic' computing procedures which involve more and more states of mind. Let's leave aside the word 'systematic' for the moment, and see what Gödel is claiming for the mathematical powers of the human race. For definiteness, let T be the set of all sentences of L which are true of the natural numbers; this set is known as *true arithmetic*. A sentence of L is deducible from T if and only if it is in T. Recall the yes-or-no digital family which we named F_T:

> For each natural number n, $S(n)$ is the statement: 'The n-th string is a sentence in T.'

We know that this is not a Turing-computable function, thanks to Gödel's first incompleteness theorem. (For if it was Turing-computable, then T would be Turing-enumerable.)

Gödel certainly hoped (see his APS lecture) that we shall eventually have a proof or a disproof of each sentence $S(n)$, so that we can work out every value of F_T. His idea was that although no mechanical method will give all the right answers, we might reach the answers by some sort of concept calculus.

We have to comment straight away that this proposal only makes sense if the human race is assumed to evolve new mental powers as time proceeds. There are two reasons for this. The first is Turing's observation that if we are allowed only a fixed finite number of states of mind, then we are limited to computing mechanically computable functions.

The second reason lies in psychology rather than mathematics. The evidence overwhelmingly suggests that our powers of concept formation are hardwired into us genetically. We can turn one of Gödel's own arguments against him. In the APS lecture he says:

> If one considers the development of a child, one notices that it proceeds in two directions: it consists on the one hand in experimenting with the objects of the external world and with its own sensory and motor organs, on the other hand in coming to a better and better understanding of language, and that means - as soon as the child is beyond the most primitive designating of objects - of the basic concepts on which it rests. With respect to the development in this second direction, one can justifiably say that the child passes through states of consciousness of various heights, e.g., one can say that a higher state of consciousness is attained when the child first learns the use of words, and similarly at the moment when for the first time it understands a logical inference.
>
>

The conceptual advances which Gödel cites - learning the use of words and learning to make logical deductions - are standard examples of skills which are built into the human organism; provided the environment isn't hostile, they emerge automatically when the child reaches a certain age. It's not at all plausible that a child who can't use words learns to use them simply by 'directing its attention in a certain way', or any of the other devices of Husserl's concept world.

So Gödel's manipulation of concepts by itself won't be sufficient to allow us to compute all the values of F_T. But if genetics is allowed to play a role, maybe Gödel's manipulation of concepts won't be necessary either. Who can say?

At first sight there is another way in which we might eventually calculate each value of F_T. That is to look for proofs by some kind of random process. If it's truly random then surely it should eventually throw up a proof or disproof of each statement $S(n)$. Gödel doesn't consider this route; but on closer inspection it doesn't give us anything new. If the proofs can be written in English, then we don't need a random process; we can simply set a mechanical process to generate one by one all strings of English, and as each string is generated, we check whether it proves one of the statements $S(n)$. The problem is that a string of English doesn't tell us that $S(n)$ is true unless we can understand it (the string) as a proof of $S(n)$. There is no guarantee that our present concepts are strong enough to understand even all proofs that can be written in English. (Every teacher knows examples of person A who can't understand proof B and probably never will be able to understand B.) So we are back to the question of learning new concepts.

Also it's not clear from the start that English does contain a proof or disproof of each sentence $S(n)$. Maybe some stronger language is needed? On this point I think Gödel is right to choose optimism. English is a universal language; if anything can be said at all, then it can be said, somehow, in English. A worse possibility is that there are some truths $S(n)$ which no finite being could ever prove. But it would be crazy methodology to start out by assuming this.

Fine then, the human race might evolve so that eventually it can compute every value of F_T, and thus solve every problem of first-order arithmetic. What would count as a 'systematic method of actualizing this development', to use Gödel's words? Some genetic advance would be needed. The only way to make this a 'systematic method' would be for the (human) computer to take control of the genetics of the human race. This is more plausible now than it would have seemed when Gödel wrote in the 1970s. But to recall the original problem, would that really count as a method of 'computing'?

The notion of 'computing' normally arises in contexts where genetic alteration of the conceptual powers of the human race is not one of the allowed operations. That being so, we have some freedom to decide whether we would include that kind of genetic operation as part of a computing procedure. I'd vote No, and I suspect I am in a large majority.

If pressed for an argument to support this vote, I would say the following. One feature of computation—as we normally understand it—is that the rules are fixed in advance. In terms of Turing's analysis, this means that the rules which specify which simple operation to perform next, depending on the computer's state of mind and the symbol being inspected, must be agreed before the computation begins. Now in principle we could write down a definition which tells us these rules even if there are infinitely many states of mind. But it is hard to see how we could do this when the states of mind include some which at present we can't even conceive.

So Turing's gap is plugged, but we do it by a sensible linguistic choice, not by finding a secret method. By the same token, I think we are entitled to some freedom in what we would count as 'systematic'; that word may not add anything meaningful to Gödel's statement.

Gödel is still entitled to his optimism, but he should be a little more modest. Maybe the human race will eventually compute every value of F_T; but we in this generation don't have to lay down the procedure that will get us there. All we can hope to do is to jump over the next conceptual hurdle. If we take the matter one step at a time, there is no longer any point in asking for a 'systematic' procedure, whatever that might mean. Even Gödel precipitated only a small finite number of conceptual revolutions. We should be thankful for each one as it comes.

6.11 Further Reading

If you want to follow up this chapter with some reading, the original texts are the best if you can get them. Gödel was a very clear and careful writer. All Gödel's published and unpublished papers are now available with commentaries in [1]. The items referred to in this chapter are as follows:
Volume I:

- On formally undecidable propositions of *Principia mathematica* and related systems (1931).

Volume II:

- Remarks before the Princeton bicentennial conference on problems in mathematics (1946).
- What is Cantor's continuum problem? (1947).
- On an extension of finitary mathematics which has not yet been used (*Dialectica* paper, 1958 and 1972 versions).
- Some remarks on the undecidability results (Remark 3, 1972).

Volume III:

- The modern development of the foundations of mathematics in the light of philosophy (APS Lecture, 1961).

The Kurt Gödel Society has its web page at:

`http : //www.logic.tuwien.ac.at/kgs/home.html`

Turing's paper is found in [2]. You should also sample the Turing page on the Internet, set up by Andrew Hodges (no relation) at:

`http://www.wadham.ox.ac.uk/~ahodges/Turing.html`

You may find Husserl rather harder to make sense of, but one place to start might be the Husserl page:

`http://www.mesa.colorado.edu/~bobsand/husserl.html`

6.12 Bibliography

[1] Kurt Gödel, *Collected Works*, Volume I (Publications 1929–1936), Volume II (Publications 1938–1974), Volume III (Unpublished essays and lectures), ed. Solomon Feferman, Dawson, et al., Oxford University Press Inc., New York, 1995.

[2] A. M. Turing, *On computable numbers, with an application to the Entscheidungsproblem*, Proceedings of the London Mathematical Society, ser. 2, vol. 42 (1936–7), pp. 230–265; reprinted on pages 116–151 of *The Undecidable*, edited by Martin Davis, Raven Press, Hewlett, New York 1965.

Chapter 7

Penrose's Philosophical Error

7.1 Can Computers Think?

The United States Department of Energy's Sandia National Laboratory and Intel Corporation have built a supercomputer that on December 17, 1996 reached the one trillion-operations-per-second mark.[1] It weighs 44 tons, has 573 gigabytes of memory and 2.25 trillion bytes of disk space. In the time it takes you to blink an eye, the computer will complete 40 billion calculations. It is only 25 years since Intel introduced the first microprocessor, which could carry out 60,000 operations per second. Who can say what breakthroughs will occur in the next 25 years or 250 years? Computers with tremendous power and speed will surely be developed. Will these computers be able to think? Will they be conscious? Is there today anyone who would say that no matter what breakthroughs occur in microelectronics, no matter what developments occur in computer programming, such as learning and random elements as in neural networks, no matter how powerful computers become, a computer will never think? One such person is Roger Penrose, who has written in his book *Shadows of the Mind* [14] that:

> there must be more to human thinking than can ever be achieved by a computer. ...Consciousness, in its particular manifestation in the human quality of understanding, is doing something that mere computation cannot. I make clear that the term 'computation' includes both 'top-down' systems, which act according to specific well-understood algorithmic procedures, and 'bottom-up' systems,

[1] For more details see Intel's website

`http://www.intel.com/pressroom/archive/releases/cn121796.htm`

and Sandia's website

`http://www.sandia.gov/LabNews/LN06-20-97/teraflops_story.html`

Figure 7.1: Roger Penrose, 1996. ©Steve Green. Reproduced with permission.

> which are more loosely programmed in ways that allow them to learn by experience.

The question at issue is not a dualistic distinction between the conscious mind and the physical brain. Penrose accepts that the conscious mind arises as a functioning of the physical brain, but he does not believe this functioning could be simulated by a large number of silicon chips operating as they do in a modern computer.

John Lucas

In the 1930's Alan Turing (1912-1954) proved a theorem showing that certain types of idealized computers, operating according to fixed rules, do have limitations. Turing's theorem is of a general type first proved by Kurt Gödel (1906-1978) concerning limitations to formal mathematical systems. These limitations are discussed in Chapter 6, *Turing's Philosophical Error?*, and we shall consider them again in this chapter as well. Penrose bases his argument that the mind is not computational on these limitations uncovered by Turing and Gödel. His use of Gödel's theorem is similar to that of J.R.Lucas [13]. Lucas's article begins like this:

> Gödel's theorem seems to me to prove that Mechanism is false, that is, that minds cannot be explained as machines.

It is interesting, however, that Lucas doesn't rule out the construction of intelligent computers! At the end of his article, Lucas says:

> When we increase the complexity of our machines there may, perhaps, be surprises in store for us. [Turing] draws a parallel with a fission pile. Below a certain "critical" size, nothing much happens: but above the critical size, the sparks begin to fly. So too, perhaps, with brains and machines.... This may be so. Complexity often does introduce qualitative differences. Although it sounds implausible, it might turn out that above a certain level of complexity, a machine ceased to be predictable, even in principle, and started doing things on its own account, or, to use a very revealing phrase, it might begin to have a mind of its own. It would begin to have a mind of its own when it was no longer entirely predictable and entirely docile, but was capable of doing things which we recognized as intelligent.... But then it would cease to be a machine, within the meaning of the act.

So for Lucas, a machine cannot think *by definition of what a machine is.* Penrose certainly does not agree with Lucas's view here. On the contrary, Penrose claims that no matter how complex they become, computers will not have the capacity of human understanding.

Hilary Putnam

The results of Turing and Gödel belong to an area of mathematics known as mathematical logic. How do mathematical logicians react to Penrose's argument?[2] In his book review [16] of *Shadows of the Mind*, Hilary Putnam[3] states:

> In 1961 John Lucas - an Oxford philosopher well known for espousing controversial views - published a paper in which he purported to show that a famous theorem of mathematical logic known as

[2] In addition to Putnam's review, discussed here, an on-line version of Soloman Feferman's review of *Shadows of the Mind* can be found at

`http://psyche.cs.monash.edu.au/v2/psyche-2-07-feferman.html`

You can find various other criticisms of Penrose's approach, together with replies by Penrose, at

`http://psyche.cs.monash.edu.au/v2/psyche-2-23-penrose.html`

A large bibliography on the subject of 'mind and metamathematics' can be found at

`http://ling.ucsc.edu/~chalmers/biblio4.html#4.2`

which is *Part 4: Philosophy of Artificial Intelligence* of the Contemporary Philosophy of Mind: An Annotated Bibliography, at

`http://ling.ucsc.edu/~chalmers/biblio.html`

[3] Putnam's own view on the nature of human mentality is indicated by the following quote from [17]:

> Why *should* all truths, even all empirical truths, be discoverable by probabilistic automata (which is what I suspect we are) using a finite amount of observational material?

> Gödel's Second Incompleteness Theorem implies that human intelligence cannot be simulated by a computer. Roger Penrose is perhaps the only well-known present-day thinker to be convinced by Lucas's argument.... The right moral to draw from Lucas's argument, according to Penrose, is that noncomputational processes do somehow go on in the brain, and we need a new kind of physics to account for them.
>
> *Shadows of the mind* will be hailed as a "controversial" book, and it will no doubt sell very well even though it includes explanations of difficult concepts from quantum mechanics and computational science. And yet this reviewer regards its appearance as a sad episode in our current intellectual life. Roger Penrose ... is convinced by - and has produced this book as well as the earlier *The emperor's new mind* to defend - an argument that all experts in mathematical logic have long rejected as fallacious. The fact that the experts all reject Lucas's infamous argument counts for nothing in Penrose's eyes. He mistakenly believes that he has a philosophical disagreement with the logical community, when in fact this is a straightforward case of a mathematical fallacy.
>

Kevin Warwick

How do the robotics experts react to Penrose's argument? Some of the work of the cybernetics group at Reading University is discussed in Chapter 4, *Neural Network Control of a Simple Mobile Robot.* The leader of that group is Kevin Warwick, who has just published a book [28] subtitled *Why the New Race of Robots will Rule the World*, in which he states:

> I believe that in the next ten to twenty years some machines will become more intelligent than humans.
>
> We need to go to an appropriate level of modelling the functioning of a human brain in order to obtain a very good approximation.... Is it down, much further than the neuron level, to the atomic level? Should we look at quantum physics... as Penrose thinks, generating an atomic model of the brain's functioning that we do not yet have?... In reality I feel that we already do have sufficient basic modelling blocks.... What are those blocks? They are the neurons, the fundamental cells in the brain which cause it to operate as it does.
>
> One major problem with consciousness is that it is a state which we ourselves feel that we are in, and through this we are aware of ourselves. It can, therefore, be extremely difficult, when thinking about ourselves, to realise that this is simply a result of a specific state of the neurons in our brain. These 'private' thoughts and feelings we have, how can they possibly be simply a state of mind,

Figure 7.2: Alan Turing, 1951. By courtesy of the National Portrait Gallery, London.

> realised by a state of the neurons? Well, that is exactly what they are.

Alan Turing

Since part of Penrose's argument is based on a theorem of Turing concerning a limitation on what can be achieved by computers, it is interesting to know Turing's thoughts on these issues. Turing wrote an article [24][4] in which he gives his opinion on whether or not a computer will ever be able to think. In the article Turing states his belief that his own fundamental theorem concerning the limitations of computing machines is not an obstacle to the creation of intelligent computers.

Turing begins his article with the sentence:

> I propose to consider the question, 'Can machines think?'

The question in this form is, he believes, too vague and he restricts the machine to being a computer. He then replaces the question whether a computer can think like a human with the question whether a computer can *behave* like a human, and more specifically whether a computer can answer questions posed by an interrogator so that the interrogator cannot distinguish the replies of the

[4] All quotations from Turing's article in this chapter are reprinted by permission of Oxford University Press.

computer from those of a human. (He calls this the imitation game. It is now known as the Turing test[5].)

Turing states clearly that he believes the computer will do well at imitating human behavior and gives 50 years as the time scale when computers will have become sufficiently powerful to begin doing this. He was writing in 1950 and so we are just coming to his predicted time period. (Remember the new Sandia-Intel supercomputer!) About his beliefs he says:

> I believe further that no useful purpose is served by concealing these beliefs. The popular view that scientists proceed inexorably from well-established fact to well-established fact, never being influenced by any unproved conjecture, is quite mistaken. Provided it is made clear which are proved facts and which are conjectures, no harm can result. Conjectures are of great importance since they suggest useful lines of research.

To construct intelligent computers, Turing proposes to use the technique of computer learning:

> Instead of trying to produce a programme to simulate the adult mind, why not rather try to produce one which simulates the child's? If this were then subjected to an appropriate course of education one would obtain the adult brain.... Our hope is that there is so little mechanism in the child-brain that something like it can be easily programmed. The amount of work in the education we can assume, as a first approximation, to be much the same as for the human child.

Turing considers, and rejects, contrary views to his own. He considers in particular the mathematical objection based on Turing's theorem, the issue we will be considering in this chapter. We give his discussion of this point in its entirety:

> There are a number of results of mathematical logic which can be used to show that there are limitations to the powers of discrete-state machines. The best known of these results is known as Gödel's theorem, and shows that in any sufficiently powerful logical system statements can be formulated which can neither be proved nor disproved within the system, unless possibly the system itself is inconsistent. There are other, in some respects similar, results due to Church, Kleene, Rosser, and Turing. The latter result is the most convenient to consider, since it refers directly to machines, whereas

[5] In 1990 Dr. Hugh Loebner pledged a Grand Prize of $100,000 for the first computer whose responses were indistinguishable from a human's. Each year a prize of $2000 is awarded to the most human computer, which this year was won on May 16, 1997 by the program Converse, entered by the University of Sheffield. The homepage of the Loebner prize is

`http://acm.org/~loebner/loebner-prize.htmlx`

> the others can only be used in a comparatively indirect argument: for instance if Gödel's theorem is to be used we need in addition to have some means of describing logical systems in terms of machines, and machines in terms of logical systems. The result in question refers to a type of machine which is essentially a digital computer with an infinite capacity. It states that there are certain things that such a machine cannot do. If it is rigged up to give answers to questions as in the imitation game, there will be some questions to which it will either give a wrong answer, or fail to give an answer at all however much time is allowed for a reply. There may, of course, be many such questions, and questions which cannot be answered by one machine may be satisfactorily answered by another. We are of course supposing for the present that the questions are of the kind to which an answer 'Yes' or 'No' is appropriate, rather than questions such as 'What do you think of Picasso?' The questions that we know the machines must fail on are of this type, "Consider the machine specified as follows... Will this machine ever answer 'Yes' to any question?" The dots are to be replaced by a description of some machine in a standard form, which could be something like that used in §5. When the machine described bears a certain comparatively simple relation to the machine which is under interrogation, it can be shown that the answer is either wrong or not forthcoming. This is the mathematical result: it is argued that it proves a disability of machines to which the human intellect is not subject.
>
> The short answer to this argument is that although it is established that there are limitations to the powers of any particular machine, it has only been stated, without any sort of proof, that no such limitations apply to the human intellect. But I do not think this view can be dismissed quite so lightly. Whenever one of these machines is asked the appropriate critical question, and gives a definite answer, we know that this answer must be wrong, and this gives us a certain feeling of superiority. Is this feeling illusory? It is no doubt quite genuine, but I do not think too much importance should be attached to it. We too often give wrong answers to questions ourselves to be justified in being very pleased at such evidence of the fallibility on the part of the machines. Further, our superiority can only be felt on such an occasion in relation to the one machine over which we have scored our petty triumph. There would be no question of triumphing simultaneously over *all* machines. In short, then, there might be men cleverer than any given machine, but then again there might be other machines cleverer again, and so on.

Turing ends his article looking towards the future:

> We may hope that machines will eventually compete with men in all purely intellectual fields. But which are the best ones to start

> with? Even this is a difficult decision. Many people think that a very abstract activity, like the playing of chess[6], would be best. It can also be maintained that it is best to provide the machine with the best sense organs that money can buy, and then teach it to understand and speak English. This process could follow the normal teaching of a child. Things would be pointed out and named, etc. Again I do not know what the right answer is, but I think both approaches should be tried.
>
> We can only see a short distance ahead, but we can see plenty there that needs to be done.

Penrose's Error

We have seen that Penrose's argument, its basis and implications, is rejected by experts in the fields which it touches. So why expend effort studying it? Firstly, a discussion of Penrose's argument gives us the opportunity to consider the mathematical theorems on computation and formal systems which are interesting in their own right. Second, the conscious mind remains shrouded in mystery. If a precise mathematical theorem (and the bedrock of scientific precision is mathematics) can shed even a glimmer of understanding on the nature of consciousness, then it is worth careful study indeed. Perhaps you don't have the opportunity to invest the time and effort required to understand fully the point of Penrose's argument. As Alwyn Scott says [21] in answer to the question why he has such a stark difference of opinion with Penrose:

> One answer to this question might be that I am unable - because of my intellectual limitations, interests in other matters, general laziness, or some combination of all three - to follow Penrose's arguments through their many details. Perhaps he is right, and I just don't get it.

To understand Penrose's mathematical argument against a computational mind, it will be necessary to disentangle all the various ingredients: Turing's theorem on limitations to computations, Gödel's theorem on limitations to formal systems, the formal system used to analyze a computation, computational models, mathematical truths, and mathematical beliefs. We shall uncover in section 7.6 a category-mistake in Penrose's reasoning, a confusion of the deduction of **H believes X** with the deduction of **X** itself. By supposing that the beliefs such as **X** are theorems, Penrose has made the consistency and correctness of his deductive system dependent on the consistency and correctness of H's beliefs, and this cannot be right. The beliefs and other thoughts of a mind cannot be theorems of the deductive system composing our theory of mind. The basic structure of the theory must be consistent whether or not the beliefs of a

[6] On May 11, 1997 the IBM computer Deep Blue won a six game match against the world chess champion Garry Kasparov. For more details see

`http://www.chess.ibm.com/home/html/b.html`

particular mind are consistent. Indeed, even in our everyday reasoning about other people's beliefs, those beliefs do not become part of our reasoning. And if those beliefs are contradictory, it does not follow that our reasoning about those beliefs is contradictory. We should be able, for example, to consistently and correctly deduce that someone will perform a particular action as a result of the incorrect beliefs he holds.

But let us begin by discussing the basic mathematical results, unencumbered by possible applications to a theory of mind.

7.2 Solving Problems in Arithmetic

Suppose I give you a piece of paper with this written on it:

```
27+96=?
```

You might say, 'That's easy; 123.' Or you might say 'What do you want me to do with this piece of paper?', or 'Que voulez-vous que je fasse de ce papier?' if you're French. The point is that before you can respond to something I do you have to understand what I'm doing. We would have had to establish a means of communication so you will understand that I'm asking a question and understand what the question is before you can answer it. All sorts of conventions have been established as you have grown up and studied at school so that the initial communication hurdle is not an obstacle. Suppose you understand that I'm asking a question but you're not sure what some of the symbols on the paper represent. You might ask me 'What is that symbol 27?' Perhaps I might show you a bowl with 27 marbles in it. But you might not know what aspect of what I showed you corresponded to the 27. So I might then show you a bowl with 27 oranges. In case you still do not understand I might show you a bowl with 27 mice. You might then get the idea that 27 is the common characteristic of all the things I showed you: It must be a bowl! No, I'm sure you would understand that 27 represents the quantity of marbles, or oranges. In fact to keep everything as concrete as possible and avoid unnecessary abstractions, no harm will result if we agree that we are always going to talk about marbles. So 27 means 27 marbles. Going back to what was written on that piece of paper, we will now suppose that you understand that I'm asking you a question about quantities of marbles. There are two quantities mentioned: 27 and 96. You might guess that the symbol + represents some *operation* you are to perform on the two quantities of marbles, but which operation? I'm sure you would say 'That's easy, it means to add the two quantities of marbles together.' It's only easy because you were taught how to do addition and the symbol which represents it in school. We must start somewhere, with some things understood[7]

---

[7]The difficulty in defining each and every concept one uses is nicely put [2] by Ludwig Boltzmann (1844-1906); reprinted by kind permission of Kluwer Academic Publishers:

> Let me begin by defining my position by way of a true story. While I was still at high school my brother (long since dead) tried often and in vain to convince me how absurd was my ideal of a philosophy that clearly defined each concept at the time of introducing it. At last he succeeded as follows: during a lesson a

(for example the meaning of 27 marbles and certain basic operations which could be carried out on quantities of marbles). Then the meaning of more and more complicated operations on marbles could be defined. For example, I might hand you a piece of paper with this written on it:

$$(27 + 96) \times (5 + 8^2) =?$$

You might say, 'That's easy; 8487.'

In order to solve problems in arithmetic (which for concreteness we will take to mean questions about quantities of marbles) we must first agree on a certain set of basic instructions, or operations on the marbles, from which a rule may be given to perform more complicated operations. The rule is a program to be executed by performing the basic instructions one after the other. We might ask questions about the result of executing a program starting from some initial collection of bowls with marbles in them. If the program represents addition then we are asking a question about addition. If the program represents multiplication then we are asking a multiplication question. And so on.

Why do we choose to formulate arithmetic in terms of bowls of marbles? Because in this way arithmetic corresponds to concrete actions on concrete objects (marbles, or coins, etc.). This is how arithmetic originally developed, as a concrete activity rather than an abstract mental construct. At this stage there is no logic, no deduction or proof, just concrete actions carried out according to instructions which define addition, multiplication, and other more complicated arithmetical procedures. At a later stage we may reason about these arithmetical procedures, using rules of deduction which seem to us to be sound.

The Basic Instructions

There are a number of ways of selecting basic instructions out of which all arithmetical operations can be performed. Turing took one approach and arrived at Turing machines. We'll use a different approach here. The basic instructions are simple operations on bowls of marbles (for concreteness). We start with a sufficient supply of bowls, each of which is labeled with a symbol $0, 1, 2, \ldots$. Each operation which we will define will require a particular number of bowls, but we won't specify an upper limit to the number of bowls so that arbitrary arithmetical procedures can be carried out. So we have bowls $R_0, R_1, R_2, \ldots$ which are referred to as *registers*. The number r_k of marbles in the bowl (register) R_k is referred to as the *content* of the register. An arithmetical procedure or program consists of a finite sequence of instructions $I_0, I_1, I_2, \ldots, I_b$. Only four types of instructions are used:

---

certain philosophic work (I think by Hume) had been highly recommended to us for its outstanding consistency. At once I asked for it at the library, where my brother accompanied me. The book was available only in the original English. I was taken aback, since I knew not a word of English, but my brother objected at once: "if the work fulfils your expectations, the language surely cannot matter, for then each word must in any case be clearly defined before it is used."

Zero instructions. For each $k = 0, 1, 2, \dots$ there is a zero instruction $Z(k)$. The meaning of the instruction $Z(k)$ is to change the content r_k of the register R_k to 0, leaving all other registers unaltered. Thus, to carry out the instruction $Z(k)$ simply empty the kth bowl.

Successor instructions. For each $k = 0, 1, 2, \dots$ there is a successor instruction $S(k)$. The meaning of the instruction $S(k)$ is to change the content r_k of the register R_k to $r_k + 1$, leaving all other registers unaltered. Thus to carry out the instruction $S(k)$ simply add a marble to the kth bowl.

Transfer instructions. For each $j = 0, 1, 2, \dots$ and $k = 0, 1, 2, \dots$ there is a transfer instruction $T(j, k)$. The meaning of the instruction $T(j, k)$ is to replace the content r_k of the register R_k with the number r_j contained in R_j, leaving all other registers unaltered. Thus to carry out the instruction $T(j, k)$ take a spare bowl and put into it a quantity of marbles equal to the quantity of marbles in the jth bowl. Then replace the marbles in the kth bowl with the marbles in the spare bowl.

Jump instructions. The preceding three instructions are called arithmetical instructions, and they manipulate the quantity of marbles contained in the bowls. The final type of instruction is of a different sort and determines how you process the program. For each $j = 0, 1, 2, \dots$, $k = 0, 1, 2, \dots$, and $\ell = 0, 1, 2, \dots$ there is a jump instruction $J(j, k, \ell)$. The meaning of the instruction $J(j, k, \ell)$ is as follows[8]:

1. if $r_j = r_k$ proceed to the ℓth instruction in the program
2. if $r_j \neq r_k$ proceed to the next instruction in the program

None of the contents of the registers is altered by a jump instruction.

That's all there is to it. The program $P = I_0, I_1, I_2, \dots, I_b$ is executed by carrying out each instruction, one after the other, except in response to a jump instruction which may cause you to jump to another instruction in the program. The execution of the program P ends when the required next instruction does not exist. For example, having executed the last instruction I_b you proceed to the next instruction I_{b+1}, but there is no such instruction. Alternatively, a jump instruction $J(j, k, b + 1)$ may indicate a jump to the instruction I_{b+1}, which does not exist.

Note: A program, containing only finitely many instructions, can refer to only finitely many registers[9], called the *working registers*. The contents of the re-

[8] To determine if the number of marbles in two bowls is the same, remove one marble from each bowl, then another marble from each bowl, and so on. If both bowls become empty at the same step then they contained the same number of marbles. If only one bowl is empty then they contained different numbers. (Then replace the marbles in their respective bowls.)

[9] We do not allow instructions of the sort, for example, $S(r_1)$, which adds a marble to the bowl with number equal to the content of register R_1. In fact, although we are using numbers as names for the bowls, any symbols could be used for these names. They do not have the same quantitative significance as the content of a bowl. We also do not allow instructions of the sort $J(1, 3, r_0)$ which may cause a jump to the instruction with number equal to the

maining registers will not be changed during the execution of the program, and the values in these registers will not affect the values in the working registers.

The setup we have been considering, consisting of registers (bowls) R_k containing numbers (of marbles) r_k and programs of instructions of the previously discussed type, is called an *Unlimited Register Machine* or URM for short.[10] Using this approach you can compute any function which can be computed in any other way. In particular, you can compute the same functions as Turing machines compute.

A program P has a finite specification by the instructions $I_0, \ldots, I_b$. You can however execute the program with infinitely many possible initializations of the registers. The program thus represents a *uniform* way for you to deal with the infinitely many cases you may be presented with.

Comment

Let's make the remark that a program P has no 'purpose'and does not 'ascertain' anything. It's just a list of instructions which you carry out if you decide to execute that program. It is we who give the program a purpose according to the use we put the result we get when we complete the program.

A Sample Program

Let's show how to write a program P to add any two numbers (quantities of marbles).

$$P = J(2,1,4), S(0), S(2), J(0,0,0), Z(1), Z(2) \tag{7.1}$$

The registers are initially all set to 0 except R_0 which contains m (marbles) and R_1 which contains n (marbles). The numbers r_0 and r_2 will be increased by 1 until $r_1 = r_2$, whereupon registers R_1 and R_2 will be emptied (set to 0) and the program completed. (Note that the jump instruction $J(0,0,0)$ has the effect of causing you to jump to the first (0th) instruction, since r_0 is always equal to r_0.) You will wind up with $m+n$ (marbles) in the register R_0, and 0 (marbles) in all other registers. The way the computation goes adding 3 and 2 is shown in Figure 7.3. At each step of the program, the state of the computation is described by the configuration of the registers[11] and the next instruction to be carried out. The computation halts when the next instruction is I_6, since there is no such instruction.

content of register R_0. Again, an instruction number and a number of marbles are two different types of things. One is an *ordinal* number, representing an ordering (first, second, etc.), the other a *cardinal number*, representing an amount.

[10] The URM was invented by J.C.Shepherdson and H.E.Sturgis [23]. A nice discussion of the URM is given by N.J.Cutland [7].

[11] The working registers are R_0, R_1, R_2; the others remain unchanged and do not influence the computation.

R_0	R_1	R_2	R_3	next instruction
3	2	0	0	0
3	2	0	0	1
4	2	0	0	2
4	2	1	0	3
4	2	1	0	0
4	2	1	0	1
5	2	1	0	2
5	2	2	0	3
5	2	2	0	0
5	2	2	0	4
5	0	2	0	5
5	0	0	0	6

Figure 7.3: The computation $3 + 2$.

Another Sample Program

The only *comparison* of two numbers n and m which we use as part of our basic instructions is equality[12], $n = m$: whether or not n is the *same* as m. Another useful comparison is $n \leq m$: whether or not n is *less than or equal to* m. Here is a program to determine if $n \leq m$. Initialize the registers to $r_0 = n, r_1 = m$, and all other registers initialized to 0. When the program is completed, the register R_0 will contain 1 if n is less than or equal to m, or 0 if n is greater than m. All other registers will contain 0:

$$T(0,2), T(1,3), J(1,2,7), J(0,3,10), S(2), S(3), J(0,0,2),$$
$$Z(0), S(0), J(0,0,11), Z(0), Z(1), Z(2), Z(3)$$

The program works by first transferring a copy of n and m into registers R_2 and R_3. Then the contents of those registers are successively increased until either n is equal to some successor of m, in which case $n > m$, or m is equal to some successor of n, in which case $n < m$.[13] The register R_0 is then set appropriately to 0 or 1, and finally all other registers are set to 0.

Numbering All Programs

Programs are specified by giving their lists of instructions. When reasoning about programs and what can be achieved by using them, it is useful to specify the programs in another way, by giving a rule to *encode* all the details of the program into a single number. Thus, if I give you a list of instructions making up a program, you will be able to compute a single number which will represent

[12] Used explicitly as part of the jump instructions and implicitly as part of the transfer instructions.

[13] The case $n = m$ is dealt with by the first application of $J(1,2,7)$.

that program. Conversely, if I give you a single number which represents a program, you will be able to *decode* the number and find the program to which it refers. There are many ways this could be done, but whatever way I choose I must *explain* it to you, so you will understand which program I am referring to if I say 'Execute program number 7.' Here is one way to number the programs: in order to assign numbers to programs, we first assign numbers to instructions. The first step is to assign numbers to the instructions of each of the four sorts:

Zero instructions. Assign numbers to the zero instructions by associating the instruction $Z(k)$ with the number k.

Successor instructions. Assign numbers to the successor instructions by associating the instruction $S(k)$ with the number k.

Transfer instructions. In order to assign numbers to the transfer instructions we must be able to encode *pairs* of numbers (j, k) by a single number. We associate[14] the pair of numbers (j, k) with the number $n = f(j, k) = 2^j(2k + 1) - 1$. Conversely, we can find j and k from n by $j = f_1(n)$, $k = f_2(n)$ where $f_1(n)$ is the largest power j such that 2^j divides $n + 1$, and:

$$f_2(n) = \frac{1}{2}\left[\frac{n+1}{2^{f_1(n)}} - 1\right]$$

Then the transfer instruction $T(j, k)$ is associated with the number $n = f(j, k)$.

Jump instructions. In order to assign numbers to the jump instructions we must be able to encode *triples* of numbers (j, k, ℓ) by a single number. In fact, once we have encoded pairs of numbers, we can encode triples, quadruples, etc. We associate to the triple of numbers (j, k, ℓ) the number $m = g(j, k, \ell) = f(f(j, k), \ell)$. From m we can find j, k, ℓ by $\ell = f_2(m)$, $j = f_1(f_1(m))$, and $k = f_2(f_1(m))$.

Then the jump instruction $J(j, k, \ell)$ is associated with the number $m = g(j, k, \ell)$.

The next step is to combine the above encoding into a single encoding of all instructions:

Zero instructions. The zero instruction $Z(k)$ is associated with the number $4k$.

Successor instructions. The successor instruction $S(k)$ is associated with the number $4k + 1$.

Transfer instructions. The transfer instruction $T(j, k)$ is associated with the number $4f(j, k) + 2$.

[14] See Cutland [7] for further discussion of this and other assignments.

Jump instructions. The jump instruction $J(j,k,\ell)$ is associated with the number $4g(j,k,\ell)+3$.

In this way given any instruction I we know how to find the corresponding number n, and conversely, given any number n we know how to find the corresponding instruction I.

In the final step we assign numbers to the programs $P = I_0, I_1, \ldots, I_b$. Since we have assigned numbers to the individual instructions, we can associate with P the sequence of numbers $m_0, m_1, \ldots, m_b$, where m_j is the number assigned to the instruction I_j. Thus we must encode all such sequences of numbers into a single number. One way to do this is to associate to the numbers $m_0, m_1, \ldots, m_b$ the number:

$$\begin{aligned} h(m_0,\ldots,m_b) &= 2^{m_0} + 2^{m_0+m_1+1} + 2^{m_0+m_1+m_2+2} \\ &+ \cdots + 2^{m_0+m_1+m_2+\cdots+m_b+b} - 1 \end{aligned} \tag{7.2}$$

This works because every number ≥ 1 may be uniquely expressed as a sum of powers of 2 (which just gives the binary expression for the number). Hence if m_j is the number associated with the instruction I_j in the program P, then P is associated with the number $h(m_0,\ldots,m_b)$.

Computing Gödel Numbers

The numbers associated with instructions and programs are called Gödel numbers, since Gödel first used the coding of non-numerical quantities by numbers. Today such coding is used all the time. For example, a chess playing computer is running a program which codes the chess pieces, their positions on the board, and their allowed moves into numerical values and relations which can then be manipulated as the game progresses.

Let's find the programs corresponding to some Gödel numbers:

1. The program with Gödel number 0: Since $0 = 2^0 - 1$ we see from (7.2) that $b = 0$ and $m_0 = 0$. Thus the program has only one instruction and that instruction has number 0. The zero instruction $Z(k)$ has the number $4k$, hence the instruction $Z(0)$ has number 0. In conclusion, the program with Gödel number 0 is $P_0 = Z(0)$. What would you accomplish by executing the program P_0? Simply empty the bowl R_0.

2. The program with Gödel number 1: Since $1 = 2^1 - 1$ we see from (7.2) that $b = 0$ and $m_0 = 1$. The successor instruction $S(k)$ has the number $4k+1$, hence the instruction $S(0)$ has number 1. In conclusion, the program with Gödel number 1 is $P_1 = S(0)$. What would you accomplish by executing the program P_1? Simply add a marble to bowl R_0.

3. The program with Gödel number 2: Since $2 = 2^0 + 2^1 - 1$ we see from (7.2) that $b = 1$ and $m_0 = 0, m_1 = 0$. Then $P_2 = Z(0), Z(0)$. You will obtain precisely the same result by executing P_2 or P_0 (but you empty bowl R_0 twice!).

4. The program with Gödel number 7: Since $7 = 2^3 - 1$ we see that $b = 0$ and $m_0 = 3$. Thus the program consists of only one instruction and that instruction has number 3. This is the jump instruction $J(0,0,0)$. So $P_7 = J(0,0,0)$. If you execute this program you will never halt. You will keep repeating the first instruction (which is to perform the first instruction) but you will make no change to the contents of the registers!

Let's find the Gödel number of the program P (7.1) which adds any two numbers. First write the Gödel number of each instruction in the program:

Zero instructions. Since the Gödel number of $Z(k)$ is $4k$, the Gödel number of $Z(1)$ is 4 and the Gödel number of $Z(2)$ is 8.

Successor instructions. Since the Gödel number of $S(k)$ is $4k+1$, the Gödel number of $S(0)$ is 1 and the Gödel number of $S(2)$ is 9.

Jump instructions. Since the Gödel number of $J(j,k,\ell)$ is

$$4g(j,k,\ell) + 3 = 4f(f(j,k),\ell) + 3 \ ,$$

we compute the Gödel number of $J(0,0,0)$ as $4f(f(0,0),0) + 3$. Since $f(0,0) = 2^0(2 \cdot 0 + 1) - 1 = 0$, we get $4f(0,0) + 3 = 4 \cdot 0 + 3 = 3$. So the Gödel number of $J(0,0,0)$ is 3. Similarly, the Gödel number of $J(2,1,4)$ is $4g(2,1,4) + 3 = 4f(f(2,1),4) + 3$. Now $f(2,1) = 11$, so $g(2,1,4) = f(f(2,1),4) = 2^{11} \cdot 9 - 1 = 2048 \cdot 9 - 1 = 18431$. So the Gödel number of $J(2,1,4)$ is 73727.

The sequence of numbers corresponding to the instructions in P is thus:

$$73727, 1, 9, 3, 4, 8$$

Hence the Gödel number of the program P is:

$$h(73727, 1, 9, 3) = 2^{73727} + 2^{73729} + 2^{73739} + 2^{73743} + 2^{73748} + 2^{73757} - 1$$

A very large number! Gödel numbers can be extremely large, but we use them as a theoretical device rather than a practical one.

The Marrakech Game

Let's pause for a moment and consider an analog of this process of Gödel numbering, where non-numerical quantities are coded into numbers. Here is an amusing way to code the game tic-tac-toe (noughts and crosses) into numbers. First assign the numbers from 1 to 9 to squares of the board as shown in Figure 7.4. This is a magic square, where the numbers add to 15 along any row, column, or diagonal.

The first player begins the game by putting an X in one of the squares on the board. This is the same as choosing a number from 1 to 9. The second player now puts an O in one of the remaining squares. This is the same as choosing

8	1	6
3	5	7
4	9	2

Figure 7.4: Assigning numbers to the squares of a tic-tac-toe board.

one of the remaining numbers. Play proceeds like this until one player has marked all the squares in a row, column, or diagonal. This is the same as one player having three numbers which add to 15.

We have thus reformulated the game of tic-tac-toe as an equivalent game involving a purely numerical procedure of choosing numbers from 1 to 9, the game being won when a player gets three numbers which add to 15. It is described in Karl Fulves' *Self-Working Number Magic* [9] where it is called the *marrakech game.*

Of course, as numbers do not carry any meaning beyond their place in the succession of numbers, the marrakech game no longer has the same 'meaning' as the original game. (The marrakech game is harder to play than tic-tac-toe because the geometrical symmetries of the tic-tac-toe board and the geometry of the game play are not present in the marrakech game.)

7.3 The Barber Paradox

Imagine a small town with one barber who shaves all those and only those who do not shave themselves. Can there be such a barber? No, there can't. To see this consider whether or not the barber shaves himself. If he doesn't shave himself then he must shave himself, and if he shaves himself then he must not shave himself. This paradox can be translated into mathematics and leads to a version of Gödel's incompleteness theorem. A nice discussion of the application of this and other paradoxes to obtain limitations on what can be achieved using programs is given by Gregory Chaitin [5].

We have seen that all programs can be numbered: $P_0, P_1, P_2, \ldots$. Let's say that a number m is produced by a program P_n if there is an initial value r_0 of register R_0 such that when all other registers are initialized to 0 ($r_k = 0$ for $k \neq 0$) and the program executed, the program will be completed with m in register R_0 (and any numbers in the other registers). Let $\mathcal{R}_n$ denote all the numbers produced by P_n. We will be interested in the family of statements S_n of the form 'n is not produced by P_n', which could be expressed as $n \notin \mathcal{R}_n$. Which statements S_n can be shown to be true? For some values of n, by reasoning about the sequence of instructions in P_n you might find a way of demonstrating that S_n is true. For other values of n you might be able to demonstrate that S_n is false. For still other values of n you might have no idea whether or not S_n is true (or even whether or not it is meaningful to ask if it is true). For example:

1. The program $P_0 = Z(0)$. Hence no matter what the value r_0 initially in register R_0 this program will be completed after one step and the content of register R_0 will be 0. Thus 0 is the only number produced by P_0, i.e. $\mathcal{R}_0 = \{0\}$. Since 0 is produced by P_0 the statement S_0 is false.

2. The program $P_1 = S(0)$. Hence if r_0 is initially in register R_0 this program will be completed after one step and the content of register R_0 will be $r_0 + 1$. Thus all the numbers $1, 2, 3, \ldots$ will be produced by P_1, i.e. $\mathcal{R}_1 = \{1, 2, 3, \ldots\}$. Since 1 is produced by P_1 the statement S_1 is false.

3. The program $P_2 = Z(0), Z(0)$. This program produces only the number 0, i.e. $\mathcal{R}_2 = \{0\}$. Since 2 is not produced by P_2, the statement S_2 is true.

4. Let a be the number $2^{73727} + 2^{73729} + 2^{73739} + 2^{73743} + 2^{73748} + 2^{73757} - 1$. Then P_a is the program (7.1) which adds any two numbers. Hence if we set r_0 to any initial value and all other registers to 0 (in particular $r_1 = 0$) then the result of executing P_a will be $r_0 + r_1 = r_0$ in register R_0. Thus P_a produces all numbers, i.e. $\mathcal{R}_a = \{0, 1, 2, \ldots\}$. Since a is produced by P_a, the statement S_a is false.

Perhaps if we rummage around among our programs $P_0, P_1, \ldots$, we might come across one which just happens to produce precisely the numbers n such that S_n is true! Is this possible? Let's call a program *sound* if all the numbers n that it produces are numbers of true statements S_n. So might there be a sound program P_ℓ which actually produces the numbers of *all* the true statements? No, *there is no such program*, and the reason could be thought of as coming out of the barber paradox.

To see that no such P_ℓ exists, suppose P_ℓ produces the number ℓ. Then, given that P_ℓ is sound, it follows that S_ℓ is true, i.e. P_ℓ does not produce ℓ. This contradiction means that *a sound program cannot produce its own Gödel number*. Thus, given that P_ℓ is sound, it follows that P_ℓ does not produce ℓ. Therefore S_ℓ is *true*. Hence the sound program P_ℓ cannot produce the number ℓ of the true statement S_ℓ. Hence we have our first main result:

There is no sound program which can produce the numbers n of *all* true statements S_n.	(7.3)

Our reasoning showed more, giving a particular true statement S_ℓ whose number is not produced by P_ℓ. This is our second main result:

If the program P_ℓ is sound, then the statement S_ℓ is true and ℓ is not produced by P_ℓ. In this way, **we can go beyond any program which is known to be sound, and obtain a true statement not produced by the program.**	(7.4)

But it is not just the one statement S_ℓ which is missed by P_ℓ, but infinitely many statements. This can be seen by observing that given the sound program P_ℓ, we can find a program $P_{\ell'}$ which produces precisely the numbers produced by P_ℓ together with the number ℓ. We illustrate how to do this in the flow diagram, Figure 7.5: The registers are initialized to 0 except R_0 which is set to $r_0 = m$. First test if $m = 0$ by comparing r_0 and r_1 (since $r_1 = 0$). If $m = 0$, add[15] ℓ to r_0 and halt. If $m \neq 0$, subtract[16] 1 from r_0 and then execute the program P_ℓ.

Consequently the program $P_{\ell'}$ is also sound and furthermore produces the number ℓ, a number not produced by P_ℓ. But the previous argument applied to $P_{\ell'}$ shows that $S_{\ell'}$ is true but not produced by $P_{\ell'}$ (and hence also not by P_ℓ). This is our third main result:

For each sound program P_ℓ, there is another program $P_{\ell'}$ which is also sound. The statements produced as true by $P_{\ell'}$ are the statements produced as true by P_ℓ together with S_ℓ, which was not produced by P_ℓ. The true statement $S_{\ell'}$ is not produced by $P_{\ell'}$ (nor by P_ℓ).	(7.5)

We may now apply the preceding argument to $P_{\ell'}$ to obtain another program $P_{\ell''}$ which is again sound and which gives all the true statements given by $P_{\ell'}$ together with $S_{\ell'}$. We may repeat this process as often as desired, leading to our fourth main result:

For each sound program P_ℓ, there is a sequence $P_{\ell_1}, P_{\ell_2}, \ldots$ of programs, each of which is sound and each $P_{\ell_{j+1}}$ outputs the same as the preceding program P_{ℓ_j} together with ℓ_j. For each P_{ℓ_j} there remain infinitely many true statements not obtained by P_{ℓ_j}.	(7.6)

In all of this discussion it is absolutely essential that the initial program P_ℓ is sound. Is there a method of obtaining *all* the sound programs P_ℓ? That is, is there a program P_k which produces precisely the numbers of all sound programs? No, there isn't. Indeed, if P_k is such a program then if P_k produces ℓ then it follows from the preceding discussion that S_ℓ is true. Thus P_k is sound. It follows that P_k doesn't produce k and hence P_k cannot produce the numbers of all sound programs. Hence no such program exists. So we have

[15] Here is a program to add ℓ to any number (which is placed in register R_0, all other registers initialized to 0): $S(0), S(0), \ldots, S(0)$ [ℓ copies].

[16] Here is a program to subtract 1 from any number m (which is placed in register R_0, all other registers initialized to 0): $S(1), J(0,1,5), S(1), S(2), J(0,0,1), T(2,0), Z(1), Z(2)$. If $m \geq 1$, when you complete the execution of this program there will be $m-1$ in register R_0, all other registers containing 0. If $m = 0$, this program will not halt; i.e. you will never complete the program.

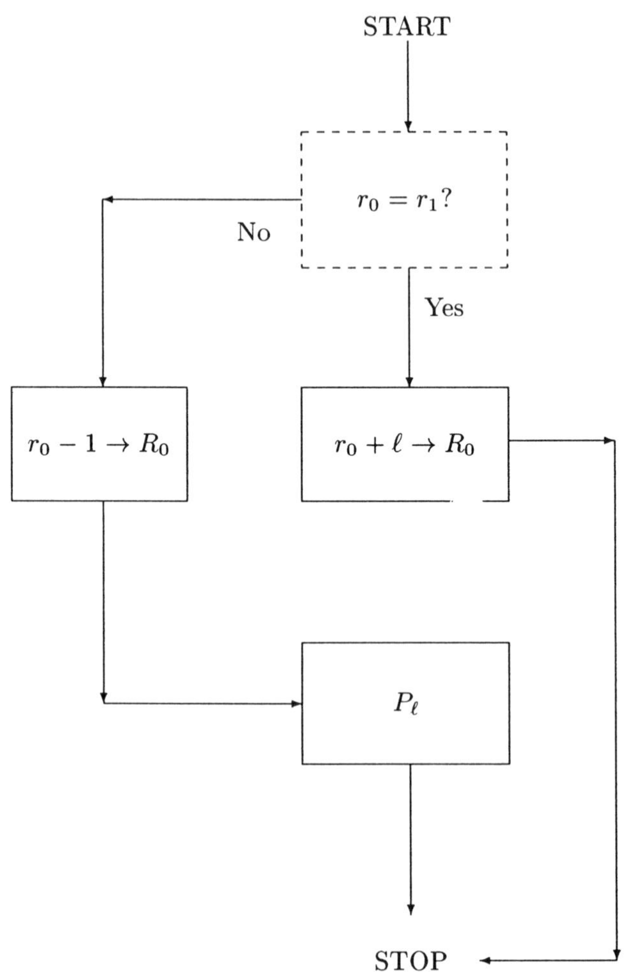

Figure 7.5: Flow diagram for the program $P_{\ell'}$.

Figure 7.6: Kurt Gödel, 1939. Reproduced by permission of the Institute for Advanced Study, Princeton, USA

obtained our fifth and final main result:

There is no program P_k which produces precisely the numbers of all sound programs.	(7.7)

Comment

Given a program P, we can execute the instructions and determine one by one the numbers produced by P. Note that the numbers produced by P depend only on the instructions listed by P. If we choose some Gödel numbering of the programs then the program P will be assigned some number, say 7, so P is P_7. Now if P doesn't produce the number 7 then S_7 is true. But suppose we choose some other Gödel numbering. Perhaps now some other program P' is assigned the number 7, so P' is now P_7. If P' produces the number 7 then S_7 is false. So whether or not the statement S_7 is true depends on our choice of Gödel numbering. Consequently, whether or not a program is sound also depends on our choice of Gödel numbering. This shows again that programs do not have an inherent 'purpose'. They just produce numbers. It is we who interpret and utilize the numbers as we wish.

7.4 Gödel's Theorem

In section 7.2 we defined the basic instructions and the programs composed of these instructions. I can explain to you very simply what you are to do when following each instruction and you may then execute the program. It is all very concrete and involves moving marbles in a specific way. A chimpanzee could be trained to execute some programs. A mechanical contraption could be built which would execute the programs, depending on how various levers were set. A present-day computer could be set up to execute any program if you type the instructions on the keyboard. There is no 'logic' or reasoning which forms part of the process of executing a program. On the other hand, in section 7.3 we have been *reasoning about programs*. This involves something different from the programs themselves. What constitutes valid reasoning? Can we be absolutely precise as to what shall consistute an acceptable proof? In what language shall the reasoning take place?[17] It was with the objective in mind of setting down a precise, consistent framework in which to conduct mathematical reasoning that formal mathematical systems were developed. A formal mathematical system is formulated with a specified language, which consists of an alphabet of symbols, together with rules for writing down formulas. In addition, there are rules[18] which determine when a finite sequence of formulas constitutes a proof, the final formula of the proof being the theorem which is proved. The most characteristic property of such systems is the mechanical nature of proof-checking, a property emphasized by David Hilbert (1862-1943). Let's decide on a method[19] to encode formulas and finite sequences of formulas into (Gödel) numbers. Then the mechanical nature of proof-checking means that there is a program P_ℓ which will produce precisely the Gödel numbers of proofs, and a program P_m which will produce the Gödel numbers of the last formula in each proof. Every proof will eventually be produced by P_ℓ and hence every theorem

[17] Even if you could carry out your reasoning without language, you could not communicate to others how you arrived at your conclusions unless a common language is agreed.

[18] What rules of deduction should be included in the formal system? Is there only one correct logical system? Consider other areas of mathematics. What is the correct geometrical theory? The classical geometry of Euclid was considered the only correct geometry for thousands of years, but now we consider the correct geometry to be an *empirical* question. The curved space-time of Einstein's general relativity theory is considered a better mathematical model than the flat Euclidean geometry. What is the correct algebraic theory? The algebraic structure of a theory of observable quantities had been supposed to satisfy $xy = yx$ (commutativity) and $x(y + z) = xy + xz$ (distributivity). But now it is considered that quantum theory, involving non-commutative observables, is a better mathematical model of atomic phenomena. Quantum theory still keeps distributivity, but if the empirical data require it, then a further modification of the algebraic structure of the theory could be made. What is the correct logic? Perhaps there is no one correct logic. It may also depend on the properties of the system being modelled. We have *empirical geometry*, *empirical algebra*, why not *empirical logic*?

[19] We could use a method similar to the one used to encode programs, since programs are finite sequences of instructions and formulas are finite sequences of symbols - and proofs are finite sequences of formulas. So first give numbers to each symbol in the alphabet of L. If ϕ is the formula $a_1 \cdots a_b$, where m_j is the Gödel number of the symbol a_j, then ϕ is assigned the number $h(m_0, \cdots, m_b)$, where the function h is given in (7.2). Similarly, if $\phi_0, \ldots, \phi_b$ is a finite sequence of formulas where m_j is the Gödel number of the formula ϕ_j, then the sequence of formulas is assigned the number $h(m_0, \cdots, m_b)$.

will eventually be produced by P_m.

When we use a formal mathematical system we have a particular interpretation in mind for the symbols and formulas. However, someone else may use the same formal system with a different interpretation.[20] The formal system gives rules for writing symbols on paper. It has no intrinsic meaning and does not come with an interpretation. For this reason, Bertrand Russell once said (quoted in chapter XI of [12]):

> Mathematics may be defined as the subject in which we never know what we are talking about nor whether what we are saying is true.

and Henri Poincarè said (quoted in chapter XII of [12]):

> Mathematics is the art of giving the same name to different things.

Once an interpretation is settled upon we can inquire as to whether or not a particular formula is *true*. You may use any interpretation you wish provided all the axioms of the formal system are true in your interpretation. The formal system is said to be *sound* if its theorems are true *no matter which interpretation you are using.*[21]

So suppose we have settled on a particular formal system to make deductions about the programs we have been considering. (In the appendix to this chapter, section 7.8, we have included an overview of the predicate calculus and the axioms used to make deductions about a program.) We interpret[22] the formulas as statements about programs. The statement S_n, discussed in the preceding section, will be expressed by a formula ϕ_n. Then if S_n is true we will say that the formula ϕ_n is true, and if S_n is false we will say that the formula ϕ_n is false.

From the program P_m which produces the Gödel numbers of all the theorems of the formal system, we can write down a program P_j which produces the number n if and only if ϕ_n is a theorem. Here's a prescription to do this. For given input, execute P_m. If P_m halts and produces the number k, then we know k is the Gödel number of a theorem. To find out if k corresponds to one of the formulas ϕ_n, we need to use the program P_M which, given input n, produces the Gödel number of ϕ_n. We execute P_M repeatedly with input 0, then 1, then 2, etc. comparing the number produced with k. If k is not the Gödel number of some ϕ_n the program will not halt. If k is the Gödel number of ϕ_n then the program produces n as output. This prescription in words must be translated into a program P_j. This is easy to do once we have the programs

[20] Different interpretations are often given to the same English sentence by different people. They may think they have a disagreement as to which statements are true, but actually they may simply have assigned a different meaning to the same words! This is called a problem of semantics.

[21] Consequently, if the formal system is sound and if there is a formula ϕ which is true in one interpretation and false in another, then neither ϕ nor its negation can be a theorem of the formal system.

[22] Someone else may interpret the formulas in a completely different way, as statements about some other activity or physical process.

P_m and P_M. A flow diagram for P_j is shown[23] in Figure 7.7. Summarizing, from the program P_m which produces the Gödel numbers of theorems of our formal system, and the program P_M which from input n produces the Gödel number of ϕ_n, we can write down a program P_j which produces the numbers n such that ϕ_n is a theorem.

Let us now suppose that our formal system is sound, i.e. the theorems are true. Thus P_m will produce only Gödel numbers of true formulas, and hence P_j will produce only numbers n such that S_n is true. Thus P_j is sound. It follows by (7.4) that the number j is not produced by P_j and S_j is true. Thus ϕ_j is true and not produced by P_m, i.e. ϕ_j is true but is not a theorem of the formal system. This conclusion holds for any sound formal mathematical system with a language sufficiently broad to express the statements S_n. We thus have the following version of Gödel's theorem:

For any sufficiently broad sound formal mathematical system, we can explicitly exhibit a true formula ϕ which cannot be proved by the formal system. Furthermore, the negation of ϕ is false and so also cannot be proved.	(7.8)

Recall that the formula which we have exhibited is ϕ_j, which expresses the statement S_j: 'j is not produced by P_j'. But by the way P_j was constructed, this is equivalent to the statement 'ϕ_j is not a theorem of $\mathfrak{F}$'. Thus the formula ϕ_j says of itself that it is not a theorem of $\mathfrak{F}$. And this is true!

How to Understand Gödel's Theorem

The version of Gödel's theorem (7.8) which we have discussed above says that the sort of formal system $\mathfrak{F}$ (supposed sound) which mathematicians use to derive truths (say, about arithmetic) will not be able to deduce *all* the true statements. Furthermore we can explicitly exhibit a true formula ϕ which cannot be proved in the formal system $\mathfrak{F}$. How are we to understand this limitation to formal systems? There are a number of viewpoints which can lead to insights into the significance of Gödel's theorem and even to new theorems concerning formal systems. Here are some ideas to think about concerning Gödel's theorem:

1. We can explicitly exhibit a formula ϕ_j which is not a theorem of the formal system but is nevertheless true *in our interpretation*. However *in another interpretation* ϕ_j will be false.[24]

   An interpretation gives *meaning* to the formal system and this meaning leads us to truths not derivable from the formal system itself. This must

[23] We suppose that when P_m halts, all its working registers other than R_0 contain 0. Similarly for P_M. This can always be arranged. In Figure 7.7 we take the working registers of P_M to be R_k for $k \leq N_M$.

[24] In our interpretation ϕ_j expresses the statement S_j, that the jth program does not produce the number j. However, in another interpretation ϕ_j no longer has this meaning. The relation between ϕ_j and S_j is lost.

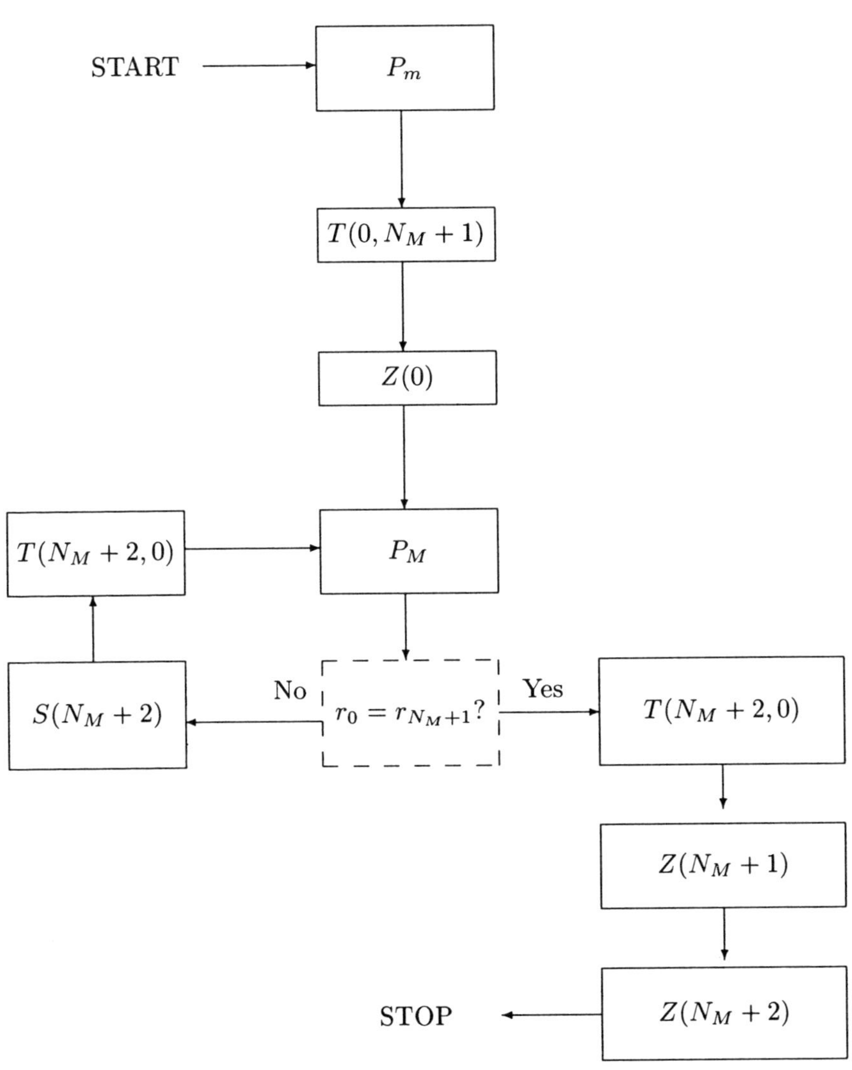

Figure 7.7: Flow diagram for the program P_j.

be expected if a formal system can be given *different* interpretations, and so different meanings and truths.

2. Gödel's theorem says that we cannot deduce as theorems *all* true formulas, from the limited set of true formulas which constitute the axioms. The axioms constitute a set of truths of limited complexity and we cannot expect to derive truths of unlimited complexity from them. Chaitin (see [5]) puts it this way:

   > The complexity of the formal system ... is a measure of the amount of information the system contains, and hence the amount of information that can be derived from it. ... Gödel's theorem does not appear to give cause for depression. Instead it seems simply to suggest that in order to progress, mathematicians, like investigators in other sciences, must search for new axioms.
   > If one has ten pounds of axioms and a twenty pound theorem, then that theorem cannot be derived from those axioms.

3. Using a formal system to deduce theorems (which are truths if the formal system is sound), you are acting as an *oracle* producing statements. Gödel's theorem describes a limitation which applies not only to computational oracles (programs or formal systems) but to any oracle, regardless of its nature. *An oracle which is known to be sound can never tell us everything we can ascertain. Thus the Gödel phenomenon cannot be overcome by searching for a non-computational oracle.* We will show how this limitation arises by analogy with our previous discussion.

   An oracle in the form of a 'black box', the interior of which we do not know, produces statements. The oracle is sound if all the statements are true. Now define[25] a device $\mathcal{D}$ which does the following: if the oracle produces the statement $\mathcal{D}$ `does not produce anything`, then $\mathcal{D}$ produces the number 0 in response. If the oracle produces any other statement, $\mathcal{D}$ does not produce anything in response.

   We may easily deduce that *if the oracle is sound*, then it will not produce the statement $\mathcal{D}$ `does not produce anything`, and furthermore it is true that $\mathcal{D}$ does not produce anything. Thus if we know the oracle is sound, then we know a truth not produced by the oracle.

7.5 Penrose

We will rephrase Penrose's argument using the particular mathematical results we have developed in sections 7.3 and 7.4. We will try to adhere to the meaning

[25]The element of self-reference in the definition is inevitable. The oracle must be able to refer to itself or to something dependent on the oracle. As we have seen in the case of formal systems, a formula can be written down which may be interpreted as stating that the self-same formula cannot be proved by the formal system.

of Penrose's argument, and thus follow the moral of the following story related in section 2.10 of *Shadows of the Mind*:

> I am reminded of a story concerning the great American physicist Richard Feynman. Apparently Feynman was explaining some idea to a student, but mis-stated it. When the student expressed puzzlement, Feynman replied: 'Don't listen to what I say; listen to what I *mean*!'

In Chapter 2, *The Gödelian case*, and Chapter 3, *The case for non-computability in mathematical thought*, Penrose argues that the human mind cannot be simulated by a computer:

> I shall shortly be giving (in Chapters 2 and 3) some very strong reasons for believing that effects of (certain kinds of) understanding cannot be properly simulated in any kind of computational terms... Thus, the human faculty of being able to 'understand' is something that must be achieved by some non-computational activity of the brain or mind... The term 'non-computational' here refers to something beyond any kind of effective simulation by means of any computer based on the logical principles that underlie all the electronic or mechanical calculating devices of today.

Penrose puts forward his main argument in section 2.5 of *Shadows of the Mind*, and then deals with various counter-arguments in Chapters 2 and 3:

> The argument I shall present in the next chapter (section 2.5) provides what I believe to be a very clear-cut argument for a non-computational ingredient in our conscious thinking... In due course (in Chapters 2 and 3), I shall be addressing, in detail, *all* the different counter-arguments that have come to my attention.

The Gödelian Case

Consider again the problem of determining the true statements S_n, that is, the numbers n such that the nth program does not produce n. According to the discussion in section 7.3 there is no sound program which will produce the numbers n of all the true statements S_n. *Can the human intellect succeed in producing all the true statements* S_n*?* There is no evidence of this, so how good is the human intellect at finding the true statements S_n? Say that a program P_ℓ *encapsulates human understanding* if every number n of true statements S_n that human mathematicians can produce is also produced by P_ℓ.

Question: Can human mathematical understanding concerning the statements S_n be encapsulated in a sound program?

According to (7.4) if the ℓth program is sound, then S_ℓ is true and ℓ is not produced by P_ℓ. Does this mean that the answer to the question is no: human mathematical understanding concerning the statements S_n cannot be encapsulated in a program? Not quite! In order to go beyond the program P_ℓ we must *know* that P_ℓ is sound. If we know a program is sound call it *knowably sound.* Then we can give an answer to the question as follows:

> **Answer:** Human mathematical understanding concerning the statements S_n cannot be encapsulated in a knowably sound program. (7.9)

This answer is given[26] by Penrose in his statement $\mathcal{G}$ in section 2.5 of *Shadows of the Mind.* This is basically all there is in section 2.5. Penrose would probably have preferred the answer:

> **Penrose's Preferred Answer:** Human mathematical understanding concerning the statements S_n cannot be encapsulated in a sound program.

but this has not been shown. Here is what Putnam says [16] about Penrose's argument so far:

> What Penrose has shown is quite compatible with the claim that a computer program could in principle successfully simulate our mathematical capacities. The possibility exists that each of the rules that a human mathematician explicitly relies on, or can be rationally persuaded to rely on, can be known to be sound and that the program generates all and only these rules but that the program itself cannot be rendered sufficiently "perspicuous" for us to know that that is what it does. Actual programs sometimes consist of thousands of lines of code, and it can happen that by the time a program has been tinkered with and debugged no one is really able to explain exactly how it works. A program which simulated the brain of an idealized mathematician might well consist of hundreds of thousands (or millions or billions) of lines of code. Imagine it given in the form of a volume the size of the New York telephone book. Then we might not be able to appreciate it in a perfectly conscious way, in the sense of understanding it or of being able to say whether it is plausible or implausible that it should output correct mathematical proofs and only correct mathematical proofs. ©1994 by The New York Times Co. Reprinted by permission.

In short, there may be a program P_ℓ which is sound and which produces precisely the same numbers n of true statements S_n that human mathematicians can produce. But for this it is necessary that human mathematicians are unable to ascertain that that is what P_ℓ does.

[26]Recall the Feynman story above.

Do Mathematicians Use Unsound Reasoning?

Are the statements S_ℓ produced by human mathematicians as true really true? Might mathematicians claim that P_n does not produce n when it actually does? If P_n produces n this fact can in principle be determined, simply by executing P_n with all possible values for r_0, one after another, until n is produced.[27] Thus human mathematicians could eventually determine that they are making mistakes. In Chapter 3 of *Shadows of the Mind* Penrose argues that this possibility is implausible:

> I cannot really see that it is plausible that mathematicians are *really* using an *unsound* formal system $\mathbb{F}$ as the basis of their mathematical understandings and beliefs. I hope the reader will indeed agree with me that whether or not such a consideration is *possible*, it is certainly not at all *plausible.*

But in fact there has been a loss of certainty in the soundness and completeness of mathematics, as Penrose acknowledges and discusses in sections 2.10 and 3.4 of *Shadows of the Mind.* Here are the comments of several influential mathematicians:

> I have told the story of this controversy in such detail, because I think that it constitutes the best caution against taking the immovable rigour of mathematics too much for granted. This happened in our own lifetime, and I know myself how humiliatingly easy my own views regarding the absolute mathematical truth changed during this episode, and how they changed three times in succession!... It is hardly possible to believe in the existence of an absolute, immutable concept of mathematical rigor, dissociated from all human experience.
>
> John von Neumann (from [26])

> Mathematics may be likened to a Promethean labor, full of life, energy and great wonder, yet containing the seed of an overwhelming self-doubt. It is good that only rarely do we pause to review the situation and to set down our thoughts on these deepest questions. During the rest of our mathematical lives we watch and perhaps partake in the glorious procession.... This is our fate, to live with doubts, to pursue a subject whose absoluteness we are not certain of, in short to realize that the only "true" science is itself of the same mortal, perhaps empirical, nature as all other human undertakings.
>
> Paul J. Cohen (from [6])

[27] On the other hand, if P_n does not produce n but a mathematician claims that it does (without any indication as to what input would produce n), it may not be possible to prove him wrong. There may be no way to determine the truth or falsity of S_n and the law of the excluded middle (every statement is either true or false) may not be applicable.

> I wanted certainty in the kind of way in which people want religious faith. I thought that certainty is more likely to be found in mathematics than elsewhere. But I discovered that many mathematical demonstrations, which my teachers expected me to accept, were full of fallacies, and that, if certainty were indeed discoverable in mathematics, it would be in a new field of mathematics, with more solid foundations than those that had hitherto been thought secure.... After some twenty years of very arduous toil, I came to the conclusion that there was nothing more that I could do in the way of making mathematical knowledge indubitable.[19]... The splendid certainty which I had always hoped to find in mathematics was lost in a bewildering maze.[20]
>
> Bertrand Russell

> Only he who recognizes that he has nothing, that he cannot possess anything, that absolute certainty is unattainable, who completely resigns himself and sacrifices all, who gives everything, who does not know anything, does not want anything and does not want to know anything, who abandons and neglects everything, he will receive all; to him the world of freedom opens, the world of painless contemplation and of - nothing.
>
> L.E.J.Brouwer (from [3])

Unassailable Mathematical Beliefs

In Chapter 3 of *Shadows of the Mind*, Penrose considers *beliefs* rather than truths. Consider for example the result (7.4). If P_ℓ is sound then S_ℓ is true and P_ℓ does not produce ℓ. Thus we know a true statement S_ℓ not produced by the program P_ℓ. But this holds only if we *know* that P_ℓ is sound. We may very well *not know for sure* that P_ℓ is sound. Then we don't know for sure that S_ℓ is true. On the other hand, if we *believe* that P_ℓ is sound (but we're not absolutely sure) then we *believe* (with the *same* level of confidence, on the basis of (7.4)) that S_ℓ is true and ℓ is not produced by P_ℓ. Then we will believe (with the same level of confidence, on the basis of (7.6)) that all the programs $P_\ell, P_{\ell_1}, P_{\ell_2}, \ldots$ are sound. Thus we may construct ever more comprehensive programs which we believe to be sound with the same level of confidence that we believe P_ℓ to be sound. In summary:

We can go beyond any program P_ℓ believed to be sound, and obtain a statement S_ℓ which we believe to be true and not produced by P_ℓ.

A similar argument can be given concerning formal mathematical systems. Take for example a formal system $\mathfrak{F}$ of the sort we have considered earlier. If $\mathfrak{F}$ is sound, so all its theorems are true, then we can exhibit a formula ϕ which is true (*in our standard interpretation*) but which is not a theorem of $\mathfrak{F}$. But again, we can only do this if we *know for sure* that $\mathfrak{F}$ is sound. Perhaps we are not certain that $\mathfrak{F}$ is sound, but we *believe* it to be. (If we are using $\mathfrak{F}$ to deduce mathematical truths we certainly will believe that we are using a sound formal system.) Then we *believe* that ϕ is true (with the same level of confidence, on the basis of (7.8)) and hence we can *add* ϕ *as a new axiom* to the formal system $\mathfrak{F}$, obtaining a new formal system $\mathfrak{F}_1$. We will then believe that $\mathfrak{F}_1$ is sound with the same level of confidence that we believe $\mathfrak{F}$ to be sound. We have not lost anything by broadening our formal system with the addition of the new axiom ϕ. We may repeat this procedure, exhibiting a formula ϕ_1 which we believe to be true with the same level of confidence that we believe $\mathfrak{F}_1$ and $\mathfrak{F}$ to be sound. Proceeding in this way we obtain a sequence of formal systems $\mathfrak{F}, \mathfrak{F}_1, \mathfrak{F}_2, \ldots$ each more comprehensive than the preceding one (having an additional axiom which is not a theorem of the preceding formal system). We believe each of the formal systems $\mathfrak{F}_j$ to be sound with the same level of confidence that we believe $\mathfrak{F}$ is sound. This procedure was first studied by Turing in a paper [25], *Systems of logic based on ordinals*, published in 1939. It is still a subject of research. In summary:

> We can go beyond any sufficiently broad formal system $\mathfrak{F}$ believed to be sound and obtain a formula which is believed true (in our standard interpretation) and not a theorem of $\mathfrak{F}$.

Penrose uses [14, 15] this result as the basis for his argument against the computational modelling of mathematical understanding. But we will show in section 7.6 that his argument is mistaken.

7.6 A Computational Model for Thought?

Do the mathematical theorems on computation and formal systems have implications concerning the computational modelling of the human intellect? Here, we mean computational modelling of the sort that is done when modelling a wide range of physical phenomena. Consider for example the steps involved in the computational modelling of the motion of a projectile:

1. Decide on which aspects of the phenomenon are to be modelled: in the case of the projectile, the position of the projectile at various moments of time.

2. Set up measuring instruments to code those aspects of the phenomenon into numerical quantities. (The meter readings serve as 'Gödel numbers' which code those aspects of the phenomenon into numbers.) In the case of the projectile, we set up rulers and clocks.

3. Set up a mathematical theory (in this case, based on Newton's Laws of Motion) to compute relationships between the instrument readings. In

the case of the projectile, write a program such that with the registers appropriately initialized, the program is completed with the register R_0 containing a number which should, if the computational model is a good one, be very close to the number obtained by reading the ruler.

Let's invent a toy model, a naive computational model for the way the brain works, and let's see if any of the mathematical theorems we have discussed could be used to rule out such a model. We follow the basic ideas set out in Chapters 1 and 2:

1. The functioning of the brain is described by the connections between neurons. The state of the brain at any time is described by the state of firing or non-firing of each of the neurons.

2. A mathematical model of the brain is constructed by associating with the jth neuron a mathematical variable $f_j(t)$ which takes the value 0 or 1, describing the state of the neuron at time t (the time being measured in discrete steps). The value 1 corresponds to the neuron firing, and the value 0 to non-firing. The configuration of all the neurons at a given time is thus modelled by all the values $f_0(t), f_1(t), \ldots, f_N(t)$. The values $f_0(t+1), f_1(t+1), \ldots, f_N(t+1)$ are determined by the values $f_0(t), f_1(t), \ldots, f_N(t)$ together with a computational rule which models the dynamical behaviour of the neural network. That is, we give a mathematical rule which enables us to calculate each $f_j(t+1)$ given all the values $f_0(t), \ldots, f_N(t)$. One such rule is described in Chapters 1 and 2, in terms of weights which correspond to neural connections. The weights w_{ji} are introduced, which take positive or negative integer values[28], and model the strength of the connection from the ith neuron to the jth neuron. Then the computational rule governing the behavior of the mathematical model is:

$$f_j(t+1) = \begin{cases} 1 & \text{if} \quad w_{j1} f_1(t) + \cdots + w_{jN} f_N(t) > M_j \\ 0 & \text{if} \quad w_{j1} f_1(t) + \cdots + w_{jN} f_N(t) \leq M_j \end{cases} \tag{7.10}$$

   where M_j is a positive integer corresponding to the threshold of the jth neuron.
   Remark. External input can be incorporated into the model by 'plugging in by hand' the values $f_j(t)$ corresponding to the sensory neurons.

3. The preceding computational rule may be expressed[29] by a program P_ℓ. The program will be constructed so that by executing the program we perform the computations expressed in (7.10). We initialize the registers with $r_j = f_j(t)$ (all others initialized to 0) and execute the program P_ℓ.

[28] Positive values of w_{ji} corresponds to exitation and negative values to inhibition. Given rational values for the weights, we can multiply through by a sufficiently large integer to rewrite the equations with integer-valued weights.

[29] Negative integers need not enter directly into the computation. Use may be made, for example, of the *difference function* $Dif(n,m) = n - m$ if $n \geq m$ and $= 0$ if $n < m$.

When we complete the program the registers will contain $r'_j = f_j(t+1)$, for $j = 0, \ldots, N$.

In this way by executing the program P_ℓ we may compute the values $f_0(t), \ldots, f_N(t)$ for all (discrete) times t, and consequently the sequence of states of the neural network.

A Computational Model for the Mind?

The computational model of the brain discussed above may or may not give a good approximation to the actual behavior of a biological neural network. Assuming we could construct a good computational model of the excitation states of the neurons in the brain, what would this model tell us about the thoughts and emotions experienced by the brain? In other words, what does the model tell us about the mind? This depends on the nature of mental representation in the brain. In his book [8] Daniel Dennett begins Chapter 3, *Brain Writing and Mind Reading*, this way:

> What are we to make of the popular notion that our brains are somehow libraries of our thoughts and beliefs? Is it *in principle* possible that brain scientists might one day know enough about the workings of our brains to be able to "crack the cerebral code" and read our minds?

Let's imagine that there is a 'dictionary' which can translate from our description $(f_0(t), \ldots, f_N(t))$ of the neuronal state of a brain to the beliefs which the mind holds at time t. Denote[30] these beliefs by $B_0(t), \ldots B_{M(t)}(t)$. These beliefs could be expressed in English (together with mathematical symbols), or in French, or in any formal language. Define some Gödel numbering so that each belief $B_j(t)$ is assigned a number and the collection of all beliefs at time t is assigned the number $b(t)$. So, we will be able to decode the number $b(t)$ and obtain all the beliefs $B_0(t), \ldots, B_{M(t)}(t)$. Suppose the dictionary is computational in the sense that there is a program P_d such that if we initialize the registers to $r_k = f_k(t)$ then when we finish executing the program P_d the register r_0 will contain the *Gödel belief number* $b(t)$.

Hence to compute the beliefs held by a brain at time t initialize the registers to $r_k = f_k(0)$, $k = 0, \ldots, N$. Then execute the program P_ℓ. When you complete the program you will have $r'_k = f_k(1)$ in the registers. Continue these computations until you have obtained $f_0(t), \ldots, f_N(t)$. Now execute the dictionary program P_d with the registers initialized to $r_k = f_k(t)$. When you complete the program P_d you will obtain a number $b(t)$ which encodes all the beliefs held by the brain at time t. The number $b(t)$ could be decoded to yield all the beliefs $B_0(t), \ldots B_N(t)$ at time t stated in English. Combining the above steps, we can write a program P so that, initializing the registers to $f_0(0), \ldots, f_N(0), t$, then upon completing the program P the Gödel belief number $b(t)$ will be in the register R_0.

[30] We shall naturally suppose that a (finite) brain can have only finitely many beliefs at any one time.

Gödel's Objection?

The above computational model of the brain is as straightforward and naive as one could imagine. But if there is no way to rule out, on the basis of the mathematical results discussed earlier, such a simple model then, ipso facto, there is no way to rule out more complex and sophisticated computational models on the basis of those mathematical results. Let's use the term 'computational mind' to denote a mind whose beliefs could be computed in the above fashion.[31] We will show:

The Gödel and Turing theorems have absolutely nothing to say about the beliefs which may be held by a computational mind. Claims to the contrary are based on a mistaken application of those theorems.

Before discussing the details, consider the fact that people often hold contradictory beliefs without realizing it. Or they may believe completely foolish things, such as 'The moon is made of green cheese.' Why should such beliefs affect the consistency and correctness of your computation of those beliefs? Some mathematicians accept the logical principle of the law of the excluded middle, others do not. Some may base their mathematical beliefs on modal logic, or on non-constructive methods, or on intuitionistic philosophy. Why should these varying sorts of beliefs affect the way the brain works at the neuronal (or any other) level or how you compute those beliefs? They don't! Concerning the disagreements among mathematicians on a sound basis for developing mathematical beliefs, consider this quotation from E.T. Bell (quoted in chapter XI of [12]):

> Experience has taught most mathematicians that much that looks solid and satisfactory to one mathematical generation stands a fair chance of dissolving into cobwebs under the steadier scrutiny of the next... Knowledge in any sense of a reasonably common agreement on the fundamentals of mathematics seems to be non-existent... The bald exhibition of the facts should suffice to establish the one point of human significance, namely, that equally competent experts have disagreed and do now disagree on the simplest aspects of any reasoning which makes the slightest claim, implicit or explicit, to universality, generality, or cogency.

The consistency, correctness, or otherwise of a person's beliefs has nothing whatever to do with the consistency or correctness of the formal system used to deduce that person's beliefs. And what about thoughts in general? Think of something crazy. Something you believe is not, and could not possibly be, true. Should that crazy thing be a part of the deductive system forming a theory of mind? Or is it only the things you believe to be true that should be theorems?

[31] What could be more computational than that?

Surely all thoughts and other aspects of mind should be treated in the same way. You can think about anything you like; it won't affect the consistency or soundness of the theory of mind.

The Mistaken Application of Gödel's Theorem

Penrose uses Gödel's theorem in various arguments against the computational modelling of mathematical understanding throughout Chapter 3 of *Shadows of the Mind*, in particular in sections 3.2, 3.3, 3.14, and 3.16. All his arguments are subject to the criticism levelled at the argument below. Here is Penrose's reasoning about Gödel's theorem in a nutshell.

Suppose a human, who we call H, has a computational mind (associated with the program P), and that H is made aware of the program P which we are using to compute the properties of his brain. We may use a formal mathematical system $\mathbb{F}$ to logically deduce the behavior of P, and hence the behavior, beliefs, and other thoughts of H. We may say that the system $\mathbb{F}$ encapsulates all the knowledge and beliefs of H. Penrose argues that *if H believes statement X, then since $\mathbb{F}$ encapsulates all of H's beliefs, it must be possible to prove X in the system $\mathbb{F}$.* Penrose then reasons that H will surely believe that the system $\mathbb{F}$ is sound. Consequently, H will believe (by Gödel's theorem) that the formula ϕ is true. But since, again by Gödel's theorem, ϕ is not a theorem of $\mathbb{F}$, it follows that $\mathbb{F}$ *cannot* after all encapsulate all of H's beliefs.[32] By this contradiction Penrose concludes that H *cannot* in fact have a computational mind.

Penrose's confusion in this line of reasoning is italicized above. For if H believes **X** (on day 1) then it is necessary that $\mathbb{F}$ deduces that **H believes X on day 1**. It is not necessary that $\mathbb{F}$ deduces **X**. If on day 2 H changes his mind and believes the negation of X, $\neg$**X**, then it is necessary that $\mathbb{F}$ deduces that **H believes $\neg$X on day 2**. It is not necessary that $\mathbb{F}$ deduces $\neg$**X**. If on day 3 H goes crazy then it is necessary that $\mathbb{F}$ deduces that **H is crazy on day 3**. It is not necessary that $\mathbb{F}$ deduces all sorts of crazy formulas.

In short, although $\mathbb{F}$ cannot deduce ϕ, there is no reason why $\mathbb{F}$ cannot deduce that **H believes ϕ**. Gödel's theorem doesn't say anything about what can be proved concerning the state of H's mind!

The essential point is that H's beliefs *do not form part of the deductive system* $\mathbb{F}$. If H believes two contradictory statements **X** and **Y**, these statements are *not* theorems of $\mathbb{F}$ and so it does not follow that $\mathbb{F}$ is inconsistent, as it would be if **X** and **Y** were theorems.[33]

Let's explain in detail the distinction between **X** and **H believes X**, using

[32] It may be that the formal system $\mathbb{F}$ does not satisfy the requirements to deduce Gödel's theorem for $\mathbb{F}$. But the point of the argument is that there are statements which H believes to be true but which are not theorems of the formal system $\mathbb{F}$. This will undoubtedly be the case for mathematical statements of a sort which cannot even be *expressed* in the particular formal system $\mathbb{F}$, and even more so for non-mathematical beliefs.

[33] If H's beliefs are not theorems of $\mathbb{F}$, what are they? The beliefs are encoded into the Gödel belief number $b(t)$, which is a *term* of the formal language.

our 'toy' computational model. The statement **X** is one of H's beliefs, say $B_k(t)$, and as such is encoded along with H's other beliefs in the Gödel belief number $b(t)$. Using the formal system $\mathbb{F}$ we can deduce a theorem which states that when the program P is run with registers initialized to $r_k = f_k(0)$ then when the program is completed $r_0 = b(t)$. There is a world of difference between the theorem $r_0 = b(t)$ and the statements $B_0(t), \ldots, B_N(t)$ encoded in the number $b(t)$. The confusion undoubtedly arose because the mathematical beliefs of H could be expressed in the same language used in the formal system $\mathbb{F}$. No one would think of incorporating a non-mathematical belief, such as 'It will rain tomorrow', into the formal system $\mathbb{F}$!

Discussion

Perhaps you do not believe that a 'mind reading' program can be found which will translate brain states into the beliefs (and other thoughts) of that brain. Consider any means which H may use to communicate his beliefs to others. He may write them down with pencil and paper, or he may say them. Does Gödel's theorem impose any restrictions on what H may write down or say? No! By computing the state of the motor neurons, using our program P_ℓ, we can determine the hand movements which H will carry out, and hence the geometrical pattern of marks on paper which will result. Any pattern whatsoever can be drawn by H. And if that pattern may be interpreted as a mathematical formula, that formula does not thereby become a theorem of the formal system $\mathbb{F}$ associated with P_ℓ. Similarly, the way H's mouth and vocal cords move, which could be computed using P_ℓ, are not restricted by Gödel's theorem. The theorem which would be proved using $\mathbb{F}$ is of the form $r_j(t) = 1$, indicating a marble in the bowl R_j and hence, by our modelling of H's brain, that the jth neuron is firing. Translating the marks on paper into, say, an English sentence, that sentence plays the role of a *term*, not the role of a theorem.

Consider a present day computer. What sort of expressions might it print out? Consider the formal system $\mathbb{F}$ associated with the program it is running. Suppose the formula ϕ is not provable in $\mathbb{F}$. Is there any reason why the formula ϕ cannot be printed out by the computer? No! Again, what comes out of the printer is not a theorem of $\mathbb{F}$.

All this becomes clear when it is realized that what is printed out, or drawn, etc., need not satisfy any grammatical rules, which are required of theorems of $\mathbb{F}$.

Related Ideas in the Literature

Our criticism of Penrose's use of Gödel's theorem rests on the idea that beliefs and other thoughts of a mind are not theorems of the theory of that mind. This idea can be found in the existing literature.

John von Neumann

In August 1955 John von Neumann was diagnosed as having bone cancer. In April 1956 he was admitted to Walter Reed Hospital which he did not leave until his death on February 8, 1957. He took with him to the hospital the manuscript of the Silliman Lectures (Yale University), which he hoped to be able to present. The title of the lectures was to be *The Computer and the Brain.* The unfinished manuscript has been published as a book [27]. The last section of the book is entitled *The Language of the Brain Not the Language of Mathematics*, and the last lines of that section and of the book are these:

> When we talk mathematics, we may be discussing a *secondary* language, built on the *primary* language truly used by the central nervous system. Thus the outward forms of *our* mathematics are not absolutely relevant from the point of view of evaluating what the mathematical or logical language *truly* used by the central nervous system is. However, the above remarks about reliability and logical and arithmetical depth prove that whatever the system is, it cannot fail to differ considerably from what we consciously and explicitly consider as mathematics.

We may interpret this quotation as stating the view that the mathematical procedures and beliefs of a mind are of a different category from the theorems of the formal system $\mathbb{F}$ associated with a computational model of the mind at the neuronal level. In short, mathematical beliefs are not theorems of $\mathbb{F}$, which is the point of our criticism of Penrose's argument.

Robert Kirk

Robert Kirk, a philosopher from Nottingham University, published an article [11] in 1986 entitled *Mental Machinery and Gödel.* He levels the same criticism at Lucas's use of Gödel's theorem that we have levelled at Penrose; reproduced by kind permission of Kluwer Academic Publishers:

> If there could be adequate mechanistic accounts which represented a person by means of a formal system, yet did not correlate beliefs, thoughts or statements with that system's theorems or 'outputs', then the argument from Gödel fails. In fact, there is an approach which appears to satisfy this requirement. It deals in terms of the organism's physical states, inputs, and outputs, all at some rather low level of specification - perhaps the neurophysiological - and in possible transitions from state to state through time.... Certainly mechanists will maintain that those tokens of mathematical sentences which Alf produces are produced 'mechanically'. However, in maintaining this they need not - and should not - say that the token sentences themselves correspond to theorems of a formal system which adequately represents Alf and his part of the world.

7.7 Can Computers Think?

Let's finally return to our original question and see if the preceding considerations have gotten us any closer to an answer. We have seen that Penrose's argument against the computational modelling of mental processes does not work, and that there is no obstruction to a computational mind, of the sort discussed in section 7.6, arising from the mathematical theorems on computation and formal systems. We conclude that these theorems have not, in fact, gotten us any closer to an answer to the question:

> Will computers of the future be intelligent?

Certainly present-day computers aren't intelligent.[34] We have quoted in section 7.1 several views that intelligent computers are a likely development in the not too distant future. Here is the vision of the mathematical physicist David Ruelle, which he paints at the beginning of his book [18]:

> Supercomputers will some day soon start competing with mathematicians and may well put them forever out of work.... Of course, present-day machines are useful mostly for rather repetitious and somewhat stupid tasks. But there is no reason why they could not become more flexible and versatile, mimicking the intellectual processes of man, with tremendously greater speed and accuracy. In this way, within fifty or a hundred years (or maybe two hundred), not only will computers help mathematicians with their work, but we shall see them take the initiative, introduce new and fruitful definitions, make conjectures, and then obtain proofs of theorems far beyond human intellectual capabilities.
>

Some Thoughts about Thinking

We close with some thoughts about thinking, which you might like to think about. Consider again our computational model for the trajectory of a projectile on page 201. By executing the program to compute the numbers read on the ruler, we are not ourselves moving along the trajectory of the projectile! And if a computer is set up to execute the program, it has not itself become a projectile. Although this is painfully obvious, the implication in connection with models of the mind is perhaps that by setting up a computer to execute a program which constitutes a 'computational model of the mind' (in the sense considered in section 7.6) the computer is not thereby *thinking*! Dennett, in Chapter 11 of his book [8], puts it this way:

> It is never to the point in computer simulation that one's model be *indistinguishable* from the modelled. Consider, for instance, a good computer simulation of a hurricane, as might be devised by

[34] They only think they are.

> meteorologists. One would not expect to get wet or wind-blown in its presence. That ludicrous expectation would be akin to a use-mention error, like cowering before the word "lion".

And Searle [22] puts it this way:

> No one would suppose that we could produce milk and sugar by running a computer simulation of the formal sequences in lactation and photosynthesis, but where the mind is concerned many people are willing to believe in such a miracle because of a deep and abiding dualism: the mind they suppose is a matter of formal processes and is independent of quite specific material causes in the way that milk and sugar are not.

How might we build a computer which *does* think?

How to Build a Robot

Let's try to build a thinking robot using our URM. Our robot will have:

- eyes made of video cameras
- ears made of microphones
- a speaker for a voice box
- chemical sensors for a nose
- mechanical legs for moving about
- mechanical arms for touching and moving things

Have we left anything out? Oh yes, a *brain*. That's the crucial bit!

To make the brain, we'll use the URM. Take all the bowls and marbles, and your program P_ℓ (page 202) which defines a good computational model of some particular brain, and sit down inside the robot. Now suppose that when certain bowls have a marble in them (modelling an excited neuron) an action is performed by the robot (a small arm movement, for example). Suppose also that input from the video cameras (or microphones, etc.) leads to certain other bowls having a marble dropped in them (corresponding to the excitation of sensory neurons). You now execute the program P_ℓ. Then, during each time step $t \to t+1$, the marbles in the bowls change (modelling the varying excitation states of the neurons in the brain) causing the robot to react to external stimuli, to speak through the speaker, and so on. Of course, you must imagine yourself working faster and faster, and the bowls of marbles very small. In this way, the collection of bowls of marbles with you executing the program P_ℓ becomes the brain of the robot.

The Mind Demon

Let's summarize the above discussion this way. A mechanical robot has a brain composed of a very large number of very small bowls containing marbles[35], with a tiny very fast working 'mind demon' executing a program P_ℓ, changing the numbers of marbles in the bowls. The demon knows *only* how to execute the basic instructions of the URM. He understands nothing else.

Among the bowls are 'input bowls' whose contents reflect properties of the outside world, and 'output bowls' whose contents determine actions by the robot.

The sequence of bowl contents $r_j(t)$ at times $t = 0, 1, 2, \ldots$, for $j = 0, \ldots, N$ corresponds to the sequence of neuron firings of the jth neuron in the brain which is modelled by the program P_ℓ. The program P_ℓ corresponds to the connectivity of the neural network (through the weights w_{ij}). The demon, executing the program P_ℓ, corresponds to the electro-chemical forces which induce the temporal development of the neural network (its varying neuronal firings).

> **Question.** You know that *you* think[36]; how close does this robot come to doing the same?

Searle's Chinese Room

John Searle [22] argues that although such a robot may listen to questions and stories and make sensible replies, there is no understanding:

> Whatever purely formal principles you put into the computer, they will not be sufficient for understanding, since a human will be able to follow the formal principles without understanding anything.... I will argue that in the literal sense the programmed computer understands what the car and the adding machine understand, namely, exactly nothing. The computer understanding is not just (like my understanding of German) partial or incomplete; it is zero.

For 'human' in this quote, substitute our 'demon', whom we have agreed does not understand *anything* except the basic instructions required in the execution of the program P_ℓ. Remembering that we said the demon corresponds[37] to the electro-chemical forces responsible for the temporal development of the neural network, we certainly would not ascribe 'understanding' to those forces, nor to the demon. But although the demon does not understand anything of the outside world, does it follow that the robot has no understanding?

Imagine the robot in China, and the program P_ℓ giving a good computational model of the brain of some Chinese person. The robot will function in

[35] To the reader with such a brain: 'Don't lose your marbles!'

[36] Well, maybe.

[37] According to the construction of present-day computers, we could also say that the demon corresponds to the CPU (central processing unit) of the computer, which, while executing a program, manipulates the numbers (on/off transistor patterns) stored in its registers and RAM memory locations.

China in the same way that the Chinese person does. It will make replies, in Chinese, to questions asked of it by Chinese people. Searle imagines himself taking the part of the demon. He then compares his excellent understanding of English with his total lack of understanding of Chinese, and concludes that also the robot could not 'understand' Chinese. But again, the demon, or Searle-in-the-robot, is not where understanding is expected to reside. Searle's reference to the understanding of English which Searle-in-the-robot possesses involves a category-mistake, as pointed out by Margaret Boden [1]:

> Searle's description of the robot's pseudo-brain (that is, of Searle-in-the-robot) as understanding English involves a category-mistake comparable to treating the brain as the bearer - as opposed to the causal basis - of intelligence.

The 'Feel' of Thinking

Can our robot with its 'bowls-of-marbles + demon' brain see a sunset the same way that we see it? Does it really think and understand in the same way that we do? As we don't expect to build a brain in this way (where would we find the demon?) the question is not very important. Suppose, instead of the bowls of marbles, we use silicon transistors, and instead of the demon, we use electromagnetic forces as they are exerted on the transistors of a silicon chip. Suppose, in other words, that we make a copy of a brain 'in silicon', in such a way that the sequence of patterns of 'on/off' transistors induced by electromagnetic forces is the same as the sequence of patterns of 'firing/non-firing' neurons induced by electrochemical forces. Would our robot, with such a silicon chip central processing unit for a brain, see and think in the same way that we do? How would it be for the robot: Would it 'feel' the same? Searle [22] doesn't think so:

> The problem with the brain simulator is that it is simulating the wrong things about the brain. As long as it simulates only the formal structure of the sequence of neurone firings at the synapses, it won't have simulated what matters about the brain, namely its causal properties, its ability to produce intentional states.

But recall Turing's test based on *behavior* (page 175). The physicist Niels Bohr echoes Turing's ideas this way [10]:

> I am quite prepared to talk of the spiritual life of an electronic computer; to say that it is considering or that it is in a bad mood. What really matters is the unambiguous description of its behaviour, which is what we observe. The question as to whether the machine *really* feels, or whether it merely looks as though it did, is absolutely as meaningless as to ask whether light is "in reality" waves or particles. We must never forget that "reality" too is a human word just like "wave" or "consciousness." Our task is to learn to use these words correctly - that is, unambiguously and consistently.

Be that as it may, the 'feel' of the functioning brain could very well depend[38] on the detailed structure of the individual neurons (discussed in Chapter 3), much as the tone of a violin depends on the quality of wood used and the manufacturing techniques employed. So a computer's silicon brain might 'feel' different from our biological brain, but nevertheless, *it would be a thinking computer.*

[38]That the 'feel' of the functioning brain depends on the detailed behavior of the neurons might perhaps be surmised from the effects of drugs on the brain, as mentioned by Susan Greenfield in Chapter 11 (page 279).

7.8 Appendix: Filling in Some Details

What sort of formal system $\mathbb{F}$ might we use to make deductions about the properties of a program? The basic framework used for all mathematical systems is the *predicate calculus.* We give here a quick overview of the predicate calculus, starting with the simpler *propositional calculus.* A nice introduction to the predicate calculus may be found in [4].

The Propositional Calculus

Alphabet.

1. *propositions:* p^n for $n = 0, 1, 2, \ldots$
2. *logical symbols:* $\neg$ $\rightarrow$
3. *punctuation:* ( )

Well-formed Formulas.

1. The propositions p^n are well-formed formulas.
2. If ϕ is a well-formed formula then $\neg(\phi)$ is a well-formed formula.
3. If ϕ and ψ are well-formed formulas then $(\phi) \rightarrow (\psi)$ is a well-formed formula.

Truth.
A truth value, *true* or *false*, is assigned to all well-formed formulas as follows.

1. A truth value is assigned arbitrarily to each proposition p^n.
2. If ϕ is true then $\neg(\phi)$ is false, and if ϕ is false then $\neg(\phi)$ is true.
3. The well-formed formula $(\phi) \rightarrow (\psi)$ is false when ϕ is true and ψ is false. It is true for all other truth values of ϕ and ψ.

Propositional Tautology.
A well-formed formula ϕ is a propositional tautology if it is true no matter what truth values are assigned to the propositions occurring in ϕ.

The well-formed formulas of the propositional calculus are built up from the basic propositions p^n, which are unanalyzed. The truth value assigned to a well-formed formula depends only on the truth values assigned to the propositions from which the formula is built up. An important role is played by the propositional tautologies, which are always true no matter what truth value is assigned to the propositions from which they are built up. Here are some examples of propositional tautologies:

1. $(p^0) \rightarrow (p^0)$
2. $(p^0) \rightarrow ((p^1) \rightarrow (p^0))$
3. $(p^0) \rightarrow ((\neg(p^0)) \rightarrow (p^1))$
4. $((p^0) \rightarrow (p^1)) \rightarrow ((\neg(p^1)) \rightarrow (\neg(p^0)))$

The Predicate Calculus

Alphabet.

1. *constants:* c^n for $n = 0, 1, 2, \ldots$
2. *variables:* x^n for $n = 0, 1, 2, \ldots$
3. *n-place predicates:* R_n^m for $m = 0, 1, 2, \ldots$ and $n = 1, 2, \ldots$
4. *n-place function symbols:* f_n^m for $m = 0, 1, 2, \ldots$ and $n = 1, 2, \ldots$
5. *logical symbols:* $\neg \;\rightarrow\; \forall$
6. *punctuation:* ( ) ,

Terms.

1. The constants c^n are terms.
2. The variables x^n are terms.
3. If $t_1, \ldots, t_n$ are terms then $f_n^m(t_1, t_2, \ldots, t_n)$ is a term.

Well-formed Formulas.

1. If $t_1, \ldots, t_n$ are terms, then the *atomic formula* $R_n^m(t_1, t_2, \ldots, t_n)$ is a well-formed formula.
2. If ϕ is a well-formed formula then $\neg(\phi)$ is a well-formed formula.
3. If ϕ and ψ are well-formed formulas then $(\phi) \rightarrow (\psi)$ is a well-formed formula.
4. If ϕ is a well-formed formula then $\forall x^n(\phi)$ is a well-formed formula, for any variable x^n.

Free and Bound Variables.

1. In an atomic formula, all the variables occurring are *free*.
2. In a well-formed formula $\neg(\phi)$, the free and bound occurrences of any variable are the same as for ϕ.
3. In a well-formed formula $(\phi) \rightarrow (\psi)$, the free and bound occurrences of any variable are the same as for ϕ and ψ.
4. In a well-formed formula $\forall x^k(\phi)$, the free and bound occurrences of any variable other than x^k are the same as for ϕ, and all occurrences of x^k are *bound*.
5. The term t is *free for the variable* x^k in the well-formed formula ϕ if, when it is substituted for every free occurrence of x^k, no variables in t become bound.

Tautologies. Let ϕ be a propositional tautology (of the propositional calculus), and let $p^{n_1}, \ldots, p^{n_\ell}$ be the propositions occurring in ϕ. Let $\sigma_1, \ldots, \sigma_\ell$ be any well-formed formulas of the predicate calculus. Let ψ be the well-formed formula of the predicate calculus obtained by substituted σ_k for each occurrence of p^{n_k} in ϕ. Then ψ is a *tautology* of the predicate calculus.

The Predicate Calculus (continued)

Axioms.

1. *Tautology:* If the well-formed formula ϕ is a tautology then it is an axiom.
2. *Specialization:* For any well-formed formula ϕ and any term t which is free for the variable x in ϕ, the well-formed formula $(\forall x(\phi)) \to (\phi[t/x])$ is an axiom, where the expression $\phi[t/x^k]$ denotes the well-formed formula obtained by substituting the term t for every free occurrence of x^k in ϕ.
3. *Quantified Implication:* For any well-formed formulas ϕ and ψ, if the variable x^k does not occur free in ϕ, then the well-formed formula

$$(\forall x^k((\phi) \to (\psi))) \to ((\phi) \to (\forall x^k(\psi)))$$

is an axiom.

Rules of Deduction.

1. *Modus Ponens:* From ϕ and $(\phi) \to (\psi)$, deduce ψ.
2. *Generalization:* From ϕ deduce $\forall x^n(\phi)$, for any variable x^n.

Proofs.

A proof of the well-formed formula ψ from the well-formed formulas $\psi_1, \ldots, \psi_\ell$ (called the *premises*), is a sequence of well-formed formulas $\phi_1, \ldots, \phi_N$ such that ϕ_N is ψ and each ϕ_k is either

1. an axiom
2. a premise
3. follows from two previous well-formed formulas ϕ_i, ϕ_j (where $i, j < k$) by *modus ponens*
4. follows from a previous well-formed formula ϕ_j $(j < k)$ by *generalization*

Theorems.

A theorem is a formula which is proved from no premises.

As an example of a proof in the predicate calculus, let's show that from a contradiction (a well-formed formula and its negation) we may deduce any formula. So here is a proof of the well-formed formula ψ from the premises ϕ and $\neg(\phi)$:

	ϕ , $\neg(\phi)$	premises
1.	ϕ	premise
2.	$\neg(\phi)$	premise
3.	$(\phi) \to ((\neg(\phi)) \to (\psi))$	tautology
4.	$(\neg(\phi)) \to (\psi)$	1,3: modus ponens
5.	ψ	2,4: modus ponens

Note. The tautology in line 3 is obtained by substituting ϕ for p^0 and ψ for p^1 in the propositional tautology $(p^0) \to ((\neg(p^0)) \to (p^1))$.

The 2-place predicate of *equality* plays an important role in mathematical systems. We introduce the standard abbreviation = instead of, say, R^0_2, and we write $t_1 = t_2$ instead of $R^0_2(t_1, t_2)$. There are two special axioms, the *axioms of equality*, associated with the equality predicate. The predicate calculus together with the equality predicate and the axioms of equality is the *predicate calculus with equality*.

The Predicate Calculus with Equality

Equality Axioms.

E1. For any variable x^n,

$$x^n = x^n$$

E2. For any well-formed formula ϕ and any variables x^n and x^m, if x^m is free for x^n in ϕ,

$$(x^n = x^m) \to ((\phi) \to (\phi[x^m : x^n]))$$

where the expression $\phi[x^m : x^n]$ denotes the well-formed formula obtained by substituting the variable x^m for some (possibly all, possibly none) free occurrences of x^n in ϕ.

As an example of the use of the equality axioms, we show that the well-formed formula $(x^0 = x^1) \to (x^1 = x^0)$ is a theorem of the predicate calculus with equality:

1.	$(x^0 = x^1) \to ((x^0 = x^0) \to (x^1 = x^0))$	E2
2.	$((x^0 = x^1) \to ((x^0 = x^0) \to (x^1 = x^0))) \to$ $((x^0 = x^0) \to ((x^0 = x^1) \to (x^1 = x^0)))$	tautology
3.	$(x^0 = x^0) \to ((x^0 = x^1) \to (x^1 = x^0))$	1,2: modus ponens
4.	$x^0 = x^0$	E1
5.	$(x^0 = x^1) \to (x^1 = x^0)$	4,3: modus ponens

Notes.

1. In line 1 the instance of E2 is obtained taking ϕ to be $x^0 = x^0$ and then substituting x^1 for the occurrence of x^0 to the left of = in ϕ.

2. The propositional tautology used in line 2 is:

$$((p^0) \to ((p^1) \to (p^2))) \to ((p^1) \to ((p^0) \to (p^2)))$$

Numerals

In order to make deductions about arithmetical procedures such as programs, we need *names* for the numbers (of marbles), and these names are called *numerals*. We will need one constant symbol corresponding to the number 0. Instead of using the 'official' symbol c^0 of the predicate calculus, we will use the symbol $\overline{0}$ as the numeral for 0. Any number of marbles in a bowl may be obtained by successively adding marbles, one after the other. The succession of numbers is basic to our conception of number, so we need a 1-place function symbol to denote the successor function, which when applied to a number n gives the next number $n+1$. Instead of using the 'official' function symbol f_1^0 of the predicate calculus, we'll use the symbol s. Consequently, since the symbol $\overline{0}$ is the numeral for zero, then $s(\overline{0})$ is the numeral for the successor of zero, namely one. Similarly, as the successor of one is two, the numeral for two is $s(s(\overline{0}))$. In the same way, the numeral for three is $s(s(s(\overline{0})))$ and the numeral for four is $s(s(s(s(\overline{0}))))$. And so on. As a shorthand, instead of writing out a numeral completely, with all the s's and brackets, we'll use the abbreviation $\overline{n}$ to denote the numeral for the number n. For example, we'll use the abbreviation $\overline{20}$ for the numeral for twenty, instead of writing out the actual numeral for twenty, which is:

$$s(s(s(s(s(s(s(s(s(s(s(s(s(s(s(s(s(s(s(s(\overline{0}))))))))))))))))))))$$

We may summarize our discussion as follows:

Definition. Numerals are defined by these rules:

1. The numeral for zero is $\overline{0}$.
2. $\overline{n+1}$ is $s(\overline{n})$

Our variable symbols x^n will be taken to denote general unspecified numbers. Whatever number x denotes, $s(x)$ denotes the successor of that number. Further successors are denoted by $s(s(s(x)))$, and so on.

Number Axioms

The variable and constant symbols are now interpreted as representing numbers. We need some axioms corresponding to the basic properties of numbers:

N1. $(s(x^0) = s(x^1)) \rightarrow (x^0 = x^1)$

N2. $\neg(\overline{0} = s(x^0))$

N3. $(\neg(x^0 = \overline{0})) \rightarrow (\exists x^1(x^0 = s(x^1)))$

N4. $((\phi[\overline{0}/x^n]) \rightarrow (\forall x^n((\phi) \rightarrow (\phi[s(x^n)/x^n])))) \rightarrow (\forall x^n(\phi))$ for any well-formed formula ϕ

Remark. The axiom N2 expresses the fact that 0 is not the successor of any number. The axiom N4 is the induction axiom.

The Formal System Associated with a Program

What we have discussed so far concerns solely *numbers* (of marbles). But we want to make deductions about the course of a computation as a program is executed. We want to be able to describe the state of the computation after each step: the configuration of the registers and the next instruction to be carried out. Suppose the program is:

$$P = I_0, I_1, \ldots, I_b$$

At the start, the next instruction is I_0, and at the finish the next instruction is I_{b+1} (there being no such instruction). (For the example of the computation of $3 + 2$, we want to describe the states the computation goes through as illustrated in Figure 7.3.) Suppose the working registers of the program are $R_0, R_1, \ldots, R_N$. The contents of the other registers do not change and they do not influence the contents of the working registers. The computation determined by the program P depends on the initial contents of the registers $R_0, R_1, \ldots, R_N$. We introduce an $N + 3$-place function symbol[39] $\mathfrak{r}$, and we interpret:

$$\mathfrak{r}(x, y, x^0, x^1, \ldots, x^N)$$

as the content of the xth register after y steps[40] of the computation have been completed, with the registers initialized to $x^0, \ldots, x^N$. Thus, in the case $N = 2$, $\mathfrak{r}(\overline{2}, \overline{5}, \overline{1}, \overline{8}, \overline{3})$ is interpreted as the content of register R_2 after 5 steps of the computation have been completed, with R_0 initialized to 1, R_1 initialized to 8, and R_2 initialized to 3.

We also introduce the N+2-place function symbol $\imath$, and we interpret:

$$\imath(y, x^0, x^1, \ldots, x^N)$$

as the number in the program of the next instruction to be carried out after y steps of the computation have been completed, with the registers initialized to $x^0, \ldots, x^N$.[41] Thus, again in the case $N = 2$, $\imath(\overline{5}, \overline{1}, \overline{8}, \overline{3})$ is interpreted as the number of the next instruction to be carried out after 5 steps of the computation have been completed, with the registers initialized as above.

Remark. Note that $\imath$ refers to the number i of the instruction I_i in the program $I_0, I_1, \ldots, I_b$, and *not* to the Gödel number of I_i. For example, suppose we are executing the program (7.1) to add two numbers. If after 2 steps the next instruction to be executed is $S(2)$ then the number of the next instruction to be carried out is 2, since $S(2)$ is the instruction I_2. It is not the Gödel number of $S(2)$ which is 9.

Summarizing,

---

[39] Instead of using the 'official' function symbol f^0_{N+3} we use the simpler $\mathfrak{r}$.

[40] Instead of the 'official' variables x^{N+1} and x^{N+2} we'll use the simpler x and y.

[41] The next instruction to be carried out depends on the initialization of the registers since jump instructions involve a comparison of the contents of registers, and these will depend on the initialization of the registers.

Let the working registers of the program P be $R_0, \ldots, R_N$.

1. There is one 2-place predicate, the equality predicate $=$.
2. There is one 1-place function symbol s, representing the successor function.
3. There is one $N+3$-place function symbol $\mathfrak{r}(x, y, x^0, \ldots, x^N)$, representing the content of the xth register after y steps of the computation have been carried out, with registers initialized to $r_j = x^j$, $j = 0, \ldots, N$.
4. There is one $N+2$-place function symbol $\imath(y, x^0, \ldots, x^N)$, representing the next instruction number in the program to be carried out after y steps of the computation have been carried out.
5. There are no other function symbols or predicates.

Let's consider some examples of statements about the program P which we may express in our formal system, and let's take the simplest case where there is only one working register, $N = 0$. The formula:

$$\mathfrak{r}(\overline{0}, \overline{3}, \overline{2}) = \overline{4} \tag{7.11}$$

is interpreted as the statement:

> `If the content of register` R_0 `is initialized to 2, then after 3 steps of the computation, the content of register` R_0 `is 4.`

Writing out the numerals in full, the formula (7.11) is:

$$\mathfrak{r}(\overline{0}, s(s(s(\overline{0}))), s(s(\overline{0}))) = s(s(s(s(\overline{0}))))$$

The formula:

$$\neg(\mathfrak{r}(\overline{0}, \overline{3}, \overline{2}) = \overline{4}) \tag{7.12}$$

is interpreted as the statement:

> `If the content of register` R_0 `is initialized to 2, then after 3 steps of the computation, the content of register` R_0 `is not 4.`

The formula:

$$\forall x^0(\neg(\mathfrak{r}(\overline{0}, \overline{3}, x^0) = \overline{4}))$$

is interpreted as the statement:

> `For all initial values of the register` R_0`, after 3 steps of the computation, the content of register` R_0 `is not 4.`

And if we want to write a formula which is interpreted as:

> `There exits an initial value of the register` R_0 `such that, after 3 steps of the computation, the content of register` R_0 `is 4.`

we use the *existential quantifier* $\exists$:

$$\exists x^0(\mathfrak{r}(\overline{0}, \overline{3}, x^0) = \overline{4})$$

But, wait a minute! We do not have the symbol $\exists$ in our predicate calculus. This is not a problem since: the statement

> `There exists an` x `such that ...`

means the same as:

> `Not for all` x `not ...`

Hence $\exists x^n(\phi)$ is taken to be an abbreviation for the correct formula $\neg(\forall x^n(\neg(\phi)))$.

How would we write a formula which is interpreted as the following statement?

> `For every initialization of the register` R_0`, the program eventually halts.`

That is, no matter what the content of the register R_0 initially, you will eventually complete the execution of the program. Suppose the program consists of the instructions $I_0, I_1, \ldots, I_5$. Then you will complete the program when the number of the next instruction is 6 (since there is no instruction I_6). Thus the previous statement could be expressed as:

> `For every initialization of the register` R_0`, there exists a step of the computation such that the next instruction number is 6.`

We express this statement with the formula[42]:

$$\forall x^0(\exists y(\imath(y, x^0) = \overline{6}))$$

To express in our formal system the statement:

> `For all initial values of the register` R_0`, if after 3 steps of the computation the content of the register` R_0 `is 5, then after 6 steps of the computation the content of the register` R_0 `is 7.`

we write the formula:

$$\forall x^0(\mathfrak{r}((\overline{0}, \overline{3}, x^0) = \overline{5}) \to (\mathfrak{r}(\overline{0}, \overline{6}, x^0) = \overline{7}))$$

[42] Notice that this formula does *not* have the same interpretation as the formula:

$$\exists y(\forall x^0(\imath(y, x^0) = \overline{6}))$$

The Proper Axioms for a Program

The preceding discussion developed a general framework for analyzing any program. But now we must explain how to introduce axioms which encapsulate the particular program we wish to analyze. We'll illustrate how it is done for the program (7.1). You will then be able to write the axioms for any program by using the same technique. For each instruction in the program, we write axioms which express the transformation in the state of the computation which results from carrying out the instruction. For the program (7.1) the instructions are $I_0 = J(2,1,4), I_1 = S(0), I_2 = S(2), I_3 = J(0,0,0), I_4 = Z(1), I_5 = Z(2)$. The working registers are thus R_0, R_1, R_2. The axioms corresponding to the instruction I_4 are:

$$\begin{aligned}
(\imath(y,x^0,x^1,x^2)=\bar{4}) &\rightarrow (\mathfrak{r}(\bar{1},s(y),x^0,x^1,x^2)=\bar{0}) \\
(\imath(y,x^0,x^1,x^2)=\bar{4}) &\rightarrow ((\neg(x=\bar{1}))\rightarrow(\mathfrak{r}(x,s(y),x^0,x^1,x^2)=\mathfrak{r}(x,y,x^0,x^1,x^2))) \\
(\imath(y,x^0,x^1,x^2)=\bar{4}) &\rightarrow (\imath(s(y),x^0,x^1,x^2)=\bar{5})
\end{aligned}$$

The axioms corresponding to the instruction I_1 are:

$$\begin{aligned}
(\imath(y,x^0,x^1,x^2)=\bar{1}) &\rightarrow (\mathfrak{r}(\bar{0},s(y),x^0,x^1,x^2)=s(\mathfrak{r}(\bar{0},y,x^0,x^1,x^2))) \\
(\imath(y,x^0,x^1,x^2)=\bar{1}) &\rightarrow ((\neg(x=\bar{0}))\rightarrow(\mathfrak{r}(x,s(y),x^0,x^1,x^2)=\mathfrak{r}(x,y,x^0,x^1,x^2))) \\
(\imath(y,x^0,x^1,x^2)=\bar{1}) &\rightarrow (\imath(s(y),x^0,x^1,x^2)=\bar{2})
\end{aligned}$$

The axioms corresponding to the instruction I_0 are:

$$\begin{aligned}
(\imath(y,x^0,x^1,x^2)=\bar{0}) &\rightarrow (\mathfrak{r}(x,s(y),x^0,x^1,x^2)=\mathfrak{r}(x,y,x^0,x^1,x^2)) \\
(\imath(y,x^0,x^1,x^2)=\bar{0}) &\rightarrow ((\mathfrak{r}(\bar{2},y,x^0,x^1,x^2)=\mathfrak{r}(\bar{1},y,x^0,x^1,x^2))\rightarrow \\
&\qquad (\imath(s(y),x^0,x^1,x^2)=\bar{4})) \\
(\imath(y,x^0,x^1,x^2)=\bar{0}) &\rightarrow ((\neg(\mathfrak{r}(\bar{2},y,x^0,x^1,x^2)=\mathfrak{r}(\bar{1},y,x^0,x^1,x^2)))\rightarrow \\
&\qquad (\imath(s(y),x^0,x^1,x^2)=\bar{1}))
\end{aligned}$$

The remaining instructions are treated similarly. Since there are no transfer instructions in the program (7.1), let's add the instruction $T(0,1)$ at the end of program (7.1) so that it becomes instruction I_6. The axioms for this instruction are:

$$\begin{aligned}
(\imath(y,x^0,x^1,x^2)=\bar{6}) &\rightarrow (\mathfrak{r}(\bar{1},s(y),x^0,x^1,x^2)=\mathfrak{r}(\bar{0},y,x^0,x^1,x^2)) \\
(\imath(y,x^0,x^1,x^2)=\bar{6}) &\rightarrow ((\neg(x=\bar{1}))\rightarrow(\mathfrak{r}(x,s(y),x^0,x^1,x^2)=\mathfrak{r}(x,y,x^0,x^1,x^2))) \\
(\imath(y,x^0,x^1,x^2)=\bar{6}) &\rightarrow (\imath(s(y),x^0,x^1,x^2)=\bar{7})
\end{aligned}$$

Finally, we should add an axiom which corresponds to the fact that the first instruction to be carried out is I_0:

$$\imath(\overline{0}, x^0, x^1, x^2) = \overline{0}$$

and axioms which correspond to the fact that the contents of the non-working registers do not change:

$$\mathfrak{r}(\overline{n}, y, x^0, x^1, x^2) = \mathfrak{r}(\overline{n}, \overline{0}, x^0, x^1, x^2) \quad \text{for each } n \geq 3$$

Remark. You may feel that the axioms should be stated *for all* x^0, x^1, x^2. However, by generalization we may always deduce the formula $\forall x^n(\phi)$ from ϕ. Conversely, by the axiom of specialization and modus ponens we may deduce ϕ from $\forall x^n(\phi)$. Consequently it makes no difference if the axioms are stated with or without the universal quantifiers.

7.9 Bibliography

[1] M.A. Boden, *Escaping from the Chinese Room*, in *The Philosophy of Artificial Intelligence*, ed. M.A. Boden, (Oxford University Press, 1990), pp.40-66

[2] L. Boltzmann and B. McGuinness, *Theoretical Physics and Philosophical Problems*, vol.5 (D. Reidel Publishing Company, 1974) p.57

[3] L.E.J. Brouwer, *Life, Art and Mysticism 1905 exerpts*; in Collected Works 1, ed. A. Heyting, (North-Holland, 1975)

[4] J.N. Crossley, et.al., *What is Mathematical Logic?*, (Oxford University Press, 1978)

[5] G. Chaitin, *Gödel's Theorem and Information*, International Journal of Theoretical Physics **22** (Plenum Publishing Corp., 1982) pp.941-954; reprinted in G.J. Chaitin, *Information, Randomness, and Incompleteness* (World Scientific,1987)

[6] P.J. Cohen, *Comments on the Foundations of Set Theory*, in Proceedings of the Symposia in Pure Mathematics Volume XIII Part I, (American Mathematical Society, 1971)

[7] N.J. Cutland, *Computability*, (Cambridge University Press, 1980)

[8] D.C. Dennett, *Brainstorms*, (Harvester Wheatsheaf, 1981)

[9] K. Fulves, *Self-Working Number Magic* (Dover, 1983)

[10] J. Kalckar, *Niels Bohr and his youngest disciples*; in S. Rozental, ed., *Niels Bohr: His life and work as seen by his friends and colleagues*, (North-Holland, 1967)

[11] R. Kirk, *Mental Machinery and Gödel*, Synthese **66** (D. Reidel Publishing Company, 1986) pp.437-452

[12] M. Kline, *Mathematics: The Loss of Certainty*, (Oxford University Press, 1982)

[13] J. Lucas, *Minds, Machines and Gödel*, Philosophy, Vol.XXXVI (Cambridge University Press, 1961); reprinted in *Minds and Machines*, ed. A.R. Anderson, (Prentice Hall, 1964)

[14] R. Penrose, *Shadows of the Mind*, (Vintage, 1995)

[15] R. Penrose, *Psyche*, Volume 2, Number 1 (1996) pp.89-129

[16] H. Putnam, *The Best of all Possible Brains* (Review of Shadows of the Mind by R. Penrose), New York Times, 20 November 1994.

[17] H. Putnam, *Mathematics without foundations*, Journal of Philosophy **64** (Journal of Philosophy Inc., 1967) pp.5-22; reprinted in *Philosophy of Mathematics*, edited by P. Benacerraf and H. Putnam, (Cambridge University Press, 1985)

[18] D. Ruelle, *Chance and Chaos*, (Princeton University Press, 1991)

[19] B. Russell, Portraits from Memory, Bertrand Russell Peace Foundation (George Allen and Unwin, 1956)

[20] B. Russell, My Philosophical Development (Unwin, 1975).

[21] A. Scott, *Stairway to the Mind*, (Copernicus, 1995)

[22] J.R. Searle, *Minds, Brains, and Programs*, Behavioral and Brain Sciences **3** (Cambridge University Press, 1980) 417-24, reprinted in *The Philosophy of Artificial Intelligence*, ed. M.A. Boden, (Oxford University Press, 1990), pp.67-88

[23] J.C. Shepherdson and H.E. Sturgis, 'Computability of Recursive Functions', *J. Assoc. Computing Machinery* **10** (1963) 217-255

[24] A. Turing, *Computing Machinery and Intelligence*, Mind **59** No.236 (Oxford University Press, 1950) 433-460, reprinted in *The Philosophy of Artificial Intelligence*, ed. M.A. Boden, (Oxford University Press, 1990), pp.40-66

[25] A. Turing, *Systems of Logic Based on Ordinals*, P. Lond. Math. Soc. (2) **45** (1939) 161-228

[26] J. von Neumann, *The Mathematician. "Works of the Mind."* in Collected Works Volume 1, ed. A.H. Taub, (Pergamon Press, 1961)

[27] J. von Neumann, *The Computer and the Brain*, (Yale University Press, 1958)

[28] K. Warwick, *March of the Machines*, (Century, 1997)

Chapter 8

Attentional Modulation in Visual Pathways

8.1 Introduction

Within the sphere of consciousness we can selectively attend to different things. However, events in the periphery of attentional focus still have access to conscious awareness. An example might illustrate this point: while performing a highly automated task, like driving a car, we are still able to think about other things or listen to music. Although we can attend to an auditory process we are still driving consciously (Searle, 1994 [33]). On the other hand if a dangerous situation presents itself we can shift our attention entirely to the process of driving. Another example from Heidegger, frequently quoted by Dreyfus (1991) [8], is a skilled carpenter hammering. The carpenter's girlfriend might be the centre of his attention whilst hammering. But it is wrong to say that the automatic process (ie. hammering) is performed unconsciously. In this chapter we will concentrate on attention as a selective process in the context of conscious awareness of a phenomenal type. As an example we use attention to visual motion.

Driving at night, reduces the amount of information available about the car ahead to two red rear lights. The ensuing radial motion is a 2-dimensional representation of motion in depth. The velocity of the two - radially moving - dots allows one to estimate how fast one is approaching the other car. The depth of processing depends, however, on whether the observer is the driver or a passenger. Both observe the same scene but the driver has to pay attention to the rear lights and to the optical flow engendered, while the passenger might attend to the radio. Both the driver and passenger are aware of the visual scene but with a different focus of attention (visual motion vs. radio). This emphasises the fact that the sensorium can be attended to, or simply perceived, both being conscious processes.

Electrophysiological studies have identified a motion sensitive area (V5) in the middle temporal region in the Macaque (Dubner and Zeki, 1971 [9]). A homologous area (human V5) can be defined in man by functional imaging using positron emission tomography (PET) (Zeki et al. 1991 [40]) or functional magnetic resonance imaging (fMRI) (Tootell et al. 1995 [36]). Primate studies (Treue and Maunsell 1996 [38]) and neuroimaging data (O'Craven and Savoy 1995 [30]; Corbetta et al. 1991 [7]) have demonstrated that activity in these functionally specialised extrastriate visual areas can be modulated by attention.

These results suggest the responses of extrastriate areas processing signals are modulated according to task demands; this may provide a mechanism where individual stimulus attributes such as colour, and motion can be selected by attention. The locus of this modulation appears to be early in the sensory processing pathways, compatible with 'early selection' theories of selective attention (Treisman & Gelade 1980 [37]). Such modulation presupposes the interaction of sensory input and modulatory 'top down' influences from other cortical areas. In this framework, the expression of an early selection process, operating in primary and secondary sensory cortex, cannot be isolated from top-down influences exerted by higher (i.e. parietal and frontal) association areas (Rees et al. 1997 [32]).

Functional neuroimaging has been used to infer modulation of extrastriate cortical areas by attentional processes (Corbetta et al. 1991 [7]; O'Craven and Savoy 1995 [30]). Electrophysiological and fMRI studies have identified changes due mainly to attentional processing at the level of the posterior parietal cortex (Assad and Maunsell 1995 [2]; Bushnell et al. 1981 [6]; Mesulam 1981 [27]; Mountcastle et al. 1981 [29]). Recent results (Treue and Maunsell 1996 [38]) demonstrate this modulation at the level of V5 in primates. Inferences about modulation in the aforementioned studies were based on changes in activity (i.e. BOLD contrast in fMRI or neuronal activity in single cell electrophysiology). Increased responsiveness of an area to input from another area can be construed as an increase in the influence that one area exerts over another, or the connectivity between them. Therefore attentional modulation can be characterised directly in terms of changes in effective connectivity among cortical areas.

The notion that attention can be expressed as a modulation or change in effective connectivity is also captured in a quotation by La Berge (1995) [22]: "The expression of attention in a brain area appears to be described effectively as an enhancement of activity in the attended set of pathways relative to the unattended set of pathways".

The concept of effective connectivity was originated in the analysis of separable spike trains obtained from multi-unit electrode recordings (Aertsen and Preissl 1991 [1]; Gerstein et al. 1989 [17]; Gerstein and Perkel 1969 [18]; Gochin et al. 1991 [19]). Effective connectivity is closer to the intuitive notion of a connection than functional connectivity and can be defined as the influence one neural system exerts over another (Friston et al. 1995e [16]), either at a synaptic (c.f. synaptic efficacy) or cortical level. Functional connectivity is operational and is simply the covariances or correlations among activities at different parts of the brain.

One method used to estimate effective connectivity in neuroimaging is structural equation modelling. This technique combines an anatomical (constraining) model and the inter-regional covariances of activity. The ensuing functional model represents the influence of regions on each other through the putative anatomical connections. The indices of these influences represent a measure of effective connectivity.

In this chapter we present fMRI data from one subject, scanned under identical stimulus conditions, whilst changing only the attentional component of the tasks employed. In the first stage of the analysis we identify regions that show differential activations in relation to attentional set. In the second stage changes in effective connectivity between these areas are assessed using structural equation modelling (McIntosh et al. 1994 [24]). In the final stage we show how these attention-dependent changes in effective connectivity can be explained by modulation, of the dorsal visual pathway, by frontal cortical areas. These effects are characterised by extending standard structural equation modelling to include non-linear interaction terms or moderators (Kenny and Judd 1984 [21]). We also demonstrate the regional specificity of the interaction (i.e. modulation) and give an intuitive illustration of modulation using simple regression analysis.

8.2 Structural Equation Modelling

Structural equation modelling or path analysis is a technique developed in economics, psychology and the social sciences. The basic idea differs from the usual statistical approach of modelling individual observations. Instead of considering variables individually, the emphasis is on the covariance structure. The covariance matrix can also be seen as the functional connectivity matrix. Parameters are estimated in structural equation modelling by minimising the difference between the observed covariances and these implied by a structural or path model. The parameters of the path model are connection strengths or path coefficients and correspond to an estimate of effective connectivity.

How are path coefficients interpreted? The path coefficient represents the response of the dependent variable for a unit change in an explanatory variable, whilst the other variables in the model are held constant (Bollen 1989 [4]).

In terms of neural systems a measure of covariance represents the degree to which the activities of two or more regions are related. The study of covariance structures in neuroimaging has a unique advantage compared to applications in other fields: The interconnections among the dependent variables (regional activity of brain areas) are anatomically determined and the activation of each region can be directly measured. This is in contrast to 'classical' structural equation modelling in the behaviourial sciences, where sometimes models are hypothetical or 'latent' and cannot be assessed directly.

Mathematical implementation

As mentioned above structural equation modelling minimises the difference between the observed S and implied covariance matrix Σ. The variance-covariance structure S of the observed variables is given by:

$$S = \frac{1}{N-1} y^T \cdot y \tag{8.1}$$

where y is a N by p matrix of deviation (from the mean) scores of the p observed variables with N observations and y^T is y transpose. The matrix S is square and symmetric with the sample variances down its main diagonal and the covariances off the diagonal.

Consider a model, where variables y are 'caused' by a set of independent variables z (Figure 8.1). This could also be construed as a set of variables y with residual influences z (outside the model). In addition the variables y may cause each other.

Algebraically, the model for y is:

$$y \cdot I = y \cdot B + z \tag{8.2}$$

where B is a matrix of unidirectional path coefficients, I is the identity matrix. Here x appears on both sides of the equation. This reduces to:

$$y = z \cdot (I - B)^{-1} \tag{8.3}$$

Looking at the variance-covariance structure implied by the model and omitting the denominator $1/(N-1)$ we have:

$$\begin{aligned} y^T \cdot y &= \Sigma = (z \cdot (I - B)^{-1})^T \cdot (z \cdot (I - B)^{-1}) \\ &= (I - B)^{-1T} \cdot C \cdot (I - B)^{-1} \end{aligned} \tag{8.4}$$

where $C = z^T.z$. B is not symmetric because of asymmetric connections. C (variance- covariance structure of z) is a diagonal matrix and contains the residual variances. If interactions among the residual influences were to be incorporated into the model, their covariances would appear (symmetrically) off the leading diagonal in C (not shown in Figure 8.2). $y^T.y$ is the implied variance- covariance structure Σ. Parameters in the matrices C and B are called free parameters. The free parameters are estimated by minimising a function of S and Σ. To date the most widely used objective function for structural equation modelling is the maximum likelihood (ML) function:

$$F_{ML} = \log|\Sigma| + \mathrm{tr}(S.\mathrm{inv}(\Sigma)) - \log|S| - p \tag{8.5}$$

where tr(.) is the trace of the matrix and p is the number of free parameters. The Newton-Raphson or other gradient descent methods are used to estimate the parameters.

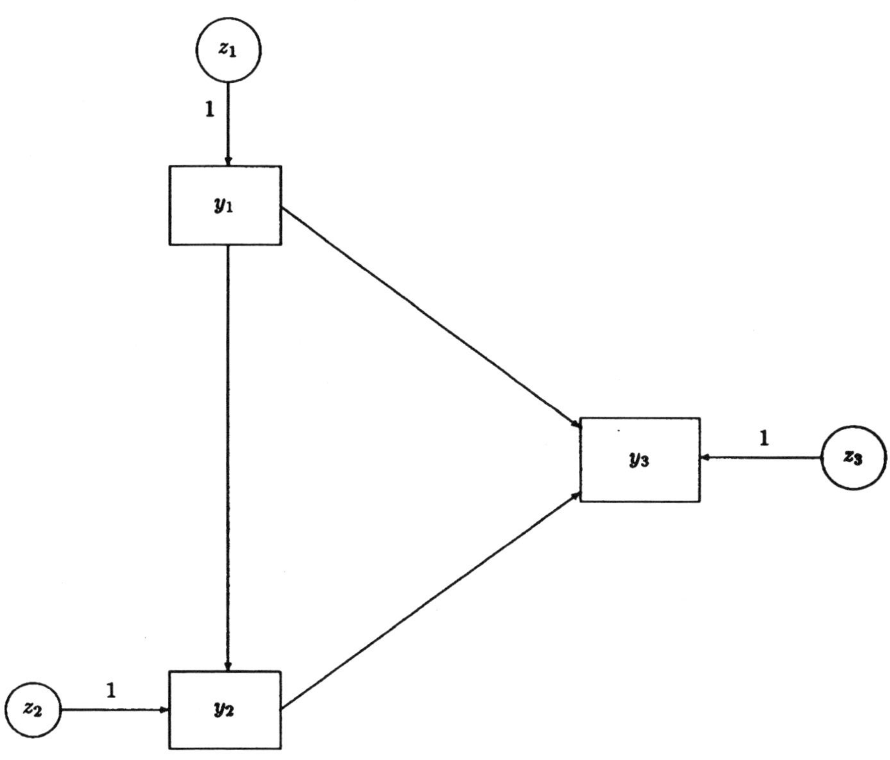

Figure 8.1: Simple path model to demonstrate the mathematical background of structural equation modelling. y are observed variables, the (uni-directional) path coefficients between variables y constitute matrix B. The 1 at each path between z and the corresponding variable y denotes the fact that z are residual variances, which are not explained by the model. The path-coefficients for the connections from z to y are therefore 1, which is simply the identity matrix (see equation 8.4).

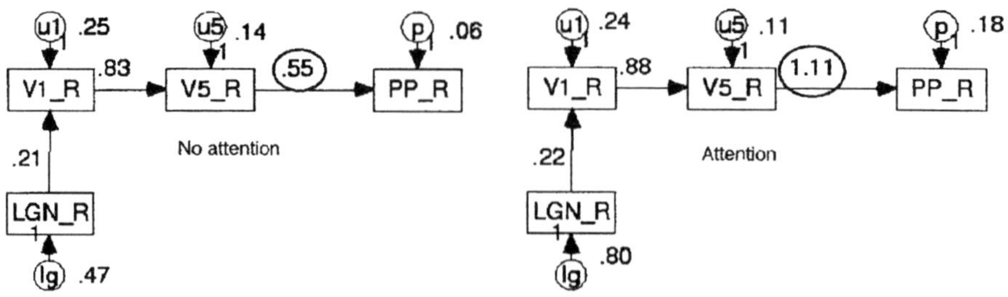

Figure 8.2: Structural equation model for the dorsal visual pathway. The parameters are evaluated for "attention" and "no attention" separately. It can be seen that the connectivity between V5 and PP is significantly larger during "attention" than during "no attention".

Anatomical model

An important issue in structural equation modelling is the determination of the underlying anatomical model. This model comprises regions and connections between those regions. Different methods can be combined to identify important regions: categorical comparisons between different conditions and eigenimages highlighting structures of functional connectivity in conjunction with results from primate electrophysiology have been used (Grafton et al. 1994 [20]; McIntosh and Gonzalez-Lima 1991 [24]). The connectivity between the identified regions is generally based on neuroanatomical tracer studies in primates.

A model is always a simplification of reality. In the context of effective connectivity one has to find a compromise between complexity, anatomical accuracy and interpretability. There are also mathematical constraints on the model. If the number of free parameters (unknowns) exceeds the number of observed covariances the system is underdetermined and no single solution exists.

Statistical inference

Statistical inference in structural equation modelling can address two points. The goodness of the overall fit of the model, i.e. how significantly different are the implied and observed covariance structures and the difference between alternative models (e.g. different conditions, subjects etc.). In the context of multivariate normally distributed variables the minimum of the maximum likelihood fit function times the number of observations minus one, follows a χ^2-distribution with $(q/2).(q+1)-p$ degrees of freedom (Bollen 1989 [4]). p is the number of free parameters and q is the number of observed variables. The hypothesis is that the model is not able to reproduce the observed variance covariance structure. That implies that a non- significant chi-square value is desired, if one wants to infer the validity of the hypothesis of no difference between model and data. This is problematic because, from the point of view of the χ^2-statistic, this is like trying to confirm the null hypothesis (of no differences). In contrast to 'classical' path analysis, the validation of models used in the context of functional brain imaging relies on the neuroanatomy. Therefore goodness of fit estimation plays only a minor role (McIntosh and Gonzalez-Lima 1994 [25]). However, the goodness of fit measure is useful, when comparing different models with each other. Here the alternative hypothesis of a difference is interesting. This so called 'stacked model' approach can be used to compare different models (e.g. data from different groups or conditions) in the context of structural equation modelling (Grafton et al. 1994 [20]). A so called 'null-model' is constructed where the estimates of some parameters (i.e. path coefficients) are constrained to be zero or equal for both groups. The alternative model allows these parameters to differ between groups. The significance of the differences between the fit of both models is expressed by the difference in the chi-squared goodness of fit indicator. This chi-squared statistic has n degrees of freedom, where n is the difference of the degrees of freedom between the null-model and the one in question. For example, if the null-model constrains one parameter to be equal between groups, the resulting degree of freedom for the chi-squared statistic would be one.

Non-linear interaction terms

Current applications of structural equation modelling generally use linear models. However it is possible, to incorporate additional variables containing a non-linear function (e.g. $f(x) = x^2$) of the original variables (Kenny and Judd 1984 [21]). Interactions between variables can be incorporated in a similar fashion; wherein a new variable, containing the product of two interacting variables, is introduced as an additional influence.

Even if the variables are multinormally distributed, the product of two such variables is not. Under these circumstances, maximum likelihood estimators (ML) are still consistent, but the chi-squared fit index and tests of statistical significance may not be valid (Bollen 1989 [4]). To overcome this problem, parameters can be estimated by weighted least squares (WLS) (Browne 1984 [5]; Kenny and Judd 1984 [21]). As we use non-linear interaction terms in the second part of our analysis, the parameters are estimated using WLS, whereas the path coefficients in our linear models are estimated by ML.

8.3 Design and Image Acquisition

The experiment was performed on a 2 Tesla whole body MRI system equipped with a head volume coil. Contiguous multislice $T2^*$ weighted fMRI images (TE=40ms; 90 ms/image; 64x64 pixels [19.2 cm x 19.2 cm]) were obtained with echo-planar imaging (EPI) using an axial slice orientation. A $T2^*$ weighted sequence was chosen to enhance blood oxygenation level dependent (BOLD) contrast. The effective repetition time was 3.22 seconds. The subject was scanned during 3 different conditions: "fixation", "attention" and "no attention". Each condition lasted 32.2 seconds giving 10 multislice volumes per condition. We acquired a total of 260 images.

During all conditions the subjects looked at a fixation point in the middle of a transparent screen. In conditions with visual motion ("attention" and "no attention") 250 white dots (size $0.1°$) moved radially from the fixation point in random directions towards the border of the screen, at a constant speed of $4.7°$ per second. The difference between "attention" and "no attention" lay in the explicit command given to the subject shortly before the condition: "just look" indicated "no attention" and "detect changes" the "attention" condition. Both visual motion conditions were interleaved with "fixation". No response was necessary.

Before scanning, subjects were exposed to five 30 second trials of the stimulus. The speed of the moving dots was changed 5 times during each trial. Subjects were asked to indicate any change in speed. Changes in speed were gradually reduced over the 5 trials, until a 1% change was presented on the last occasion. Once in the scanner, changes in speed were completely eliminated (without the knowledge of the subject), so that identical visual stimuli were shown in all motion conditions.

Regions of interest were defined by comparing "attention" and "no attention" and comparing "no attention" and "fixation". As predicted with a stimulus consisting of radially moving dots we found activation of primary visual cortex (V1), V5 and the posterior parietal complex. For the subsequent analysis of effective connectivity, we defined regions of interest (ROI) with a diameter of 8 mm, centred around the most significant ($p < 0.05$, corrected) voxel as revealed by the categorical comparison. A single time-series, representative of this region, was defined by the first eigenvector of all the voxels in the ROI.

8.4 Image Analysis and Categorical Comparisons

Image processing and statistical analysis were carried out using SPM96 (Friston et al. 1996 [13]; Friston et al. 1995c [14]; Worsley and Friston 1995 [39]). All volumes were realigned, coregistered to a structural T1 weighted image, spatially normalised (Friston et al. 1995a [11]) to a standard template (Evans et al. 1993 [10]; Talairach and Tournoux 1988 [35]) and smoothed using a 6 mm isotropic Gaussian kernel full width at half maximum. Data analysis was performed by modelling the different conditions ("attention", "no attention" and "fixation") as reference waveforms in the context of the general linear model as employed by SPM96 (Friston et al. 1995b [12]). Specific effects were tested with appropriate linear contrasts of the parameter estimates for each condition, resulting in a t-statistic for each and every voxel (Friston et al. 1995d [15]). These t-statistics (transformed to Z-statistics) constitute a statistical parametric map (SPM). These SPMs are then interpreted by referring to the probabilistic behaviour of Gaussian random fields. Data were analysed for each subject individually. The threshold adopted was $p < 0.05$ (corrected for multiple comparisons).

8.5 Modelling of the Posterior Visual Pathway

Our model included the LGN, primary visual cortex (V1), V5 and the posterior parietal complex (PP). Although connections between regions are generally reciprocal, due to mathematical restrictions (the relative numbers of known and unknown variables and stability of the model), we only included unidirectional paths.

To assess effective connectivity in a condition-specific fashion, we used time-series that comprised observations during the condition in question. Path coefficients for both conditions ("attention" and "no attention") were estimated using a maximum likelihood function with the software package AMOS (Amos for Windows, Version 3.5, SmallWaters Corp., Chicago). To test for the impact of changes in effective connectivity between "attention" and "no attention", we defined a free model (allowing for different path coefficients between V5 and PP for "attention" and "no attention") and a constrained model (constraining the V5 $\rightarrow$ PP coefficient to be equal). This analysis revealed a highly significant difference ($\chi^2 = 26$, $p < 0.01$) in model fit in favour of the free model, indicating a significant change (ie. increase) in effective connectivity for "attention" (Figure 8.2). This finding is in accordance with primate electrophysiology and human imaging studies, showing attentional modulation of responses at the level of V5 (Beauchamp and DeYoe 1996 [3]; O'Craven and Savoy 1995 [30]; Treue and Maunsell 1996 [38]) and the posterior parietal complex (PP).

8.6 Modelling Modulation by Interaction Terms

The linear path model of the previous section comparing "attention" and "no attention" revealed increased effective connectivity in the dorsal visual pathway in relation to attention. The question that arises is which part of the brain is capable of modulating this pathway? Based on lesion studies (Lawler and Cowey, 1987 [23]) and on the system for directed attention as described by Mesulam (1990 [28]), the dorsolateral prefrontal cortex or the anterior cingulate were candidates for such a modulatory role (only the right dorsolateral prefrontal cortex showed a significant activation during "attention" relative to "no attention").

The right prefrontal region was included in our model as a moderator variable modulating the posterior visual pathway. Given the neuroanatomical projection from PFC to PP, we included a modulation of the pathway between V5 and PP. Assuming a non-linear modulation of this connection, we constructed a new variable "V5PFC" in our analysis. This variable, mediating the interaction, is simply the time-series of region V5 multiplied by the time-series of the right prefrontal cortex. Although the time series are normalised before multiplication, the product of 2 variables (say V5 and PFC) will still show some correlation with the individual terms. Therefore we residualised the interaction term (V5 × PFC) by least squares (ie. remove any components of V5 × PFC that could be predicted by a linear combination of V5 and PFC):

$$V5PFC = V5 \times PFC - [V5\,PFC] \cdot \mathrm{pinv}([V5\,PFC]) \cdot V5 \times PFC \qquad (8.6)$$

where $pinv(.)$ is the pseudoinverse. This procedure can be compared to partial correlation analysis, where the effect of the established predictors is removed from a new predictor to assess the improvement in fit.

The influence of this new variable on PP corresponds to the influence of the prefrontal cortex on the connection between V5 and PP. The interaction model is shown in Figure 8.3. Because the non-linear model could accommodate changes in connectivity between "attention" and "no attention" the entire time-series was analysed (i.e. attention specific changes are explicitly modelled).

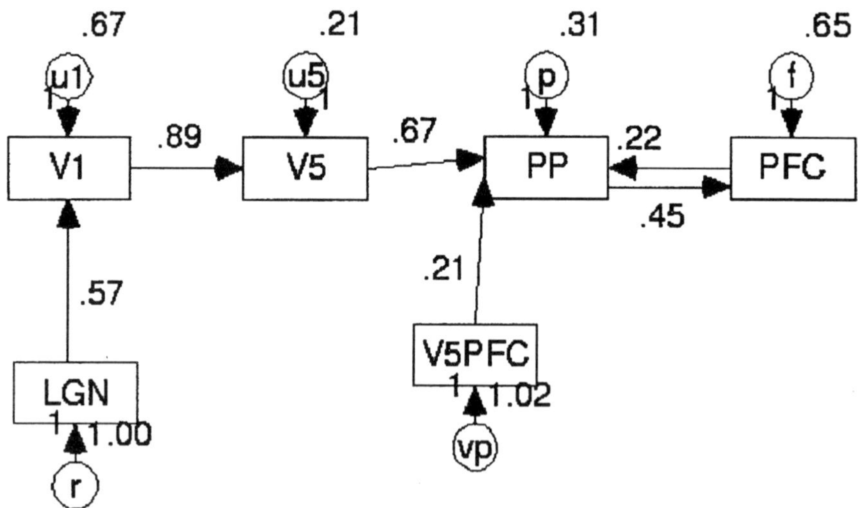

Figure 8.3: A sketch of the extended model of the dorsal visual pathway with the modulatory influence of the prefrontal cortex (PFC) on the connection between V5 and PP. The main effect of PFC is also included to show whether the interaction is significant in the presence of the main effect.

As described in the linear model, we tested for the significance of the interaction effect by comparing a restricted and free model. In the restricted model the interaction term (i.e. path from V5PFC to PP) was set to zero. It is important to note that testing for an interaction is only valid in the presence of the main effects (i.e. paths from V5 and PFC to PP). Although The path coefficient for the interaction term was 0.21, omitting this term led to a significantly reduced model fit ($\chi^2 = 34$, $p < 0.01$).

The prefrontal cortex is known to play a special role in regulating the interplay of cortical regions (Shallice 1988 [34]). This has also been shown for attention in patients with prefrontal lesions (Pierrot Deseilligny et al. 1986 [31]). In terms of effective connectivity, McIntosh and colleagues demonstrated feedback from Brodmann area 46 to the dorsal visual pathway during spatial vision (McIntosh et al. 1994 [26]). It is also interesting to note that in their model, which included interhemispheric connections, the influence of the right prefrontal region often dominated over the left prefrontal region.

8.7 Interaction Effects Demonstrated with Regression Analysis

The presence of an interaction effect of the PFC on the connection between V5 and PP can also be illustrated by a simple regression analysis. If PFC shows a positive modulatory influence on the path between V5 and PP, the influence of V5 on PP should depend on the activity of PFC. This can be tested, by splitting the observations into two sets, one containing observations in which PFC activity is high and another one in which PFC activity is low. It is now possible to perform separate regressions of PP on V5 for both sets. If the hypothesis of positive modulation is true, the slope of the regression of PP on V5 should be steeper under high values of PFC. Figure 8.4 shows exactly this and provides the regression coefficients and the F-statistic for the difference in slope.

8.8 Regional Specificity

We also addressed the question of the regional specificity of the interaction. In general this can be seen as the contribution of the interactions between two areas (V5 and PFC) in explaining the variation in activity in a third (PP). In this context a modulatory effect of PFC on the efferent projections from V5 would be expressed as a contribution from V5, that depended on activity in PFC. If the connection between V5 and PP is modulated by the PFC, the time-series of V5PFC (the orthogonalised product of the V5 and PFC time-series) should predict a component of the activity in the PP. To test this hypothesis we used a SPM analysis of BOLD signal in each and every voxel of the brain, using the interaction term (V5PFC) as a regressor or explanatory variable.

If the modulatory effect is regionally specific then we should see this second order contribution effect only in the posterior parietal complex. The ensuing t-statistics for positive regression slopes were assembled into an SPM$\{Z\}$ (Figure 8.5). The regional specificity of the interaction for the bilateral posterior parietal cortex is remarkable.

An important aspect of this analysis, is that there is an entirely equivalent and symmetric interpretation of the physiological interactions above; namely that they reflect a modulation of PFC $\rightarrow$ PP connections by V5 activity. This is because an interaction can be construed as either a modulation of the effects

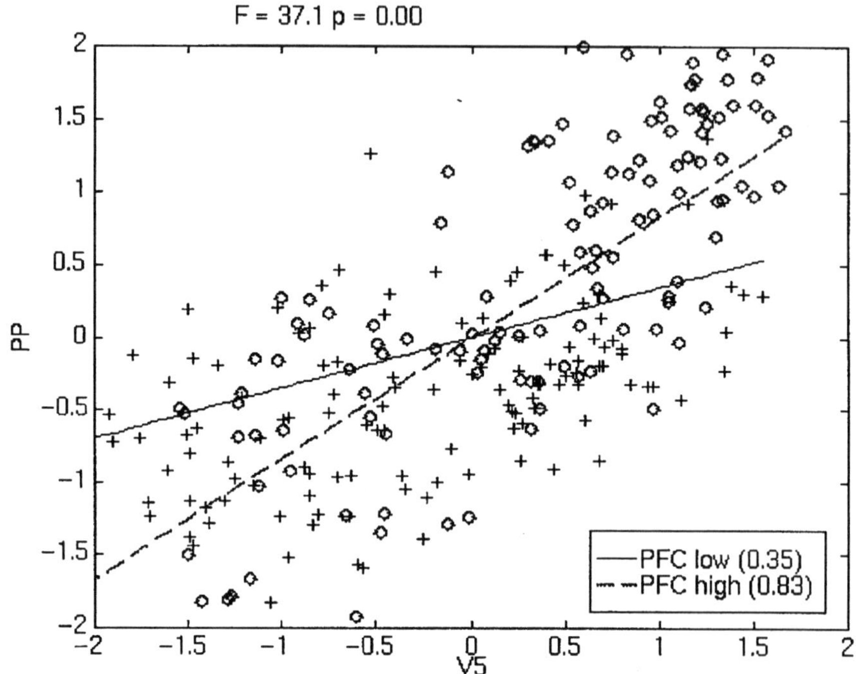

Figure 8.4: An alternative way of showing the modulatory effect of the PFC on the connection between V5 and PP. All observations were divided into two groups, one with observations in which PFC activity is high, one in which PFC activity is low. The graph shows separate regression curves for both groups and the significance for the differences in slope for the right hemisphere.

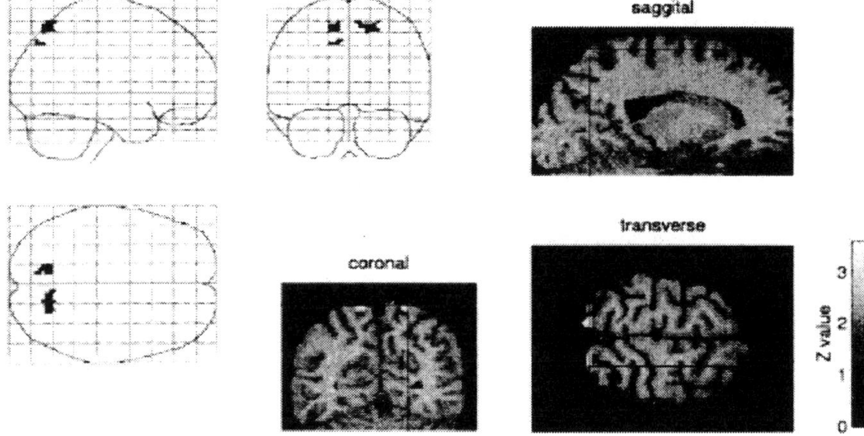

Figure 8.5: SPM(Z) using the interaction term V5PFC as a covariate of interest. The activations are shown as a maximum intesity projection, looking at the brain from the left, back and top. The right side of the figure shows the activations superimposed on a structural MRI. The significant voxels show a positive correlation with the interaction term. Note the regional specificity of the interaction term for the posterior parietal corex (PP).

of the first factor, by the second, or equivalently a modulation of the second's effects by the first. There is no formal distinction between what is an effect and what is a modulatory factor. However our explanation that PFC modulates the connection between V5 and PP as an example of top down modulation is more likely, given data that support an executive role for the prefrontal cortex (Pierrot Deseilligny et al. 1986 [31]; Shallice 1988 [34]).

8.9 Conclusion

Structural equation modelling applied to fMRI data revealed marked changes in effective connectivity in the posterior visual pathway in relation to attentional set. Further the introduction of non-linear interaction terms allowed us to model a modulatory influence from the dorsolateral prefrontal cortex on this pathway. These results were not evident from categorical comparisons. This type of analysis will be useful in a variety of imaging experiments in which one hypothesises changes in effective connectivity under different experimental conditions.

8.10 Bibliography

[1] Aertsen A., Preissl H. (1991) Dynamics of activity and connectivity in physiological neuronal Networks. New York : VCH publishers Inc

[2] Assad J. A., Maunsell J. H. (1995) Neuronal correlates of inferred motion in primate posterior parietal cortex. Nature 373: 518-21

[3] Beauchamp M. S., DeYoe E. A. (1996) Brain areas for processing motion and their modulation by selective attention. NeuroImage 3 (Supplement): S245

[4] Bollen K. A. (1989) Structural Equations with Latent Variables. New York : John Wiley & Sons

[5] Browne M. W. (1984) Asymptotic distribution free methods in analysis of covariance structures. British Journal of Mathematical and Statistical Psychology 37: 62-83

[6] Bushnell M. C., Goldberg M. E., Robinson D. L. (1981) Behavioral enhancement of visual responses in monkey cerebral cortex. I. Modulation in posterior parietal cortex related to selective visual attention. J Neurophysiol 46: 755-72

[7] Corbetta M., Miezin F. M., Dobmeyer S., Shulman G. L., E. P. S. (1991) Selective and divided attention during visual discrimination of shape, color, and speed: Functional anatomy by positron emission tomography. Journal of Neuroscience 13: 1202- 1226

[8] Dreyfus H. L. (1991) Being-In-The-World: A commentary on Heidegger's Being and Time, Division I. Cambridge, USA: MIT Press.

[9] Dubner R., Zeki S. M. (1971) Response properties and receptive fields of cells in an anatomically defined region of the superior temporal sulcus in the monkey. Brain Res, 35: 528-32

[10] Evans A. C., Collins D. L., Mills S. R., Brown E. D., Kelly R. L., Peters T. M. (1993) 3D statistical neuroanatomical models from 305 MRI volumes. Proc. IEEE-Nuclear Science Symposium and Medical Imaging 1813-17

[11] Friston K. J., Ashburner J., Frith C. D., Poline J.-B., Heather J. D., Frackowiak R. S. J. (1995a) Spatial registration and normalization of images. Human Brain Mapping 2: 1-25

[12] Friston K. J., Frith C. D., Turner R., Frackowiak R. S. J. (1995b) Characterizing evoked hemodynamics with fMRI. NeuroImage 2: 157-65

[13] Friston K. J., Holmes A. P., Ashburner J., Poline J.-B. (1996) SPM96. World Wide Web http://www.fil.ion.ucl.ac.uk/spm

[14] Friston K. J., Holmes A. P., Poline J.-B., Grasby P. J., Williams S. C. R., Frackowiak R. S. J., Turner R. (1995c) Analysis of fMRI time-series revisited. NeuroImage 2: 45-53

[15] Friston K. J., Holmes A. P., Worsley K. P.,Poline J.-B., Frith C. D., Frackowiak R. S. J. (1995d) Statistical parametric maps in functional imaging: A general linear approach. Human Brain Mapping 2: 189-210

[16] Friston K. J., Ungerleider L. G., Jezzard P., Turner R. (1995e) Characterizing modulatory interactions between V1 and V2 in human cortex with fMRI. Human Brain Mapping 2: 211-224

[17] Gerstein G. L., Bedenbaugh P., Aertsen A. (1989) Neuronal assemblies. IEEE Trans. on Biomed. Engineering 36: 4-14

[18] Gerstein G. L., Perkel D. H. (1969) Simultaneously recorded trains of action potentials: Analysis and functional interpretation. Science 164: 828-830

[19] Gochin P. M., Miller E. K., Gross C. G., Gerstein G. L. (1991) Functional interactions among neurons in inferior temporal cortex of the awake macaque. Exp. Brain Res 84: 505-516

[20] Grafton S. T., Sutton J., Couldwell W., Lew M., Waters C. (1994) Network analysis of motor system connectivity in Parkinson's disease: Modulation of thalamocortical interactions after pallidotomy. Human Brain Mapping 2: 45-55

[21] Kenny D. A., Judd C. M. (1984) Estimating nonlinear and interactive effects of latent variables. Psychol Bull 96: 201- 210

[22] La Berge D. (1995) Attentional processing. Cambridge, Mass. : Harvard University Press

[23] Lawler K. A., Cowey A. (1987) On the role of posterior parietal and prefrontal cortex in visuo-spatial perception and attention. Exp Brain Res 65: 695-8

[24] McIntosh A. R., Gonzalez-Lima F. (1991) Structural modelling of functional neural pathways mapped with 2-deoxyglucose: Effects of acoustic startle habituation on the auditory system. Brain Res 547: 295-302

[25] McIntosh A. R., Gonzalez-Lima F. (1994) Structural equation modelling and its application to network analysis in functional brain imaging. Human Brain Mapping 2: 2-22

[26] McIntosh A. R., Grady C. L., Ungerleider L. G., Haxby J. V., Rapoport S. I., Horwitz B. (1994) Network analysis of cortical visual pathways mapped with PET. J Neurosci 14: 655-666

[27] Mesulam M. M. (1981) A cortical network for directed attention and unilateral neglect. Ann-Neurol 10: 309-25

[28] Mesulam M. M. (1990) Large-scale neurocognitive networks and distributed processing for attention, language, and memory. Ann- Neurol 28: 597-613

[29] Mountcastle V. B., Andersen R. A., Motter B. C. (1981) The influence of attentive fixation upon the excitability of the light-sensitive neurons of the posterior parietal cortex. J Neurosci 1: 1218-25

[30] O'Craven K. M., Savoy R. L. (1995) Voluntary attention can modulate fMRI activity in human MT/MST. Invest. Ophthalmol. Vis. Sci. (suppl.) 36: S856

[31] Pierrot Deseilligny C., Gray F., Brunet P. (1986) Infarcts of both inferior parietal lobules with impairment of visually guided eye movements, peripheral visual inattention and optic ataxia. Brain 109: 81-97

[32] Rees G. E., Frackowiak R.S.J., Frith C.D. (1997) Two modulatory effects of attention that mediate object categorisation in human cortex. Science 275: 835-8

[33] Searle J. R. (1994) The Rediscovery of the Mind. Cambridge, USA: MIT Press.

[34] Shallice T. (1988) From neuropsychology to mental structure. Cambridge : Cambridge University Press

[35] Talairach P., Tournoux J. (1988) A Stereotactic coplanar atlas of the human brain. Stuttgart : Thieme

[36] Tootell R. B., Reppas J. B., Kwong K. K., Malach R., Born R. T., Brady T. J., Rosen B. R., Belliveau J. W. (1995) Functional analysis of human MT and related visual cortical areas using magnetic resonance imaging. J Neurosci 15: 3215-30

[37] Treisman A. M. & Gelade G. (1980) A feature-integration theory of attention. Cognitive Psychology 12: 97-136

[38] Treue S., Maunsell H. R. (1996) Attentional modulation of visual motion processing in cortical areas MT and MST. Nature 382: 539- 41

[39] Worsley K. J., Friston K. J. (1995) Analysis of fMRI time-series revisited - again. NeuroImage 2: 173-181

[40] Zeki S., Watson J. D., Lueck C. J., Friston K. J., Kennard C., Frackowiak R. S. (1991) A direct demonstration of functional specialization in human visual cortex. J Neurosci 11: 641-9

Chapter 9

Neural Networks and the Mind

9.1 Introduction

The brain still presents many enigmas to the working neurophysiologist. The general flow of input from sensory receptors to feature detectors to object storage and recognition modules to motor action programs to motor output hides a vast number of highly subtle processing activities which still need to be delineated. There is also important feedback from higher areas to lower ones, as well as continued activity of specific neurons holding 'activity traces' of previous activity [18], thereby forming a so-called 'working memory' in which temporally extended neural activity is crucially involved. Moreover the results coming from the use of non-invasive instruments (PET, fMRI, EEG and MEG, which measure activity inside the brain without having to open it up) indicate that there is great subtlety and variety in the manner in which different modules of the brain combine together both spatially and temporally to solve the information processing tasks by which the brain finds itself surrounded. All of these features are far from understood, but there is increasing data, at a variety of levels of detail, as to how the brain functions under a wide range of conditions.

In order to make progress in bringing order to this important material, and allow it to be used in answering the deep questions raised above, it seems necessary to take a high level view of the information processing performed by the brain which may support consciousness. This would avoid solving many of the detailed problems posed about the brain modules' methods of operation while allowing the development of a broad framework into which the more detailed activities of these brain modules, at a local level, may be able to be fitted. In tackling the problem of the function of the various modules of the brain without such an overarching plan it is as if one is trying to put together the pieces of a jigsaw puzzle without knowing the picture it is supposed to resemble. That is a difficult task.

Let us avoid some very hard problems by keeping to the high ground and searching for the general organisational plan used by the brain in its information processing. However, one still is faced with the question as to why the general flow of information in the brain described at such a high level requires consciousness in any concomitant manner. What more must be put into a flow chart description of brain activity in order to represent subjective experience?

One initial possibility is based on the Relational Mind model [44, 45]. The basic idea behind this approach is that consciousness arises from the active comparison of ongoing brain activity, stemming from external inputs in the various modalities, with somewhat similar past activity stored in suitable memory sites. The mental content of an experience is therefore the set of relations of that experience to stored memories of relevant past experiences. Thus the consciousness of the blue of the sky, as seen now, is determined by the stored memories of one's past experience of blue skies, say on hillsides, at the seaside stretched flat on the sands, or in one's garden sitting in a chair. Not only episodic memory (that involving oneself in specific events) need be involved; there may also be preprocessing and semantic memory of objects of a visual or other scene. (Preprocessing is the term denoting what happens at an early stage to inputs in the brain, such as the changes to them brought about at the retina. By the term semantic memory is meant general knowledge, such as that London is the capital of the United Kingdom.) The semantic memories thereby involved may also lead to further episodic memory activation, in which further relations to past experience are accessed. Other forms of preprocessing also occur in other modalities, as is being increasingly probed in vision, with orientation, colour and motion detectors being used at an early stage.

It will be explained how, by use of the Relational Mind model, one can begin to construct a more detailed model of the Mind. In terms of this model it is then possible to begin to make sense of the experimental results now becoming available from the non-invasive machines and to design tests of the model at a deeper level. Some of these aspects will be developed.

We will start with a more detailed analysis of the Relational Mind model itself so as to help focus better on the most appropriate aspects. This will lead to a set of Principles which we will use for further guidance, and to the two stage Relational Mind model which will be explored as to the emergence of consciousness at the lowest level, that of phenomenal experience. The way in which this can explain such experience will be discussed briefly before we close with some predictions of the model.

9.2 The Relational Mind

The main thesis of the Relational Mind model is that:

> The conscious content of a mental experience is determined by the evocation and intermingling of suitable past memories evoked (sometimes unconsciously) by the input giving rise to the experience.

Thus the basic idea of the relational approach to the mind is that *consciousness arises from the similarity of ongoing brain activity, stemming from external inputs to various modalities, to past activity stored away in suitable memory receptacles.* These past activities could include either preprocessing, semantic or episodic memories, where the former two involve knowledge for which the particular details of when or where the learning experience occurred has not been recorded. Episodic memory, on the other hand, always involves a record of the moment of experience, so that 'I' or 'me' is always present in such records. Both semantic and episodic forms of memory are declarative, being able to be made explicitly conscious. The preprocessing memory structures are not declarative in the same manner but their resultant neural activity is phenomenally experienced as what are called 'qualia' or 'raw feels', and possessing the ineffability or impossibility of explanation (and other features) discussed extensively by philosophers, and to which we will return later. There are also memory structures of a non-declarative form which are excited by a given input, such as value memory associated with earlier encounters with the objects involved in the input.

In all, then, *the mental content of an experience is suggested as comprising the set of relations of that experience to stored memories of relevant past experiences.* Thus the consciousness of the blue of the sky, as seen now, is determined by the stored memories of one's past experience of blue skies. Similarly the consciousness of sights, sounds or smells may not only be of the simple records of such past experiences but also of the emotions involved in them. Such emotional content may have powerful effects, even though the emotional memory is not itself consciously experienced.

In general there is strong evidence for the thesis that past experiences, however stored, influence present behaviour. Will there be a concomitant effect on present consciousness? To answer that question it appears useful to consider the function of consciousness. That has been debated with considerable energy for a long time, but with no clear conclusion. However there is experimental support for the rather minimal thesis that consciousness has at least a veto effect on developing behaviour [25], although it may not necessarily be used in determining all the details of responses. Even such a minimal interaction indicates that consciousness is expected to be influenced by past experiences, as they themselves influence behaviour and lead to a choice of responses needing conscious intervention at appropriate times.

There is strong support from the community of cognitive scientists for the Relational Mind model when it is rephrased in the form that consciousness arises from the interaction of 'top down' with 'bottom up' activities. To a cognitive scientist the thesis in this form is obviously true - he/she would ask "what else can there be as the source of conscious experience?" - and they therefore do not need any further proof. Moreover the thesis is well supported by a host of evidence already presented in numerous textbooks. However it is felt necessary to develop some independent justification for it here since it is the basis of the present approach and will be used to build a more detailed framework with which to build a neural model of consciousness; it therefore needs a strong grounding.

9.3 Exploring the Relational Mind Model

To explore the Relational Mind model more fully, let us return to the basic statement of the model given earlier:

> The conscious content of a mental experience is determined by the evocation and intermingling of suitable past memories evoked by the input giving rise to the experience.

The key phrases to focus on in this statement are:

- input
- evocation
- intermingling
- suitable memories

In other words analysis must be carried out of the manner in which input is processed so as to excite or evoke one or more memories (of various sorts) relevant to the encoded input, and so being seen as suitable. These memories are then combined (intermingled) in some manner with the input. The whole process is shown in Figure 9.1. In that figure the input is encoded up to a certain (non-conscious) level in neural modules SMA, SMB. A transformed or encoded result then excites a set of memories from a long-term memory store E. These memories, together with the encoded input, are then 'intermingled' or combined in some further manner in separate neural working memory modules WMA, WMB. The labels SM and E are used to denote preprocessing/semantic and episodic memory respectively. The essence of the relational structure in this more explicit version of the Relational Mind model is in two places:

1. the excitation of the memories from E,
2. the working memory modules WMA, WMB,... ..

These modules all depend on memory of various forms, so it is appropriate to consider what these forms might be.

9.4 Memory in the Relational Mind Model

Memory in some form or other is quite extensive throughout the brain. Earlier distinction was made between declarative (that which you explicitly know) and non-declarative (that which is implicit, such as a skill) memory; the former was divided into semantic and episodic memory. At the same time early input is transformed by preprocessing modules, which should also be included as part of the total memory structure (which includes genetic effects). Non-declarative

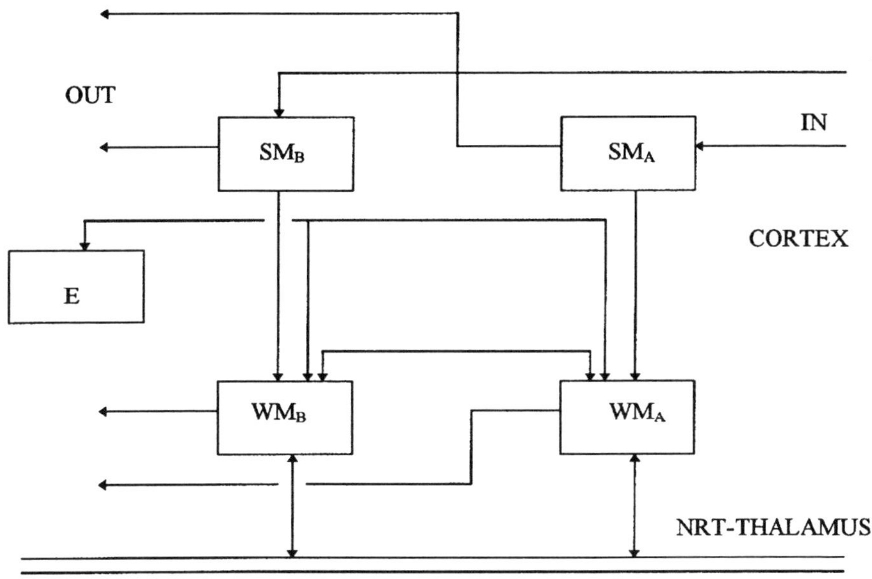

Figure 9.1: The main structures involved in the Relational Mind model; input goes to semantic/preprocessing modules SMA and SMB. The processed outputs of these modules is then sent to the respective working memory modules WMA and WMB. There is then a competition run between these sites by means of the NRT/Thalamus system. Episodic memories in the module E may also be involved in supporting activities on one or other of the working memory modules.

memory includes at least that of skills and of value, although the latter can become declarative, as when one says "I like that (object)". Emotions and effect are important features which will be beyond our present scope to consider in detail; it is to be accepted that value memory is used to make decisions and should be included in E. Thus the memory structures initially to be used in Figure 9.1 are:

SM : preprocessing/semantic memories
(in a given code, such as colour or motion)
E : episodic and value memory

The above identifications do not seem to solve the binding problem (as to how the different codes for an object, in general activated in different regions of the cortex, are combined together to give the unified experience of the object), and in particular leave open the manner in which the outputs of the preprocessing/semantic memories across the given codes (say of colour, shape, texture and motion, for objects) are integrated to give a satisfactory 'bound' representation of an object to be used as input to the episodic and value memory stores E. This latter is itself initially assumed to be independent of the separate codes, but based on unified objects representations. The binding problem will not be discussed in any detail here, but one potential manner in which such binding might occur arises from two further features: the temporal character of memory, and the nature of the decision system itself.

For the first of these, the manner in which time enters in memory, it is now clear that there are two sorts of memory. One, that of buffering, involves continued activity, without necessarily any concomitant structural changes (such as increase or decrease of synaptic weights). Such activity has been observed in various parts of the brain, but especially and most predominantly in the multi-modal regions of the cortex (inferotemporal., parietal and frontal) and in related sub-cortical structures. These are where nerve cells are found which have a relatively persistent response to inputs in a number of different modalities. The length of time such activity can persist itself seems modifiable, especially in frontal lobe (up to about 30 seconds) as experiments on monkeys have shown. On the other hand the posteriorly sited phonological store, the working memory part of the phonological loop [11, 10], can hold activity for about 1-2 seconds but does not seem variable or adaptive so as to change that length of time. There is good experimental evidence for the existence of both the variable-duration frontal lobe memories (called 'active' by Fuster [17]) and the more fixed length and more posteriorly placed 'working' memories; the latter of these will be discussed in more detail in sections 9.6,9.7 & 9.8 (for the former see [54]).

The other sort of memory involves adaptive change in the ease of re-activation or reduction of error rate of recall for the associated memory. In other words it is based on structural change of the brain and not on continued neural activity. The lifetime for such changes, once they have been established in the cortex, is well known to be for many years (as our own autobiographical or episodic memories clearly show). On the other hand there may be shorter term 'priming' memories, brought about by a single exposure to an object, which could help increase the speed or ease of access to long-term memories. This increased access may only persist for a limited time, of the order of hours or days. Such 'priming memory' may give crucial increased support to earlier experiences so as to give content to consciousness, as some of the experiments mentioned earlier in support of the Relational Mind Model indicate. Such effects would seem to involve exaggerating the strengths of ouptut from the neural sites of preprocessing/semantic memory SM, shown in Figure 9.1, for recently experienced inputs.

The above discussion of the nature of memory is a very brief account of the structures that many neuroscientists are attempting to elucidate. The problem of untangling memory is proving very hard, partly because memory seems to be so ubiquitous in the brain. If there is any hard part of the easy problems of modeling the brain other than consciousness, it is clear that memory should be included in it.

The main principles of the Relational Mind model will now be stated (a detailed justification of them is contained in [54]):

Principle 1. For each input code A there exists a related pair of preprocessing/semantic and working memory modules, denoted SMA, WMA.

Principle 2. There is a competition, on a given WMA, between neural activities representing different interpretations of inputs in the preceding second or so.

Principle 3. A competition is run between the activities on different working memories (WMs), the winner gaining access to consciousness.

Principle 4. Feedback from episodic memory is involved in the competition between the different preprocessing-semantic memory/working memory pairs.

Principle 5. Upgrading of episodic memories occurs from the output of the winning working memory, denoted WMwin.

Principle 6. The output of the winning working memory (or memories, if they are correlated) is stored in the hippocampus and amygdala as a buffer.

Having stated these principles we will now begin to explore the neural modules which may be able to support them. We will start with Principle 3, and consider what sort of system would provide uniqueness to conscious experience.

9.5 A Neural Candidate for Global Control

Consciousness appears on the face of it to be a unique phenomenon. There is assumed to be but one winner of the overall competition considered in Principle 3. Yet there are numerous sites of working memory spread throughout the brain, as we will discuss in more detail shortly. One would therefore expect there to be some 'control structures' in the brain which achieve global correlation between these various components. In particular these structures would support competition between various inputs, when suitably encoded, and only allow certain, most relevant memories to be activated and related to the corresponding winning input. Such structures may also be involved in assessing the level of discrepancy of new incoming input with that predicted from later parts of activated stored pattern sequences. These aspects were considered [46, 47], in terms of possible networks which could perform pattern matching and sustain competition. One of these involved the nucleus reticularis thalami (NRT), a sheet of mutually inhibitory neurons interposed between thalamus and cortex

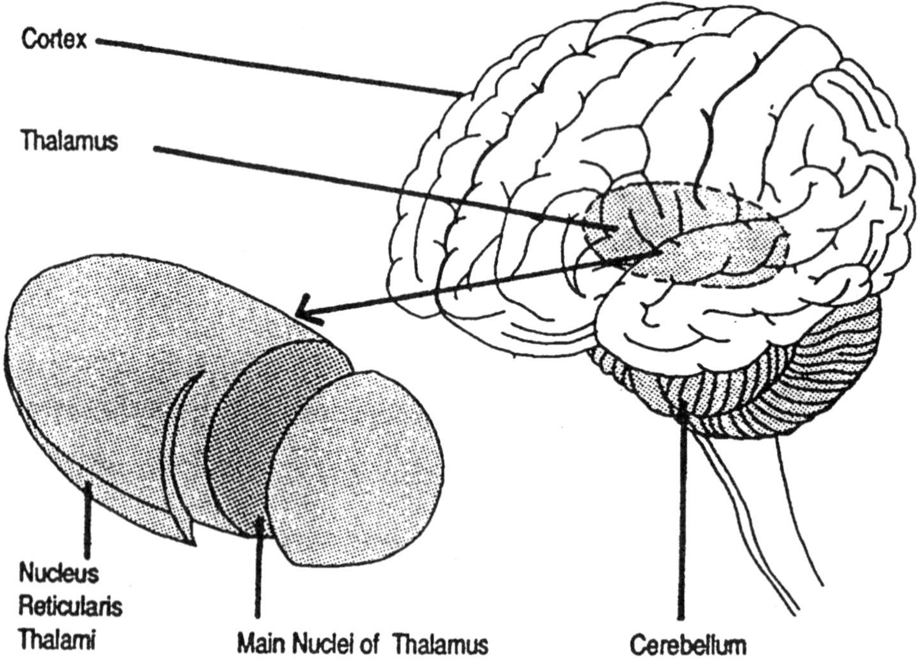

Figure 9.2: The nucleus reticularis (NRT), a sheet of inhibitory cells surrounding the thalamus, and suggested as acting as a global 'gate' to control cortical activity and support global competition between sites of lasting activity (working memories).

It was suggested that the thalamus-nucleus reticularis thalami(NRT)-cortex complex may support such activities. This is due to the fact that the NRT is composed almost entirely of inhibitory neurons which are fed by activity coming from thalamus up to cortex and also by reciprocal cortico-thalamic connections. Since NRT also sends inhibitory inputs down onto the thalamus, it is clear that such inhibition (which could also function as a release from inhibition if the NRT targets inhibitory interneurons in the thalamus more effectively than the excitatory relay cells to the cortex) could exert a powerful control influence on cortical activity. This has been shown experimentally to be the case, with global effects of the NRT sheet especially being observed in the manner in which NRT controls the nature of cortical patterns of activity in sleep. There is also some evidence for a similar global form of control by NRT on allowed cortical activity in the non-sleep states [40, 56].

It is possible, in a general manner, to understand this global control achieved by NRT of cortical activity in the following manner. Since any localised activity on the NRT will try to damp down activity on it elsewhere by lateral inhibition, the NRT can sustain 'bunched' spatially inhomogeneous activity, in which competition between neighbouring thalamic or cortical inputs onto it is occurring. This spatially structured activity may occur globally over the whole NRT sheet if it is well enough connected laterally, as is seen to be the case in spindle generation in sleep [42]. In this manner the NRT may function as a global controller of cortical activity. As such it appears of great relevance to include in models of the control circuitry for consciousness [46, 47, 49]; [1, 2, 3]; [7, 8]; [19]; [21]. The structure of the neural network used in the simulation is shown in Figure 9.3, and the wave-like activity on NRT arising from a simulation is shown in Figure 9.4.

We now turn to the question of modelling the more detailed emergence of consciousness at a local level. That requires us to explore more fully the working memory sites which have been mentioned earlier. We will present evidence for the existence and positioning of these sites and then turn to a possible neural model explaining some of their features.

9.6 Consciousness and Sites of Working Memory

9.6.1 Psychological Bases for Working Memory

Phenomenal awareness is proposed as initially arising in the relevant sites of working memory in a given modality and code. These sites are composed of modules with activity persisting over seconds of time. The modules can be considered as made up of neurons each with slow decay (through leakage of activity from its surface), although we will later consider a model in which local groups of neurons with strong recurrent connections have extended activity caused by the recurrence. In either view, competitions between different interpretations of inputs, such as might arise in vision when different inputs are presented separately to the two eyes (so-called binocular rivalry, in which the two inputs lead to a succession of switches between experience of the corresponding concepts), are able to run their course due to this extended activity. Such extended activity will also allow a suitable level of context to be able to be included in the processing, in agreement with the values obtained for these sites by psychological methods.

The existence of such working memory sites was first postulated over 20 years ago [11] on the basis of psychological evidence. Short term list learning,

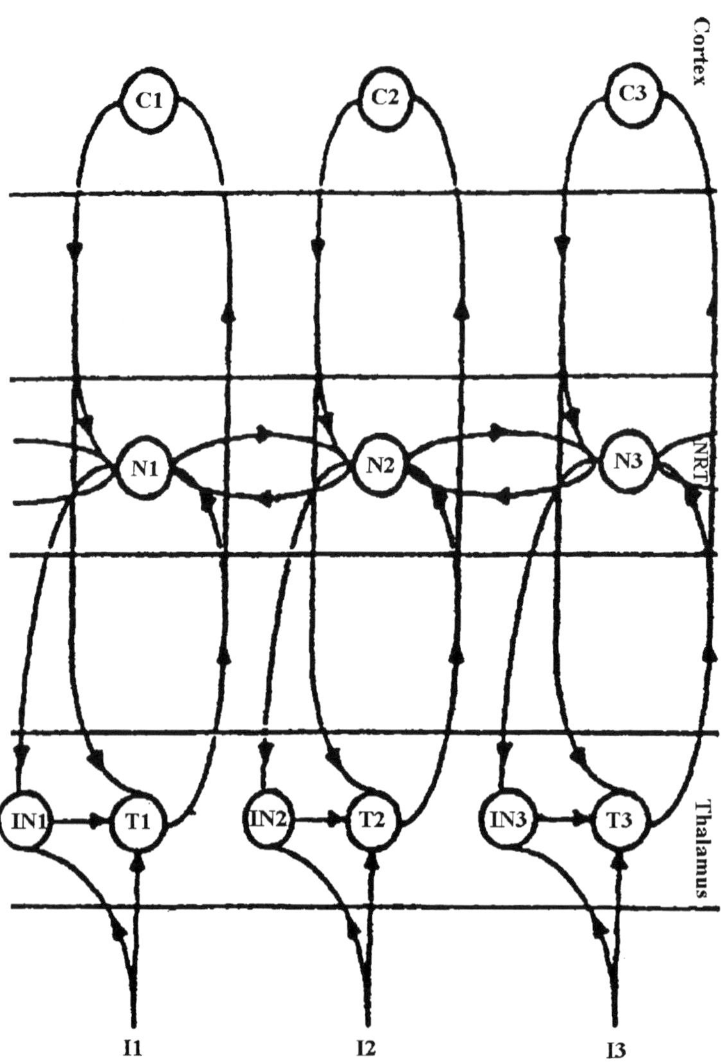

Figure 9.3: The detailed connectivity of the cortical/ NRT/ thalamic network used to simulate various aspects of the competition between different regions of cortex. Input enters thalamus at I, feeding directly onto the thalamic cells T and indirectly to the inihibitory interneurons IN. The outputs of the thalamic relay cells are then sent to the cortical cells C, with collaterals to the NRT; cortico-thalamic axons also give collaterals to the NRT.

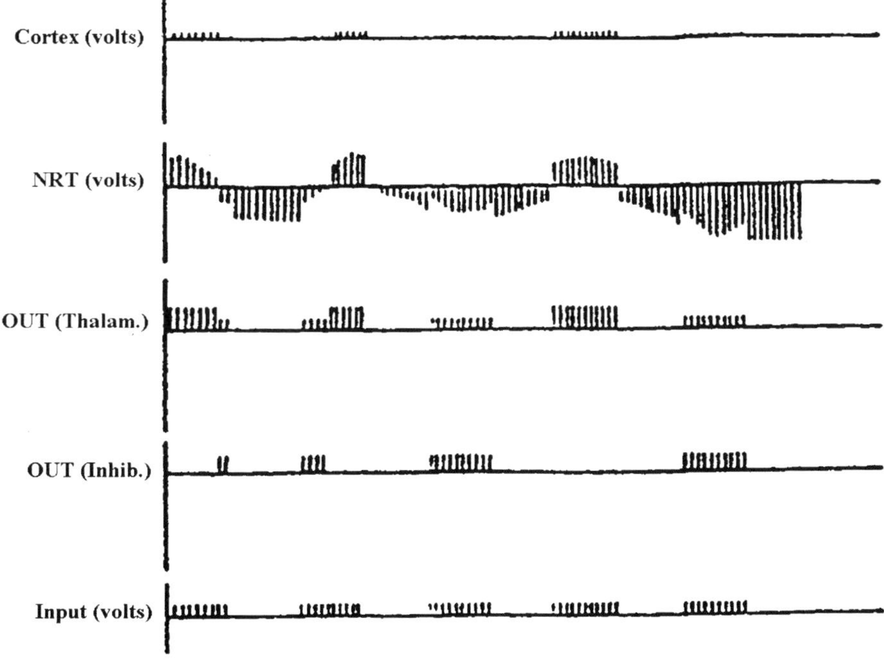

Figure 9.4: Results of a simulation of a line of a hundred cells connected as shown in Figure 9.3. Input enters the thalamic relay cells at I, and produces a global wave-like pattern leading to global influences over the whole of NRT, and so over cortex.

for example, indicated a short term memory span of about 2 seconds. In the case of auditory word input, it was suggested as being first processed up to a semantic level in Wernicke's area (BA 22), and then sent to the phonological store which acted as a buffer. The presence of such a mode of action of a dedicated site in a particular code has been put on an even surer psychological basis since then [10, 9]. The nature of the working memory site required the persistence of activity for about 2 seconds, in the case of the phonological store for words, when rehearsal by the executive component (presumed to be in the frontal lobe, and very likely identifiable at least in part with Broca's area) was disallowed by requiring a subject to perform another task such as repeating the word 'the'.

A similar site of extended activity was also postulated as existing in vision, and termed the visuo-spatial sketchpad. Psychological evidence for this has also been presented in [10, 9], although there were hints that there might be more than one such working memory site, with one for objects and one for spatial tracking. These sites were all shown to have temporally limited duration for holding their content when rehearsal was prevented by loading other tasks in the same modality. The possible dissociation of the visual stream is now better understood, so that the possibility of multiple working memories in vision must be accepted.

After the input has been processed up to a suitable level it accesses the appropriate working memory. It is then subjected to further processing on the working memory module, in particular [48] undergoing a suggested process of competition locally on the working memory site with activations from earlier inputs. The local competition itself is so constructed as to enable the most appropriate interpretation (in terms of neural representations activated most strongly), if the input is ambiguous to be singled out by means of past activity traces held on the module. Such a process will lead to competition between neural activity representing contradictory inputs. Thereafter the results of such competition would be available for use as the basis for reportability and related processes in other parts of the brain. In other words they would gain access to the postulated 'global workspace' of [6], like skaters gliding out onto the smooth ice of the ice-rink after they have waddled in an ungainly manner to the rink. They have acquired their content as part of preprocessing and of winning the competition on working memory to gain access to the smooth arena of the global workspace.

The initial stage of entering the global workspace is clearly of great interest. Indeed the present approach allows a better understanding to be obtained of the global workspace itself. For these working memory sites may be identified with the areas of so-called heteromodal cortex [32], which receive input from several modalities and which are known to be well connected to other similar sites to achieve such global broadcasting, and are to be regarded as being at the topmost level in the cortical processing hierarchy from the input. Thus it is possible to begin to identify the global workspace with the well-connected heteromodal regions of the cortex. It will be argued shortly that such sites are those with highest density of cells in the infra-granular layers 2/3 of the cortex;

these, it is proposed, are the areas where the 'raw feels' at the basis of conscious experience are expected to emerge.

9.6.2 PET Studies of Working Memory

Non-invasive instruments (EEG, MEG, PET and fMRI) are throwing new light on the working of the brain [39]. In particular, for working memory, PET results [16, 41, 35] on the positions of these sites are summarised in Table 9.1. This justifies their identification with the working memory sites introduced on a psychological basis, as discussed in the previous sub-section.

Task	Brain Area	Task/Code
Spatial	Left 40 (posterior parietal)[41]	Object shape discrimination
	Right 40/19 (occipito/parietal)[41]	
	Right 19 (occipital)[16]	Orientation discrimination
Object	Left 40 (posterior parietal)[41]	Object discrimination
	Left 37 (inferotemporal)[41]	
Phonemes/Words	Left 40 (posterior parietal)[35]	Word disambiguation

Table 9.1: PET results on posterior sites of working memory.

The PET studies of [41] showing dissociation of spatial and object short term memory was achieved by a number of different experimental paradigms. In one experiment the positions of a collection of dots on a screen had to be remembered for a period of 3 seconds, to be checked against the position of a circular probe at a later time. This was contrasted with the use of a set of geometrical objects, whose shapes had to be held in short term memory for 3 seconds before being compared with a further object presented at the centre of fixation. As always suitable control tasks were used which involved passive processing at the earlier stages; the resulting PET activity was then subtracted from the results of the short term memory trials.

The other PET study of spatial working memory [16] referred to in Table 1 considered a discrimination made by the subject between an earlier square-wave circularily shaped grating and one presented 350 msecs later. The most active region (after removal of suitable control activity) was found to be in the right superior occipital area 19 (laterally). This is suggested to contain "the additional computation involving the short- term memory and matching required by the TSD (temporal same-difference) task". It is also noted that there is a dissociation of the areas involved in such processing for different attributes or codes.

As cited in Table 9.1 there is good evidence for the existence of the phonological store in area 40 in the left cerebral hemisphere [35]. This fits in well with the psychological evidence on working memory in word processing described briefly above. In these experiments stimuli were presented visually and the subjects did not speak. Two experimental situations were used so as to dissociate the total articulatory loop (which includes the rehearsal system and not just the short term buffer) into the rehearsal system and the short term buffer, the phonological store. First a string of letters was remembered, engaging all of the articulatory loop; this activity had subtracted from it a second task involving a rhyming judgement of letter strings (which engages the subvocal rehearsal system but not the phonological store). The result gave area BA40 unequivocally as the site of the phonological store.

Table 9.1 is seen to support the notion of separate working memories in various visual codes or attributes, at least for orientation, for spatial position and for object shape. Such separation of working memories is also valid across modalities, as is also clear from Table 9.1.

9.7 Awareness of Words

The psychological and neurophysiological support for the existence of a localised site for the emergence of primary awareness in the case of word processing has been given above, the site itself being the phonological store. It is certainly the case that the evidence given so far does not prove that consciousness actually emerges in that (or any other) working memory site, although it is supportive of such a claim. In order to progress towards further justification of the claim it is necessary to develop a model of the processing on such a specialised site and compare the features of the model with those purported to obtain for phenomenal awareness itself. It is the purpose of this section to describe a possible model of this type for the emergence of consciousness of words.

In order to develop the model it is necessary to consider experiments which explore the manner in which earlier inputs may affect the response to later ones. In particular, manipulation of these earlier inputs, so that they could be either perceived or not, allows for various aspects of the emergence of awareness of the word inputs to be uncovered.

A natural class of experiences to probe are those termed 'subliminal', meaning they are below the level of consciousness but yet with information having been perceived at a nonconscious level. There has been considerable dispute about the existence of subliminal perception [20] but there are various experiments which have shown that subliminal perception does indeed occur [31]. The relevant experiments determine how subliminally processed data may affect later processing at a conscious level, and so allow the actual emergence of consciousness at the later time to be analysed experimentally..

An indication of how such awareness might emerge may be gleaned from the analysis of experiments on subliminally processed lexical decision data [30]. A subject had to make a decision as to the word or non-word character of a lexical string. A sequence of three such strings constituted a given trial; the second letter of the string was manipulated, by masking, so as to be either at conscious or at subliminal level or so as to be completely invisible to the subject. It was found that reaction time to the decision as to the lexical character of the third word was affected by the semantic relationship to the earlier words in such a manner that apparently all possible semantic codes were accessed by the second word. However these were not at a conscious level. Only one meaning of a polysemous word (when presented second in order) was accessed at conscious level, since it was then found to delay response to a word of opposite semantic character.

There is also support for the result that a number of interpretations of ambiguous words are activated non-consciously from lexical priming experiments, such as those of [29]. Thus, in what is called 'cross-modal priming', where an auditorily presented sentence containing an ambiguous word is paired with a visual probe word, there is significant facilitation in responding to targets related to either alternative meaning. This speed-up only happens within 200 msecs or so of the time of presentation of the ambiguous word. After that only the contextually appropriate meaning of the ambiguous word facilitates response to probe words. As they state, "comprehenders briefly activate multiple senses of ambiguous words even in clearly disambiguating circumstances". Moreover such activation occurs at a non-conscious level.

A model of such processing was suggested [50], based on the fact that all possible semantic meanings of a word are accessed at unconscious semantic level (as emphasised by [30] and supported by the work reported in [29]) in a semantic memory module. Only in a later working memory site is there a competitive process, assumed present so as to prevent activation of contextually incongruous word meanings and their emergence into awareness. It is here that the longer persistence of activity in working memory comes crucially into its own; previous activity still held on the working memory is used to give constraints on present winners. The competition arises from a small amount of lateral inhibition between word nodes for words with semantically opposed meanings, giving one or the other a competitive edge. On the other hand there was also lateral excitation between a consciously processed earlier word and a later semantically related one.

In detail, the model was composed of two modules, one for semantic encoding at a preconscious level and the other, the working memory module, for consciousness to emerge through excitation of the appropriate area of neuronal activity above a criterial threshold level. The nodes on each of the modules were assumed identical and, in the detailed mathematical analysis, taken to be dedicated, one to a word meaning. However, similar results are to be expected if the coding in each of the modules was more distributed but yet still topographically organised, with separate areas of activation for semantically different words. Once a node was activated on the semantic net it fed excitation forward to its identically coded partner on the working memory net. There was also feedforward excitation from the semantic memory module to the working memory module between nodes coding for semantically similar words. Only on the working memory net could any competition arise through lateral inhibition, that being from an active (firing) node to one which is semantically contradictory. On the other hand there was assumed to exist lateral excitation between nodes which are semantically similar. The lateral interactions appear as if they could arise from a 'Mexican hat' type of function. It was through the excitatory lateral interactions (both feedforward from the semantic module to the working memory one and laterally between nodes on the working memory module) that reduction of the reaction time took place due to subliminally processed competing second words in the three word string observed by Marcel [30]. Slowing down of the reaction time from a consciously processed second word arose, in the model, from the inhibitory lateral interactions between nodes on the working memory net.

9.8 Activity 'Bubbles' in Cortex

9.8.1 General Discussion

Some hint as to the neurophysiological features possessed by modules which hold activity over a time considerably longer than expected can be developed by considering some of the details of the relevant cortical microcircuitry. This may be discussed in terms of recent attempts to understand the origin of orientation or direction sensitivity of cells in early visual cortex in terms of the canonical microcircuit (Douglas and Martin, 1990). Recurrent excitation (and associated inhibition) was suggested as amplifying much weaker orientation sensitivity arising from thalamic input. The principle behind this amplification was already explored 20 years ago [4], in terms of the formation of bubbles of activity in local cortical regions due to the recurrence of neural activity. In the original model of Amari these bubbles persisted until disturbed by later competing input. It is more natural to assume that single cell adaptation (increasing threshold for cell firing due to fatigue effects and from suitable inhibitory ionic currents activated after cell firing), and direct inhibition from local circuit neurons, causes these bubbles to turn off after a specific time; this effect was noted explicitly in the model of Suarez et al [43].

The relevance of this use of recurrence in cortico-cortical circuits to understand increased effective time constants of 'working memory' neurons is as follows. Recent analyses of the cyto-architecture of the cortex indicates that the higher regions, in the sense of Mesulam [32], have an increased density of cells in layers 2/3; in particular the results of Barbas and Pandya [13] give a clear indication of this for frontal cortex, with the areas 8, 9 and 46 having highest density for these layers. Similar results are known for the multi- modal associative regions of parietal and temporal lobe. It is in these latter regions that the working memories of Table 9.1 are sited.

It can be seen by a simple neural network argument that the longest persistence of neural activity due to recurrence of activity will arise from regions with a highest cell density. Cortical cells mainly have synapses which arrive from other cortical cells, and of those the highest proportion is of a local circuit kind. Neglecting the 10% or so inhibitory cells, then the higher the density of cells the higher the number of synapses onto a given one (assuming all cells contact each other with the same probability). Even if the total number of synapses on cells is limited (due, say, to no more cell surface being available to new synapses) there will still be a greater probability of activity being circulated around in cortical regions with greater cell density. On both counts it is expected that the lifetime of recurrent activity will be longer the more nerve cells are packed into the cortex.

To conclude, the lifetime of the bubbles produced in the cortex by input are seen to be largest for those areas of cortex which have a highest density of recurrent connections, and therefore, one can conclude, for those areas with the highest density of cells in layers 2/3. As stated earlier those areas are the multimodal ones in the temporal-parietal, the inferotemporal and the frontal areas. In particular, area 40, including the phonological store, has already been shown to be the site of one such working memory store [35]. Others have been searched for in vision, and at least the visuo-spatial sketchpad detected [41]. There is also evidence of the increase of effective time constants of the neural circuitry as one ascends the processing hierarchy from the primary to the associative auditory cortex [28].

9.8.2 Technical Results

The general program of Continuum Neural Field Theory (CNFT) produced important insights from the work of Amari and his colleagues (as summarised in [5]). In particular the nature of the 'bubbles' in their dependence on parameters of the module on which they are created and in their stability was carefully discussed; learning of afferent synapses was shown, by way of input-dependent bubbles, to lead to a topographic map of the inputs onto the neurons of the module. Further extensions of that work have occurred [57].

These ideas have been applied to umasking of somatosensory inputs [36] and the modification of somatosensory topography [37]. Applications have also been made to the manner in which Nitric Oxide may be used in the context of CNFT [23]. There has also been an attempt to attack the problem of visual processing by means of CNFT [53], in which a variety of features of visual processing were considered from this viewpoint.

One of the most interesting of these was that of apparent motion, in which a spot of light shone briefly on one point on a screen and then very soon after (within 150 msecs) on a nearby one (within 5 or so visual degrees) causes the experience in the viewer of a continuous track of light from the position of the first spot to that of the second. There appears to be physical continuity of the track of the spot in the cortex since a spot of one shape is experienced as being continuously altered along the observed but non-existent track. That has been duplicated by simulation by means of a suitably connected module, as is shown in Figure 9.5 [38].

Further analysis of a number of visual phenomena, in particular the stabilised images and especially their colored version, and of transparency and binocular rivalry [53] allow for aspects of higher level visual coding to be analysed by means of CNFT. In this manner it is hoped that an analytic framework for the many features of visual processing can be obtained.

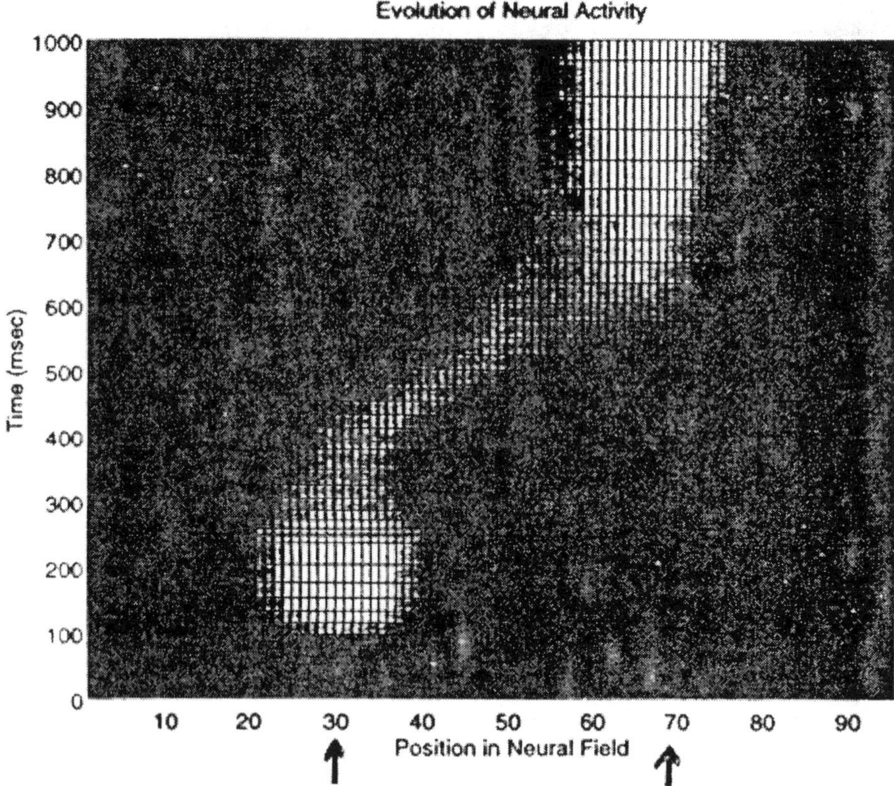

Figure 9.5: Results of a simulation of apparent motion effect, using a continuum neural field with Mexican hat lateral connections. A localised region of activity centered at position 30 is set up at t = 0. This is removed at t = 300, when an other region of activity is set up at position 70. A continuous track of activation is seen to spread from the first to the second spot of activity, so corresponding to an apparent motion effect.

9.9 The Emergence of Qualia

Let us restate the model presented so far for the emergence of phenomenal consciousness. The nature of primary awareness of the various modalities is suggested to occur after coding at a lower order feature analysis level. There may be a number of such stages, both serial and in parallel. Consciousness only then arises in those modules which possess sufficiently long time constants (from the recurrent excitation) and suitably strong inhibitory/excitatory connections for intra-module competition so as to allow a unique winner to emerge, by its activity being above some criterial level. The model has been suggested by the empirical fit to Marcel's [30] data for words, but can be used to give a more qualitative explanation of the emergence of visual awareness, as explored through experiments on binocular rivalry [27] and on opponent processing [22], as discussed in more detail in Taylor [51].

At this point the model may be regarded as promising, yet there are numerous unresolved questions which must be explored further to help flesh it out. The most important one is how the properties of qualia can be seen to arise from the specific process of neuronal computation in the model. Especially why should consciousness arise from a simple thresholding of activity on a working memory module? There must be more to consciousness than that! In particular how may the properties of qualia, such as those of transparency, ineffability and presence [33] be expected to occur?

These properties for qualia ([14]; see also [12]) are presented in Table 9.2, and will be discussed in turn. The first of these is 'presence', and includes the sense of the subjective present. The involvement of self is not part of our discussion in this section, but the temporal component of phenomenal awareness is. The latter has three characteristics, as delineated in Table 9.2: persistency, latency and seamlessness. The first of these, that of the temporal extension of phenomenal experience, is a very important component of the sense of the 'subjective now'. Such persistence is expected to be an intrinsic part of the process of winning a competition. This is the opposite side of the coin to latency (the delay of the onset of consciousness after input occurs); until another winner arises, the earlier one is still 'king'. In the competitive model, latency will arise as the time needed for a competition to be won on a working memory site. There is also the further property that we will call 'seamlessness': that the transition from one content of awareness to another happens rapidly enough to provide a sense of continuity in awareness.

Both latency and seamlessness were probed by Libet [24, 25, 26]. In the former it was shown that about 500 msec is needed to cause the 'artificial' arousal of consciousness by direct electrical stimulation of somatosensory cortex. It was also noted by Libet [26] that the change-over process of consciousness was brief, occurring in less than a tenth of the time it took for the total process of changeover. That this could occur, along with the detailed features of the dependency of the length of the latency on the stimulation parameters, was shown in a neural system by the simulation of a simplified version of the Thalamus/NRT/Cortex system ([2, 3]; [49]). Thus seamlessness arises in the com-

Qualia		Working Memory Activity
1. Presence	* persistence * latency * seamlessness	* persistence * latency * rapid change-over
2. Transparency	* can 'look through them' * fully interpreted	* smooth broadcasting to each other * at highest coding level
3. Ineffable	* infinitely distant * intrinsic	* one-way creation * at highest level, with no visible back-up
4. Uniqueness		* one winner
5. Bound		* correlation between different codes at different levels
6. Perspectival		

Table 9.2: Comparison of properties of qualia with those of working memory activities.

petitive model provided the competition, once started, is concluded suitably rapidly.

The next notion, of transparency, may be seen as arising from the well-connected system of working memories, for which there must be easy transfer of correlated information once a winner has occurred. Such transfer would also carry along with it correlated preconscious material across different codes, since there are good connections between modules at the same level of complexity, defined earlier in terms of either transience of activity or cell density in layers 2/3 [32].

It is important to follow up the feature that is posited as being played by anterior attention in the above emergence into awareness of activity on sites of working memory. If such attention is not being used then the working memory activity will not have the back-up of its related underlying preconscious activity which produced it in the first place. Both the working memory and the preconscious activity are posited as being needed for laying down a permanent memory of the event, so that it can be recalled in suitable detail and at various scales. Such encoding may be unavailable if only working memory itself is available, without its back-up; this may explain the lack of memory of any awareness experienced whilst performing so-called automatic acts, such as driving a car, in spite of there being phenomenal experience itself during the activity.

There is also the 'fully interpreted' character of qualia claimed to be part of transparency [33]. Such a property could arise from the fact that the working memory sites are at the highest level of the posterior processing hierarchy. Thus all the coding and 'interpretation', in informational terms, has been achieved when the competition has been won and incompatibilities ironed out. Indeed it would seem that such competition is indeed the final step in the interpretation process, achieved, it is claimed in the model presented here, as part of awareness being reached.

The third property of qualia listed in Table 9.2 is ineffability, that qualia are impossible to probe and 'get behind'. They are 'atomic' or intrinsic. This property is well mirrored in the one-way character of activation on working memory sites posited in the model. The properties of uniqueness and boundedness have already been briefly discussed , and have been included in Table 9.2 to make it more complete. Perspectivalness is beyond the scope of the present discussion, but is considered in some detail in [54].

Thus in conclusion all of the characteristics of qualia suggested by Metzinger (except perspectivalness) have been seen to be mirrored in the properties of activity emerging onto the well-coupled working memory system of heteromodal cortical sites in the posterior cortex. This gives support to the suggestion that the model is indeed one for the emergence of phenomenal awareness. In particular it may help towards answering the 'what is it like to be?' question for animals with no sense of self.

The most important feature of the technical analysis is the possiblity of proving that the lifetime of adapting neural activity in a recurrently connected module increases as the cell density is increased, other parameters being held constant. This analysis can be achieved mathematically, and leads to a prediction of the particular cell density which produces adapting, but input independent, bubbles. As noted earlier there is observed to be an increase of cell density, especially in layers 2/3, in higher cortical regions. The main prediction of the analysis is thus in support of the main thesis of this section:

Consciousness emerges where continued activity is supported to occur

This is not a new thesis, but has been backed up both by data and by exploration of the model at a deeper level than before so as to give an initial physical explanation of how to bridge the gap between brain activity and mental experience. Moreover it identifies the sites of such emergence explicitly in the brain as being those of working memory posited in earlier sections as being the site of consciousness. Our results reported here therefore support and amplify this thesis, in a very detailed neural manner, in terms of CNFT.

9.10 Discussion

The competitive relational model of the mind is testable in a variety of ways. This was already indicated earlier, with observable localised sites of continued activity being observable when conscious processing is occurring as compared to much lower activity in these sites when it is not. These sites should coincide with those of highest neuronal cell density in layers 2/3, and it should be possible by means of single cell recording to determine the parameters in local circuits so as to compute (by means of the approach of CNFT discussed in section 9.6) the time of temporal duration of neural activity there; this should coincide with the observed duration by non-invasive techniques.

Activity should also be able to be observed as part of a competition being run between these different sites of activity. There should be evidence of this in terms of correlations between the activities on the two (or more) sites of working memory involved, in particular with the concurrent (time-delayed by about 3 msecs) activity on the relevant part of NRT (so close to thalamic sites). This correlation would be expected to occur until one or other of the sites of working memory has its activity reduced below a critical value; at that point the site would no longer be supporting consciousness.

The possibility of being able to perform some of these measurements using combined fMRI, EEG and MEG techniques is now becoming more realistic [34]. It is also testable in the phenomenon of neglect. Non-invasive MEG measurements should ultimately allow details of the essential circuitry for consciousness to be ascertained. The possibility of such use of non-invasive instruments allows one to propose a program of experiments to crucially test the ideas developed here as part of the Relational Mind model. This involves experiments designed to compare conscious with nonconscious (subliminal) processing activities in various modalities. A set of experiments has been selected for this program [55], and various parts of it are under way. We are entering an exciting period in brain research in thinking that we may be able to probe our very intrinsic nature.

9.11 Bibliography

[1] Alavi F and Taylor J G, "A Simulation of the Gated Thalamo-Cortical Model", pp 929-932, in *Artificial Neural Networks 2*, ed Aleksander I and Taylor J G, North Holland, Amsterdam (1992).

[2] Alavi F and Taylor J G, "Mathematical Analysis of a Global Network for Attention", in *Mathematical Approaches to Neural Networks* ed J G Taylor, Elsevier (1993).

[3] Alavi F and Taylor JG "A Global Competitive Neural Network", Biol Cybernetics 72, (1995) 233-248.

[4] Amari S-I 'Dynamics of pattern formation in lateral-inhibition type neural fields' Biol Cybernetics 27 (1977) 77-87.

[5] Amari S 'Dynamical study of formation of cortical maps' pp 15-34 in *Dynamic Interactions in Neural Networks: Models and Data*, MA Arbib and S Amari, eds, New York: Springer-Verlag (1989).

[6] Baars B J, A Cognitive Theory of Consciousness, Cambridge University Press (1988)

[7] Baars BJ and Newman JB 'A Neural Attentional Model for Access to Consciousness : A Global Workspace Perspective', Concepts in Neuroscience 4 (1993) 255-290.

[8] Baars BJ and Newman JB 'A Neurobiological Interpretation of Global Workspace Theory' pp 211-226 in Consciousness in Philosophy and Cognitive Neuroscience, ed A Revonsuo and M Kamppinen, Hillsdale NJ : Lawrence Erlbaum (1994)

[9] Baddeley A D, "Working Memory; The interface between memory and cognition", J Cogn. Neurosci. 4 (1992) 281-288

[10] Baddeley AD, Working Memory, Oxford : Oxford University Press (1986)

[11] Baddeley A D and Hitch G, "Working memory" in Bower G A (ed) The Psychology of learning and motivation, 8 (1974) 47-89, Academic Press, New York.

[12] Bailey A, 'The baby food definition of 'qualia", Abstract # 29, Consciousness Research Abstracts, Towards a Science of Consciousness, Tucson, and preprint, Univ of Calgary (1996)

[13] Barbas H and Pandya DN, 'Patterns of Connections of the Prefrontal Cortex in the Rhesus Monkey Associated with Cortical Architecture' ch 2 in Frontal Function and Dysfunction, eds HS Levin, HM Eisenberg and AL Benton, Oxford : Oxford University (1992) Press.

[14] Dennett DC, 'Quining Qualia' in Consciousness in Contemporary Science, Marcel AJ and Bisiach E (eds), Oxford : Oxford University Press (1988)

[15] Douglas RJ and Martin Kac, 'A Functional Microcircuit for Cat Visual Cortex', J Physiol. 440 (1990) 735-69.

[16] Dupont P, Orban GA, Vogels R, Bormans G, Nuyts J, Schiepers C, De Roo M and Mortlemans L, 'Different perceptual tasks performed with the same visual stimulus attribute activate different regions of the human brain', Proc Natl Acad Sci USA 90 (1993) 10927-10931.

[17] Fuster J, 'Frontal Lobes', Current Biology **3** (1993) 160-165

[18] Goldman-Rakic P, 'Working memory and the Mind', Scientific American (1992) 73-79.

[19] Harth E, 'The Sketchpad Model', Consciousness and Cognition 4 (1995) 346-368.

[20] Holender D, 'Semantic activation without conscious identification in dichotic listening, parafoveal vision and visual masking: A survey and appraisal', Behavioural and Brain Sciences 9 (1986) 1-23.

[21] Kilmer, Neural Networks (1996) (in press).

[22] Kolb FC and Braun J, 'Blindsight in normal observers' Nature 377 (1995) 336-8.

[23] Krekelberg B and Taylor JG, 'Nitric oxide in cortical map formation' J Chem Neuroanatomy 10 (1996) 191-196.

[24] Libet B,Alberts W W, Wright Jr E W, Delattre D L, Levin G & Feinstein B, "Production of Threshold Levels of Conscious Sensation by Electrical Stimulation of Human Somato-Sensory Cortex", J Neurophysiol 27 (1964) 546-578.

[25] Libet B, Wright Jr E W, Feinstein B and Pearl D K, "Subjective Referral of the Timing for a Conscious Sensory Experience", Brain 102 (1979) 193-224.

[26] Libet, private communication (1994)

[27] Leopold D and Logothetis N, Nature 379 (1996) 549-553.

[28] Lu Z-L, Williamson SL and Kaufman L, 'Human auditory primary and association cortex have differing lifetimes for activation traces', Brain Research 572 (1992) 236-241

[29] MacDonald MC, Pearlmutter NJ and Seidenberg MS, 'Lexical Nature of Syntactic Ambiguity Resolution', Psychological Review 4 (1994) 676-703

[30] Marcel A J, "Conscious and preconscious recognition on polysemous words: Locating selective effects of prior verbal contexts" in Nickerson RS ed, *Attention and Performance VIII*, Erlbaum, Hillsdale NJ (1980)

[31] Merikle PM, 'Perception without Awareness' American Psychologist 47 (1992) 792-5

[32] Mesulam MM, 'Patterns in Behavioural Neuroanatomy : Association areas, the limbic system and hemispheric specialisation', Ann Neurol. 10 (1981) 309-325.

[33] Metzinger T, 'The Problem of Consciousness' pp3-40 in *Conscious Experience*, ed Metzinger T, Thorverton UK: Imprint Academic (1995)

[34] Muller-Gartner H, private communication (1996)

[35] Paulesu E, Frith C and Frakowiak RSJ, 'The neural correlates of the verbal components of working memory', Nature 362 (1993) 342-5

[36] Petersen R and Taylor JG, 'Unmasking in somatosensory cortex', Neural Information Processing (1995)

[37] Petersen R and Taylor JG, 'Stimulus-led adaptive changes in somatosensory cortex', Comp Neuro Science Proceedings (1995)

[38] Petersen R and Taylor JG, 'Apparent motion from Neural Field Theory' (in preparation) (1996)

[39] Posner MI and Raichle ME, Images of Mind, New York : WH Freeman (1994)

[40] Skinner JE and Yingling CD (1977) 'Central gating mechanisms that regulate even-related potentials and behaviour' pp 30-69 in Desmedt JE ed, *Progress in clinical neurophysiology : Attention, voluntary contraction and event -related potentials*, Klinger I. (1977)

[41] Smith EE and Jonides J, 'Working memory in Humans : Neurophysiological Evidence', ch 66, pp1009-1020 in *The Cognitive Neurosciences*, ed M Gazzaniga, Cambridge MA : MIT Press (1995)

[42] Steriade M, Gloor P, Llinas R R, Lopes da Silva F H and Mesulam M M, "Basic Mechanisms of Cerebral Rythmic Activities", Electroenc. and Clin. Neurophysiol, 76 (1990) 481- 508.

[43] Suarez H, Koch C and Douglas RJ, 'Modeling Direction Selectivity of Simple Cells in Striate Visual Cortex within the Framework of the Canonical Microcircuit', J Neuroscience 15 (1995) 6700-19.

[44] Taylor J G, 'A model of thinking neural networks', Seminar, Institute for Cybernetics, Univ of Tübingen (1973)

[45] Taylor J G, 'Can neural networks ever be made to think?' Neural Network World 1 (1991) 4-12.

[46] Taylor J G, 'Towards a neural network model of the mind', Neural Network World 2 (1992) 797-812.

[47] Taylor JG, 'A Global Gating Model of Attention and Consciousness', in Neurodynamics and Psychology, ed. Oaksford M and Brown G, Academic Press, New York (1993)

[48] Taylor JG, 'The Relational Mind' in *Neural Networks*, ed Browne A, Inst of Physics Press (1996)

[49] Taylor J G, 'A Global Competition for Consciousness?', Neurcomputing 11 (1996) 271- 296

[50] Taylor J G, 'Breakthrough to Awareness' Biol Cybernetics 75 (1996) 59-72

[51] Taylor JG, 'Where and how does consciousness emerge?', King's College preprint (1996)

[52] Taylor J G, 'Modelling what it is like to be', in Proc Arizona Conf on Scientific Basis of Consciousness, ed Hammeroff S , Cambridge MA : MIT Press (1996)

[53] Taylor JG, 'Modeling Cortical Processing', invited talk, WCNN96, San Diego (1996)

[54] Taylor JG, 'The Race for Consciousness', Boston: MIT Press (1997)

[55] Taylor JG and Muller-Gartner H (1997), 'Experiments for Consciousness', in *Special Issue on Consciousness*, Neural Networks (to appear)

[56] Villa A, 'Influence de L'Ecorce Cerebrale sur L'Activite Spontanee et Evoque du Thalamus auditif du Chat', These, Univ de Lausanne (1988)

[57] Zhang J, 'Dynamics and Formation of Self-Organising Maps', Neural Computation 3 (1991) 54-66

Chapter 10

Confusions about Consciousness

10.1 Consciousness Studies

Much of the confusion about consciousness is generated by lack of clarity on the issue of *dualism*. The majority of scientists who are caught up in the current excitement about consciousness studies would probably deny that they are dualists, if the question were put to them explicitly. But at the same time I think that many of them are closet dualists. They strive to resist the temptations of dualist thinking, but as soon as their guard drops they slip back into the old dualist ways. The very language in which they normally pose the problem of consciousness gives the game away. "How can brain states 'give rise' to conscious feelings?" "How are conscious states 'generated' by neural activity?" The way these questions are phrased makes it clear that consciousness is being thought of as something *extra* to the material brain, even if the official doctrine is to deny this.

To help make the point clear, let me move away from the mental realm for a moment, and consider two contrasting analogies from purely physical science, the theory of electromagnetism, and the theory of heat. These two theories work rather differently. Think of how they relate heat and electromagnetism to other physical processes. The theory of the electromagnetic field is a theory of an *extra* physical entity, of something additional to other physical goings-on, such as the movement of charged particles. The charged particles are one thing, and the field they produce something further. But the theory of heat does not explain heat in a similar way. Heat is not something *extra* to the kinetic energy of moving particles. Rather, talk of the heat in a body is just another way of referring to the kinetic energy of the particles in it. There aren't two entities here, the moving particles and the heat. It's not as if a 'heat field' arises when the particles move. Heat is nothing but the movement of the particles, described in other terms.

Now, which of these is the better model for the relation between conscious feelings and brain activity? That is, should we expect a successful 'theory of consciousness' to show us how certain brain activities generate certain *extra* entities, the conscious feelings, on the model of the electromagnetic field? Or should we rather expect such a theory to show us how conscious feelings are *nothing but* certain brain activities, described in other terms, on the model of heat. When Francis Crick, for example, says that consciousness is associated with 40-Hertz neuronal oscillations in the visual cortex, or indeed when any scientist equates consciousness with any feature of brain activity, are we to understand them as saying that some *extra* conscious field is generated by the brain activity, or rather that consciousness is *nothing but* that brain activity, described in other terms?

We can call a theory of the former kind a *dualist* theory, and a theory of the latter kind a *materialist* theory. I suspect that much work in consciousness studies simply hasn't decided whether it is aiming at a dualist theory or a materialist theory. The indecision matters because it can lend such work an air of spurious excitement. This is because a dualist theory of consciousness, while it would certainly be exciting, is a highly implausible prospect. A materialist theory, by contrast, while it is plausible enough, is not going to yield any exciting secrets. So, by fudging the issue between these two kinds of hypothesis, theorists of consciousness can have their cake and eat it. They can present their work as sharing the excitement of a dualist breakthrough, yet at the same time denying that its claims are any more surprising than a materialist hypothesis. If we are seriously to assess their theories, however, we need to be told whether they are intended in the dualist or materialist mode.

To further clarify this issue, let me turn to another analogy, this time with the 'theory of *life*'. About a hundred and fifty years ago, scientists used to be excited about life in roughly the way that they are now excited about consciousness. While they were of course clear enough about which systems are alive and which not, they were much perturbed by further questions. *Why* are those systems alive? What mysterious power animates them? And why is this power present in certain cases, such as trees and oysters, and not in others, like volcanos and clouds?

These questions have now disappeared from active debate. Biology textbooks sometimes begin with a few perfunctory paragraphs about the distinguishing characteristics of their subject matter. But the nature of life is no longer a topic of serious theoretical controversy. Everybody now agrees that the difference between living and non-living systems is simply having a certain kind of physical organization (roughly, we would now say, the kind of physical organization which fosters survival and reproduction).

The reason for the nineteenth-century debate, and its subsequent disappearance, is that scientists used to be dualists about life, and aren't any longer. That is, they used to think that living systems are animated by the presence of a special substance, a vital spirit, or *elan vital*, which was postulated to account for those features of living systems, such as generation and development, which were thought to be beyond physical explanation. And of course, when they did believe in these vital forces, they then faced any number of exciting questions, such as why they arise in certain circumstances and not others, and what laws govern their operation.

But nobody is a dualist about life any longer. Nobody believes in vital spirits nowadays. A century and more of physiological research have persuaded scientists that the characteristic features of living systems can all be accounted for in principle in terms of normal physical forces, whithout bringing in any extra forces operating only in living bodies. With this realization all the excitement abouot the nature of life has dissolved. To be alive is just to be a physical system of a certain general physical kind. There isn't any extra property present in living systems, over and above their physical features, which distinguishes them from non-living systems. So there are no pressing questions about the mysterious nature of this extra property.

I think that this story about life carries a direct moral for the study of consciousness. If you think that there are special mental forces, over and above the familiar physical forces, then you will think that there are exciting questions which must be answered, such as why these forces arise in certain circumstances and not others, and what laws govern their operation. On the other hand, if you think that the cognitive workings of intelligent beings depend on nothing but the operation of normal physical forces, without any extra forces operating only in brains, then you will see things differently. You may begin your textbooks with a few remarks about the distinguishing characteristics of conscious systems, but once this essentially classificatory question is out of the way, you won't want to spend any more time agonising about the nature of consciousness.

As my remarks so far will no doubt have intimated, I prefer the latter, materialist view of consciousness to the former, dualist story. And the reason is the same as in the case of life. All the physiological evidence indicates that no special mental forces are needed to account for the operation of intelligent organisms. Of course the evidence isn't conclusive, and doesn't absolutely *prove* that there are no such forces. But it weighs strongly against them. If you are unpersuaded, then ask yourself this question. Are any parts of matter in your brain ever caused to accelerate by mental causes, in the absence of any other forces? That is, do we need to include purely mental causes alongside gravity, the electroweak force, and so on, in the category of fundamental forces? As I said, there is no conclusive disproof of this thought, and it has its defenders, like Sir John Eccles [2]. But I take it that it would run counter to a large body of empirical evidence. (Your colleagues in the Physics Department would certainly be very interested if such a force could be shown to exist.)

No doubt some of you will be feeling uneasy about the analogy with life. Don't we have immediate access to the nature of conscious activity, via our introspective knowledge of our own minds, in a way that we don't have access to the nature of life? And doesn't this show us directly that consciousness goings-on are distinct from any physical goings-on? When we are aware of a pain, say, or of seeing something red, don't we know directly that there is something going on in us which is quite different from any neuronal activity in our brains?

I agree that we all have strong intuitions to this effect. But they need to be handled with care. My own view is that they are illusory, and I will come to this in a moment. The other option is to take them at face value, as showing that conscious feelings really are distinct from brain activity. However, if you take this line, then you face the argument from two paragraphs back, that your colleagues in the Physics Department are going to be flabbergasted if it turns out that these extra conscious states sometimes cause bits of matter to accelerate in your brain.

One way to square this circle is to adopt *epiphenomenalism.* This is the view that extra conscious states do exist, additional to brain activity, but nevertheless have no effects of their own. David Chalmers defends this position in his recent *The Conscious Mind* [1]. Chalmers is persuaded by the intuition that conscious states must be distinct from physical states. Yet he is enough of a scientist to realize that it would fly counter to well-evidenced opinion to credit these extra conscious states with powers to influence neuronal activity. So he suggests that perhaps they are just epiphenomenal 'danglers', caused by certain kinds of brain activity, but with no power to cause anything themselves.

But this isn't a happy position either. When you pull your hand out of the fire, surely it is your conscious *pain* that causes you to do this. Yet epiphenomenalism denies this.

If we are to avoid the inefficacy of epiphenomenalism, without sliding back into the mysterious mental forces of Eccles-style dualism, we must resist the intuition that the conscious states are separate from brain states. That is, we need to say that pains and other conscious states are nothing but brain activities. Then, of course, we will have no difficulty understanding how pains affect behaviour. For there is no mystery about the causal route from brain states to behaviour.

Could conscious experiences really be the same as brain states? This seems perfectly coherent to me. This materialist position doesn't of course deny that it feels like something to have a pain. It simply identifies this with what it feels like to be in some brain state. *That* is what it is like, for beings who are in that kind of brain state. (What would you expect it to be like to be in that brain state? To be like nothing? Why?)

What about the strong direct intuition that brain states and feelings are quite different in kind? As I said, I think this is an illusion. We are so close to our own feelings that it is easy to get confused about them. The trouble is that we are able to think about our feelings in a special way - by *having* them - which is not available in the case of any other possible objects of thought. And this special way of thinking about feelings makes it difficult for us to see that the things we are thinking about - namely, by feelings - are just the same things as we can think about in other ways - namely, brain states.

Here I am only gesturing at a complex topic, which I treat in far greater detail in my book *Philosophical Naturalism* [3]. But I hope I have done something to indicate how there is room for materialists to resist contrary intuitions. More generally, I hope I have done something to demonstrate the virtues of materialism over the dualist and epiphenomenalist alternatives.

While I think that the acceptance of materialism will be good for the study of consciousness, it will be bad for 'consciousness studies'. For once we accept materialism we will recognize that there are not going to be any breakthroughs, any crucial discoveries about what 'causes' consciousness. That would be like discovering what 'causes' life. But of course there is no such thing to discover. All we can do is classify the different kinds of life, and try better to understand their mechanisms. Similarly with consciousness. We should stop getting excited about the spurious question of what 'causes' consciousness. Instead we should settle down to the serious business of classifying kinds of consciousness and exploring their mechanisms.

10.2 Bibliography

[1] Chalmers, David (1996) *The Conscious Mind: In Search of a Fundamental Theory* (Oxford University Press, Oxford)

[2] Eccles, John (1989) *Evolution of the Brain: Creation of the Self* (London, Routledge)

[3] Papineau, David (1993) *Philosophical Naturalism* (Oxford, Blackwell)

Chapter 11

Round Table Discussion

The Modelling and Understanding of Consciousness

11.1 Presentations

John Taylor:

We decided amongst ourselves in a truly democratic fashion that we would each have four minutes. There are seven of us, so that should take half an hour. Four minutes to have our response to each other's statements and talks that we heard this morning, and then the floor will be open and there will be from you hopefully comments, questions, or other things. So I feel that we should start immediately. I know that Susan has to go off reasonably quickly to get a train back to Oxford, so it's something that we should start straight away. The order is myself, Igor, Susan, Brian, then David, Christopher and Peter. So let me start off with my response to what I've heard so far, which I've enjoyed very much. I'd like to immediately say I thank all the speakers. It's been fantastic so far. It's aroused my consciousness. I hope it's aroused yours.

My responses can be summed up under four headings. First of all I certainly am not a dualist. I certainly am a hard nosed materialist and I was quite surprised that David Papineau should have got me so wrong. But then that's philosophers for you. They love stirring things up, and indeed that's their job as he said at the beginning of his talk. I would like to emphasise that consciousness is something that will not go away, in the sense he's talking about it, until it's properly understood. It won't go away of course 'cause we still will have it, so it won't go away until we've understood it. It's a wonderful control system, and if I look at it as an information processing modeller and theorist, it's the control system par exellence and if we want to get the ultimate intelligent machine then it's going to have to be conscious and possibly with much higher levels of consciousness than our own. So that at least I hope answers David that indeed I don't at all see that we are making a mistake in trying to model it - I at least am trying to model it.

Second point to Brian, can science be emotional? Well yes, there are lots of models that people are putting forward for the emotions, the neural basis for the emotions. They're putting them into computer systems. Not necessarily to have emotional P.C.'s, though sometimes when I come in the morning I think mine is too emotional. I sometimes think I should kick it. We're having a meeting for example on Friday to discuss with people from the Institute of Psychiatry and from St. George's Hospital models of various limbic regions, nucleus accumbens, anigdala, vta and so on, where dopamine and other things that Susan was talking about are originating, and that seems to me to be trying to get to the heart of the nature of emotions.

Thirdly, do we need neurons, do we need brains, to have consciousness? And this is the question I think raised by Igor, a very interesting question. Some people have suggested we could download ourselves into a PC and store it away, and go off, and let that PC work for us. Well, so far I would suggest that there are no other ways of getting the clues to the way consciousness is working, and there are no other systems we have other than those with neurons and brains and if we want to take those clues and use them and try and model and understand as for example Chris and his team are doing and as I will be spending more of my time being involved with these non-invasive instruments, then we have to follow the brain and understand it. Once we've cracked it, and I would say it is a crackable problem, as is I should suggest any aspect of the world in which we live of which we have experience, it is 'crackable' because it is quantifiable. I'm a mathematician - you can see the bias. Therefore we can attack it ultimately and solve it but through neurons and brains.

Finally, where to next? Well the facts, nothing but the facts as somebody said long ago. I go for the facts, I look for them wherever they are, and I try and incorporate them to test the model out and that's what science is about, and I hope that there will be those of you out there who are stimulated by what we've had today to try and move on with the facts. If you forget the facts, you're floating around and you're not able to do science and advance. O.K. I've had my four minutes and hopefully as a good chair I will shut up and I will now call on Igor who I will shut up in four minutes' time.

Igor Aleksander:

I'd actually like to address the audience and only partly my colleague on the left[1] who is going to be addressed a lot. I'd like to talk about one word which occurred both in Susan Greenfield's talk and in mine, but in very different ways, and this is the word automaton. The dictionary definition of an automaton is something that definitely doesn't have consciousness, a sort of zombie. And that's a very proper way of using it. I actually said that there is a branch of mathematics called automata theory with which you could handle this question of consciousness in all sorts of systems, possibly including the brain. Well

[1] David Papineau

automata theory, that sounds contradictory in some way. How can you take the theory of these zombies and then apply it to something that isn't a zombie? But automata theory is a branch of mathematics which has to do with structures of cellular systems, things made up of a lot of bits, and it relates the shape of these structures to what happens if you let these systems run. If you take photographs of what each cell is doing in time and you draw a picture of which state, which photograph, leads to another state, this gives you a structure which I refer to as a state space structure.

Now, turning to the colleague on my left, may I very respectfully suggest that in fact automata theoreticians and indeed many other scientists in other forms of engineering and control engineering have possibly stolen a march on the question of whether one is a dualist or a materialist. He said he couldn't quite work out whether I was a dualist or a materialist. This is because we probably have a view of this particular question which stems from being able to understand the duality between structure and function in engineering systems, which is in some ways parallel to the kind of dualism that one talks about in philosophy, the mind-body dualism. Now there are many philosophers who've actually taken automata theory very seriously. Fred Dretzke is one of them, and one or two others whose name I can't remember at the moment. But in some ways this question of are you a materialist or are you a dualist is to my way of thinking a slightly naive question and I think this is a point you yourself were trying to make, but rather than say that scientists get hung up on it and try and answer the wrong problem, one might consider that scientists and engineers have actually stolen a march on that one, and have a good view of what the relationship between structure and function, mind-body, might be.

So I simply end by saying that, you know, there may be a few lessons we can learn from one another, rather than just saying oh well, philosophers have got their own area of problems which are ring-fenced to scientists. Finally, I'd like to say that there's one thing that we've proved today, John, and that is that consciousness is concrete enough for you to get a belly full of it. I congratulate you. I've never been to a meeting where so many different and complex ideas have been pushed out one after another, and you just sat there and you look quite cheerful still, so congratulations.

John Taylor:

Wait till they have their turn. Susan, your turn now.

Susan Greenfield:

O.K. well, I, like my colleagues, have never ever been accused of closet dualism or dualism come to that matter. Perhaps I ought to amplify a bit why I still think there is a problem and I don't think it's adequate to say because there isn't a problem that solves the problem.

The problem for me, perhaps philosophers don't have this problem, the problem for me is I've held the human brain in my hands, I've got bits or could have got bits under my fingernail, I've seen the effects of drugs on the brain. Like a lot of people in this room I'm probably mystified why people take the drugs of abuse they do, or perhaps having seen people taking prozac - why do they do that? Now, for me, perhaps for philosophers this isn't a problem at all. For me this is near short of a miracle, or a curse, however you tend to look at it, that a physical brain can be influenced by molecules that we know about, that have actions to physiological levels we know about and can have these effects that they have, these subjective effects.

Now I know that philosophers usually accuse scientists of ignoring the subjectivity of consciousness and that's why I was quite surprised to be taken to task for actually pointing that out, or actually trying to retain the essence of consciousness which is its subjectivity instead of jumping on the scientific bandwagon as a lot of scientists do of simply weighing in and gleefully looking at the firing of neurons. I personally believe there's more to... well for me it's an exciting question to do that. So perhaps we could discuss that more fully later because it's unfair to criticize someone who can't answer back as he can't in these four minutes, so I'll just get on to the second issue.

I was taken to task for using words like 'emerge'. This again points to a difference between scientists and philosophers that I'm increasingly aware of. Scientists are very cavalier in the words they use because normally the force of our argument rests with the data - hence the sort of slides that one shows - whereas with philosophers the force of their argument rests usually on the logic of an abstract idea. So therefore they have to be very, very careful about definitions and terms and spend a lot of time defining terms. I therefore am quite happy to use a word other than 'emerge'. Like most scientists I'm easy. I'm not going to die in the ditch for a word like 'emerge'. If he wants us to use another word that's fine. What I'm trying to say is that I know that a single neuron isn't conscious but that a hundred billion are, and I want to know how that is, and that for me is a problem.

In the remaining time left I'd like to address another issue that was raised by some of the speakers and I know it intrigues people a lot and I suspect it intrigues a lot of mathematicians which is this issue of computation and how far that can be used towards consciousness. I think it's marvellous what computation can do, and I don't really want to talk about things other than how it can help consciousness. My main problem with the computational approach is that the computationalists normally concentrate on perceptual and cognitive achievements of machines which even now vastly outstrip those of very young biological brains like us as a neonate. It strikes me that that path, although it is an interesting path, won't lead to understanding consciousness 'cause for me the core of consciousness are feelings. I think that we are born with feelings. Even a one day old child has feelings, albeit not very sophisticated feelings. It's not going to get wild at the sight of the Mona Lisa or go into ecstasies over hearing Beethoven's ninth, but it has pain and it can exhibit very crude feelings that throughout life get more and more refined and more and more ideosyncratic.

And it's those feelings that to me are the core of consciousness that we have an emotional tone with us all the time even though it is only an excess that people will point to depression or ecstasy or so on and it's those feelings that of course, and this is going back towards my own bias, are manipulated by drugs and which make the brain so exciting and so different as yet from machines, because there's this added dimension of quality of mechanism, that is to say the quality of the molecules that are used as well as the quantity, the computation.

Back one final time and tying in with that the analogy of vitalism and saying that because we're no longer vitalists and we're all in this brave new scientific world we therefore ought not be interested in consciousness either, I'd like to say that molecular biologists haven't just sat back and said: "Oh well we don't believe in vitalists therefore that's all right we all know about life", they've gone on and manipulated life and are doing astounding things. I would like to think that the role for scientists with consciousness is to do the equivalent of molecular biologists, and to find out more and more how we can manipulate consciousness and thereby understand it. I know it's only a correlation. I know it's not an assumption of causality. I know it's not very glamorous, but I personally think it's a start.

John Taylor:

Thank you, Susan, thank you. So let me now turn to Brian.

Brian Josephson:

Well, it's difficult to find the right things to put into four minutes. I might just comment on this supposed analogy between vitalism and theories of consciousness. I don't believe these are equivalent but for a different reason. I think if you would have asked somebody a century or so ago: "Well you don't think that a collection of molecules could act in the way living systems do", he would say: "Well it seems to be very implausible but I don't feel I can rule it out totally". Whereas we are dealing with a different problem in the case of consciousness. It's a sort of category mismatch to regard the two as being the same kind of problem. I also personally don't feel it helps very much talking about materialism and dualism and such things because these are just labels - attempts to put people into boxes and I think thinking should be more flexible than that. I also am not quite sure how much John Taylor's statement that we study emotions with neural net models, how much that answers the question of consciousness because I also feel consciousness is more than just emotions, and that the hard problem is in any case not really solved by modelling emotions or any kind of modelling. Well, to look on a more positive side and I've been mainly critical, I have definite sympathies with Igor's approach of looking at the problem from a very abstract point of view. I think it's always helpful to do that and it may be useful to get a general model for the kind of thing that would behave as if it were conscious, or it would behave in an analogous way.

However I feel that not withstanding what the various speakers have said, there ultimately is a physics of consciousness. There's a process that makes systems conscious just as there are processes that lead to black holes. There's some kind of physical process which may allow little bubbles of consciousness to emerge and in general collapse, but those that find themselves in the right environment will become conscious agents, and perhaps we should look into the question of what kind of system is it that anything would want to be conscious in. Perhaps we just have a very good environment for something to be conscious in, in our brains, not ideal but at any rate not too bad. So I think I'll leave it there.

John Taylor:

Thank you very much Brian. David, your turn to come back to the town, stir it up a bit.

David Papineau:

I want to talk particularly to John Taylor and Susan Greenfield. Let me explain why: Dr. Büchel didn't even mention consciousness and Brian Josephson raised a number of large issues which four minutes isn't enough for. And Igor - I'm not sure if you really are a dualist. You've been much more careful in the phrases you've used than the two in the middle[2].

John Taylor:

Trust a philosopher to say that.

David Papineau:

Trust a philosopher to stir things up. I should say, you really are a dualist, and trust a mathematician not to understand what they're doing when they get away from equations. So, you say that you want to understand how qualia *arise*. How could you mean that without meaning that qualia are one thing and what they arise from is something else. You said you want to explain how consciousness has arisen in the brain. Susan, immediately after denying that you're a dualist, started insisting that the real problem that you still saw was there even if you weren't a dualist was you wanted to explain how you got subjective effects from only neurons. You said there's more to it than just the firing of neurons. You said there's an added dimension to the brain apart from the physical dimension. ...*Some squabbling*...It doesn't matter whether you use these phrases. It doesn't matter whether we call you a dualist or not. But what I would invite you to think about, is, if you really aren't a dualist,

[2]Susan Greenfield and John Taylor

why do you think that the problem of consciousness, the need for a theory of consciousness, is missing or more significant than the problem of life, the need for a theory of life? Now, I agree with you, when you responded to this analogy by saying, look, the death of vitalism didn't close down scientific studies of life, and of course it didn't, and no more do I think it should close down that kind of scientific study of consciousness. Of course you want to understand what kinds of life there are on earth. We want to understand various aspects of their working, we want to understand how to manipulate it, preserve it, etcetera. So we should do with consciousness. We want to understand what kind of consciousness there is in human beings, what its causes and effects are, how we can manipulate it, how we can ameliorate the unpleasant forms of it. Of course we should do that. But what we shouldn't do is start looking for *the* theory of consciousness, *the* key, the theoretical key to unlock the problem of consciousness. Because of course we don't look for the theoretical key to unlock the problem of life anymore, now we're not vitalists. When we were vitalists we had that kind of problem. Now we're not vitalists. We just understand that living beings are a certain complicated kind of physical system. Where you draw the line is just a matter of terminology. We want to understand the more complex ones of course, and in that sense we're understanding life. But we don't need a theory of life to do that, the same as we don't need a theory of consciousness.

John Taylor:

That's it? Thank you, David. Thank you very much. So now I will turn to our last speaker, that's Christopher, who's just finished speaking and now he's heard these new attacks. Where do you go now?

Christian Büchel:

I wasn't attacked at all.

John Taylor:

No, you were lucky.

Christian Büchel:

I was lucky, talking about attention, not about consciousness, although there is a close link. I think it is worth discussing the link between attention and consciousness later on. I'd like to make some points of the exocephalic model which I attributed to Professor Papineau at the beginning of my talk. I hope Susan Greenfield will excuse it.

Igor Aleksander:

She survived.

Christian Büchel:

And I'm really looking forward to the year 2030 when the exocephalic model can talk to us. This is a very important feature - language, mentioned in Professor Taylor's talk, and I think it's worth stressing language a bit more, because the reason Susan Greenfield mentioned on her slide we don't know whether the cat is conscious or not has something to do with the fact that a cat can't talk to us, that there's no communication, no language communication possible, although there might be other channels of communication. I think this is very important and hopefully the exocephalic magnus will talk to us soon. All the controversy about dualist or not, I'm still thinking on. I also was pleased to see that Susan in her talk used modulation of connectivity in the brain to explain consciousness where we stopped earlier and only tried to explain attention by modulation of cortical pathways, but maybe and this might be also part of the discussion what is attention and what is consciousness or how are they linked because they are closely linked in some respects and they are not closely linked in other respects.

John Taylor:

Thank you Chris. Now I'll call on our last panel member, that is Dr. Peter Fenwick. I think you have his bio in front of you. I'm delighted to welcome him with us. He's a psychiatrist, psychologist, from the Institute of Psychiatry, and has a great deal of experience in a broad range of areas associated with consciousness and its possible lack of and deficits in. So Peter, now you've only got four minutes. I know it's a bit unfair 'cause all of us have had half an hour, but it's going to be worth it. Four minutes please.

Peter Fenwick:

Thank you very much and thank you for asking me to come and it's only right and proper that the psychiatrist should end the formal presentations. Four minutes probably will be enough. ...*laughter*....because one of the things I'm going to suggest is that in fact essentially they are all wrong...*laughter*... Now they are all wrong for very specific reasons. Descartes started it. Before Descartes, the world was seen as a unity. What Descartes said is there's the extended thing which is the world and there is the mental thing. The mental thing is not part of science, the mental thing belongs to the church. So the church took on spirituality, the mental component, and the thinking thing was connected to the extended thing through the brain, by the pineal gland. Now we know that's wrong, but nevertheless we've got this split, and this split has produced our modern day science. Now modern science argues for an external objective world which you can do things to, and as you heard from all the speakers today they have been doing things to this objective external world. Now that is a cultural position. There are many other different cultures that don't think of consciousness that way. The current view in philosophy which is called postmodernism, is that our science is value-laden. Now that is a very strong statement because it is saying, when you carry out something with your science you're not looking at the nature and quality of the external world, what you're looking at is the nature and quality of the culture that you bring to bear on the external world. For some things our science is very good. It's very good indeed at asking questions about the objective world, and if you look at what Western science has developed, it is astonishing. But where it is extremely bad and where we have no models of the territory we are in now, is the whole question of consciousness. Now I'm going to go back again to this Descartian split. My own view is that it is all physical. The mental is a component of the physical. I'm going to suggest to you that the external world around you is not an external world, it is a psychological world. It has never been a physical world, it is only a psychological world and the difference between my psychological world and your psychological world is that mine is subjective and I can talk about yours because they overlap. So we're talking about a psychological reality. When Susan talks about her models she does it beautifully but they are models of brain function and they're models in our psychology. Take another culture and they'll have quite different models. So the most important point there is that we are dealing with models. Now does our scientific model of consciousness have major flaws? Well yes it has, because you cannot get from the mechanics of the brain to consciousness because consciousness is not within our science. Look up any textbook and you will see nothing about consciousness. Neurons aplenty, magnetic fields aplenty. Feelings? In terms of neural structure, yes, but there are no theories which allow us to combine the two, and so I think that the way we should look at things is something like this. First of all consider the data about consciousness. One of the things that we know from the parapsychological data is that if people or animals are in contact with things, then they can influence reality. Now there's a new idea for you, 'cause it means that you are part of this matrix. There is some lovely work coming out from France which demonstrates that chickens can influence random number generators in robots. You have to set it up by imprinting the chickens on the robots so that the chickens think the robots are their mothers and then they can alter the random number generators.

John Taylor:

Peter, I hate to say this at this point, but you've gone over your four minutes and your random generator number is up.

Peter Fenwick:

Can I just summarize it? I think, therefore, we bring our culture to bear on our science, it's without consciousness. I think mechanisms of perception within the brain are not as we think them. I think that we're dealing with a psychological reality out there and I think that you'll get part of the picture by looking at brain function, not the whole picture. Thank you.

11.2 Audience Participation

John Taylor:

Well thank you very much Peter. O.K. now that's been our turn. Obviously we're all straining at the bit to get back at each other. But that's not fair, because you've got three quarters of an hour for yourselves and I feel that while we can talk amongst each other, it's your turn, and I would like to structure this somewhat. It seems to me there are a number of topics that have come up in things that we have said and in earlier talks which we should try to cover to get ultimately more precise about the nature of consciousness or your feelings about it. First of all, one of the topics I'd like us to start with - what is Science and is it able to tackle the problem of consciousness? Secondly, this theme about consciousness being hard or not. Is there any real problem? - the one that I think David has thrown at us as a nice little bomb. And then going on from that, is it in fact the case that a number of you out there are dualists though you didn't know it all along, which, according to David, could well be the case? And then we should go on more closely to the question I think Chris raised about the nature of attention versus consciousness and then further into the deeper aspects. So I would like to have comments and questions from the floor. I think that's the way we should proceed. So, the nature of science and its problem. Can it actually attack consciousness? Is it in fact so difficult? Or is it still up to it? Yes, please, first.

From the Audience:

I wonder if anyone on the panel is religious and how would they try to explain religious revelation in terms of consciousness? Many people would think that the greatest insights accorded to mankind ever, on a par with Einstein's and other scientists' discoveries, have been religious insights, and whether those things are completely beyond the realm of science.

John Taylor:

Well can I just ask you, what insights are they that we can say were better than the special theory or the general theory of relativity?

From the Audience:

The religious insights would be less specific than a scientific theory, but perhaps more important to the people who have those experiences.

Susan Greenfield:

Let's take a crab-wise approach. So if we could avoid the question of religion - just freeze it for a second. People who take certain drugs, or people who have certain brain illnesses such as schizophrenia, will often have, I'm not saying this is the same as religion, but I'm saying the fact that, for that individual they are undergoing a very special experience, that no one else can share, that is giving them revelations onto life that are changing their perception of the world, the fact we know that can happen without invoking religious experiences makes me question why we would need to invoke religious experiences.

Igor Aleksander:

Just two very brief points. Because it is a very complex area it needs to be clarified. The first thing is that you are talking about an area of consciousness which goes under the heading of beliefs, and I think that many are interested in this question and I'm not going to tell you how this is done, but certainly it is part of the business of studying consciousness - where do beliefs come from? - and under beliefs are some blinding ones which are described as revelations. In history however there's another problem which is easily solved, which is that for a long time mind and soul were confused, and I think this is no longer done. Soul is that which remains after you are dead, and as soon as you start thinking about that it becomes part of the area of beliefs. I think one can have long discussions about that but I think they are not going to take you any further than the discussions we're actually having.

Brian Josephson:

Susan Greenfield says there's the schizophrenic experience and we doubt that and hence should doubt religious experience. I would suggest a better analogy is with mathematics, because mathematicians and people who do mathematics and get things wrong - we don't doubt the existence of mathematics because some people make a mess of it.

John Taylor:

And that is a truly religious experience. We try and impart it to our students all the time. *laughter*

Brian Josephson:

What we really want is a framework of thought which is wide enough to encompass these things. I should just mention a point which is relevant is that religious experience has been studied within the academic world - in the social sciences. For example, the Alistair Hardy research unit in Oxford studied these and made connections between experiences and life and classified them, and so on. There's also a great debate about whether near death experiences are real or caused by the brain fizzling out.

Peter Fenwick:

Just very quickly. A number of astronauts had very wide mystical experiences, and those who had it were the crew members and not the captains. From a social point of view, these people were undergoing a very strong experience and if they didn't have to retain their model of the world, which is our western scientific model, and get the capsule down, then under that influence they could produce other models of reality, and the model which they produce is in fact that of the mystical experience. You can argue that if we were in that mystical state all the time, where one saw directly through into the structure of matter, we'd have a totally different science, or no science.

John Taylor:

I'm sorry. I don't agree at all with that, Peter. I'm afraid I've worked with people who claim to have been able to see into the structure of matter. I've tested it out. I've worked with people who have been regressed back to earlier lives and tested out whether they have any extra knowledge. I've worked across the whole range of the paranormal and I found nothing there. I've worked with people who are not charlatans at least consciously. I've worked for example with children who weren't, consciously. I've worked with adults. I've worked with others who *were*, consciously, one whose name came up quite recently helping the British football team: They won, but whether that was due to him or to other influences, a good trainer etc., I'm not sure.

Peter Fenwick:

John, I didn't say that our science wouldn't produce the same sort of world. What I would say is that we would have a different science and we're talking about consciousness, and I think consciousness would have a different form.

John Taylor:

It's an intriguing question, but I think maybe at this point we should move on. Are there any other questions associated with the nature of science?

From the Audience:

What is the difference between consciousness and life?

David Papineau:

These are the people who believe there is something special about consciousness. I think this question is like: "What's the difference between an animate being and an inanimate one?" I don't think there is a specially clear line. I'm with Susan on this. For most things what it's like for them doesn't matter to them, so you wouldn't count them as conscious. Once you get beings where what it's like for them matters to them, you start thinking of them as conscious. But I don't think there is a sharp line, just as I don't think there's a sharp line between crystals alive, viruses alive, little ones, big ones ···. It's not a serious question. I don't think the point at which consciousness starts is a serious question.

Susan Greenfield:

I've got various issues, various little sort of depth-charges that have fired off when you asked your question. So I'll just say them quickly although they don't necessarily relate to each other. The first and most relevant one is persistent vegetative state, which everyone has heard of 'cause it comes up fairly often in the news now, with people being killed because they are no longer going to be conscious, which I think raises the interesting idea that if you're not conscious you might as well be dead, even though you are alive, because they just withdraw food. I think that's quite an interesting idea.

The second of course is when you're asleep. You're not conscious, but for me, as I hope I've made clear, dreaming is a form of consciousness.

And the final thing is, to me, consciousness is all tied up with movement. I think any multicellular organism that moves is animated, has an *animose*, is conscious. There's a lovely story of the tunicate, that when it's moving around as a larvae has a very very primitive nervous system, but when it becomes sessile and fastens to a rock, it eats its own brain, it becomes a vegetable, effectively. Some said it's like when you get tenure you end up like that. *laughter*

John Taylor:

Speaking as someone who has tenure, I'm afraid I disagree on that. I think there are points at which consciousness does evolve, and we come back to this question of emergence, provided you have a complex enough brain arranged in the right way. And it's that arrangement that's important, and in certain lower animals you might say it is beginning to be arranged in the right way. There is enough neuronal information processing, but again it's arranged in the right way. And I mean for David to suggest in fact that we are looking for something above and beyond the activity of the neurons, yes we are, we're looking for activity of *ensembles* of neurons, as Chris was showing and as we're finding by these non-invasive instruments. And it's a different level that we're working at now. I think Susan is right about the dangers of single cells. They don't have consciousness. But yet you embed them in an ensemble of activity with different modules, then you can begin. And we're not looking, I hope Susan isn't, but I certainly am not looking for anything more than that. 'Cause there can't be anything more than that.

David Papineau:

I'm glad to hear that.

John Taylor:

Are we resolved then?

Susan Greenfield:

Are we still dualists?

David Papineau:

No, you're not.

John Taylor:

Ah, wonderful. We're let off.

David Papineau:

There's another question which is much easier to ask and much easier to answer, but I think it's worth distinguishing from your question, which is: "At what level do things become *self-conscious*?", and then we have a whole lot of different things and it's quite easy to see that they have to have beliefs about themselves, they have to have memories of themselves, they have to have a sense of their own biography. Now I mean we can see quite easily when that arises and that's quite specific. The harder question and one that I don't think does have a very definite answer is when we go lower down than that, and we kind of have, I don't know, frogs, slugs, sea-cucumbers, and I don't know how to start addressing the question of whether they're conscious or not. I don't think it's a serious question.

John Taylor:

Once we can talk about it from a higher level maybe we can work our way down, with the models we have.

Well I think maybe we should move on at this point to the second area, and I'd like to ask *you* something, and I would wonder if we could have a show of hands. How many of you feel that consciousness *is* a hard problem? In other words, it's a very difficult one and this question of subjectivity is very hard to get inside. How many of you feel that this is the case? Oh, there is a majority out there.

Can I ask a second question, which is, how many of you feel that this is an *impossible* question? Given science, whatever it is, how many of you feel that we will never solve it, whatever way? Be honest, be honest, we've not got big brother watching you here. I don't think so. Not so many.

What about those who think it's not a real question. How many of you are conscious? I'm sorry I shouldn't have asked that. ... *laughter*

What about those who think that consciousness is impossible to solve, though it is a hard question.... So I think at this point we see that there are a group of people, non-trivial number, who do think it is impossible to handle.

Well now I'd like to ask you to make comments or ask questions about this area of hardness, versus impossibility, versus triviality.

From the Audience:

This is primarily a question directed at Professor Papineau, but I think perhaps Professor Josephson and Dr. Fenwick may deal with it too. It doesn't follow from the fact that dualism is false that materialism is true. If dualism is false, monism is true. If there are no grounds in theory, as opposed to practice, for distinguishing between me, and I take me to be a conscious non-philosopher of mind, and a non-conscious entirely material non-philosopher of mind, then there is no ground in theory for saying that they are anything but the same thing. The question then arises, is the class of things into which both the zombie and I fall a class of material objects or a class of mental objects? If monism is the right way to go about it why do you want materialism as the obvious candidate?

David Papineau:

There is a reason, but it's not an a priori philosopher's reason, it's a reason based on empirical evidence of fairly recent vintage. I'll put the argument that I see as knocking out dualism in a different way, and you'll see why it forces us I think towards materialism. Let's take, and you might want to question this, that there are some bits of matter, and in particular I'm going to be interested in the bit of matter in my arm, all right, it moves around. I take it this has got some cause of its moving around, and I take it that physics shows us that the cause of such physical events is always prior physical events, so that there'll be some physical events going up my arm into my brain that caused that. Right. I take it you'd also want to say that my decision or the pain causes my arm to move. How can that be that my arm has two quite different causes? It doesn't, clearly. The way to resolve this tension is to say the two causes are just the same cause. So you collapse the mental cause into the physical cause. The reason it's collapsed into a physical cause is because of the premise, that I invited you to believe a moment ago, that surely my arm goes up because of some physical cause. So that's the reason for materialism rather than some other kind of monism. And the evidence for that is just the scientific evidence we have, that in general all physical effects are due to physical causes.

Brian Josephson:

Well I think I'd disagree with the argument that if you have two causes for something the two causes must be the same, because you may be contemplating concepts of cause in two different ways and hence get two different causes from that point of view, so I don't think it's a logically complete argument.

John Taylor:

So there could be two even conflicting causes, one being a mental, one being a physical one, you're saying?

Brian Josephson:

Yes.

David Papineau:

Let me put it like this. Here's a reason for saying they're the same. Do you think that if the physical cause hadn't been there your arm still would have gone up anyway because of the mental cause? Do you think if you hadn't felt the pain your arm still would have gone up anyway because of the physical cause? Those are cases where you think you've got two causes, but I presume you don't want to say if I hadn't felt the pain my arm would have come out of the fire anyway. So there's strong pressure for saying there aren't two different causes here.

Brian Josephson:

Well I mean it may have done it because of force of suggestion or something.

Susan Greenfield:

So you want to put your arm up. You put your arm up. That's a mental thing surely. I'm just being devil's advocate, it's purely mental.

David Papineau:

But Brian seemed to be saying a moment ago that there were two different causes here, the mental cause and the physical cause. If he thinks that, then it would follow that even if you hadn't wanted to put your arm up it would have gone up anyway because of the physical cause. And that seems wrong.

John Taylor:

But in a sense that is what happens in the reflex response. I must say I'm sympathetic to your argument Brian at this point.

Peter Fenwick:

It's that Descartian split again. We're still looking at the mind and the body as two different things. If you just come away from Descartes, you can see the whole world as a unity. O.K., so this is monism. It was Gallileo who started this question of different sorts of stuff in the universe. There was energy stuff and there was matter stuff, and the matter stuff had shape and form and the energy stuff had different properties. Then there were subjective qualities that were not part of these categories and so didn't exist. We've got to get back again to recognizing that what we call mental causes are within the physical stuff.

John Taylor:

Very good, thank you, thank you. Maybe we should go on to the next question from the gentlemen there. Yes, please.

From the Audience:

I would like to ask you about the perception of time. Our experience of time occurs when one sensation is obliterated by its successor. It seems to me that if you just had a purely materialist description of consciousness, that your consciousness being tied to an unchanging universe, you would have conscious cells side-by-side in time, and they wouldn't actually communicate.

John Taylor:

The initial response to that is there are many clocks in the body. We know this about diurnal rhythms, but there are also clocks at the order of every twenty-five milliseconds and there are many many more. And if you take the libid experiments that I think a number of us have talked about, that libid data shows that there is a back-dating that very likely uses a thalamic level or a frontal lobe clock. People without frontal lobes have great difficulties in determining the length of time. So it would seem to me that the nature of time psychologically or physiologically is determined by these rhythms and there are many many of them.

Susan Greenfield:

I'd just like to follow on from that. I think time perception, a little bit like pain perception, is falsely simple, and one assumes it is going to be conducted in a very linear way. But I'm sure there are people in this room who have had the experience of an accident. And the common report when you have an accident is everything slows down, everything seems to move really slowly. I have a model I can't really elaborate here, but suffice it to say that I think the experience of having an accident when time slows down is comparable perhaps to children who again have very slow perception of time. You know, a kid thinks a day is forever. That ties in again with, as John touched on, the pre-frontal cortex, where we know if you have a lesion you have something called source amnesia, which is to say you lose space-time constructs of your memories, so your memories become like dreams. And for me, why I find this attractive is dreams, child, and accidents would all be associated with a small neuronal assembly. We also know in schizophrenia there is a disruption in that area. So I think that time perception is a very intriguing avenue into understanding consciousness, as pain perception is, rather than to be something that would refute materialism.

John Taylor:

Exactly. Chris, would you like to respond on this question of time, because you're interested in the flow of activity?

Christian Büchel:

Interesting thing is that Francis Crick was mentioned, and his theory is not very popular at the moment.

John Taylor:

Forty Hertz you mean.

Christian Büchel:

Exactly. He would like to say that full consciousness is due to forty Hertz, which you can say is one timing. This is obviously not true that a single timing can be due to what we like to call consciousness. And so more complex patterns might arise now there. I think the problem in time, or the perception of time itself, is not that closely linked to consciousness, or what I understand as being consciousness.

John Taylor:

Right, thank you. Can we move on? Yes, please.

From the Audience:

I'd like to say something about why I feel that anybody, including myself, would think there is something intractable about the subject at issue. I think that's because it doesn't appear to be like any other property that we know of. It seems to be the end of the line somehow. The point perhaps at which ontology runs out, and yet we have to try and explain it in terms of other things that are, exist, and so on. But we can't quite do that. So "it ain't like any other", in a certain sense. The other thing is that perhaps there are some inherent limitations for the course of science itself. Perhaps it is a culture which is going to defeat everything else and gradually eliminate all other forms of looking at the world, but perhaps not. I would think that the search for a scientific explanation of consciousness, although we are now in realms of complexity that are far advanced from what we were a few years ago, it may still be, on an analogy, like walking around a library only knowing the area of the shelves, or like a group of highly fancied droughts players who are blundering around a chess conference trying to apply their analysis to something which is actually way beyond their understanding. I'd like to hear from some members of the panel why they think that the level of complexity with which we are analyzing it goes with a chance of understanding what is going on in the most complicated object in the universe? And all of this of course would be entirely consistent with a broadly materialist point of view, but not necessarily with an anaemic one.

Brian Josephson:

I think you made some very good points. I'd like to mention in this connection some theories of John Polkinghorne, which I probably can't explain very well. But he's trying to make a connection or at least a reconciliation between science and religion which proposes that there are different levels of organization. There's a high level of organization which is to some degree the source of the organization we see in science, and so he was suggesting we got ideas as to what was fundamental a bit mixed up. And one could then have, as it were, unseen complications of the kind you suggested coming in. He was talking about it in connection with the concept of free will and showed that it might provide a way of putting free will into an apparently deterministic universe.

John Taylor:

Thank you. Any other comments?

Igor Aleksander:

Just a quick one on intractable things, which I have great difficulties with. As soon as you name one, somebody's going to try and track it down. So if you look at that as an activity you find that there is a workman-like activity which can be done in tracking things down, whereas those who believe that things are intractable can only sit there and believe they're intractable. The history of science is strewn with examples of things that were thought to be intractable until they were tracked down. But we have no way of predicting.

Susan Greenfield:

Just imagine Christopher Columbus was there in Portugal, in Lisbon, surrounded by people saying: "Oh, we don't have a chance". Of course we can't say in advance if we're going to make any progress. But we know more about the brain than we ever did. We're making huge advances. Now that might not be fruitful, but we won't know until we try.

From the Audience:

What about travel into the past?

Susan Greenfield:

You mean actually time travel? Well, that's outside my area of expertise. As physicists, I know that some people talk about that. I might as well say: "What about green martians?" We know they don't exist therefore they're not a type of consciousness. I think it's a false analogy.

John Taylor:

No, no, but wait. Wait, Susan. We can answer that directly. There are certain forms of time travel that, according to our present theories, are regarded as possible, other forms that aren't. So if you want to discuss those in detail, which are not relevant to this particular session, that can be discussed in a very detailed way. And you see, I would reply to you, since I've worked for many years in particle physics, that we have got down to very deep fundamental theories. And if you say that actually we have missed something out, and that is returning to Brian's suggestion about John Polkinghorne, that somehow there are some gaps in there, I am surprised that John Polkinghorne would say that, because he's a particle physicist too. And I was a research student with him in Cambridge and know his thoughts about this, that particle physics works extremely well. The question how you can reconcile consciousness with particle physics, if you are a religious particle physicist, is one that I would not wish to address. John Polkinghorne is not here, so he cannot address it.

David Papineau:

Can I say something, because we seem to be presenting a uniform scientific front here.

John Taylor:

Ah, the philosopher steps in.... *laughter*

David Papineau:

Since I've been coming across as the village materialist so far, I don't think, and I doubt any of my colleagues here think, that science can analyze and resolve everything that's real and important to me. I don't think there's a science of friendship or intellectual progress, or artistic beauty. I mean, I think there are some things that are important and real but outside science.

John Taylor:

For ever, for always?

David Papineau:

Yeah, I mean I just don't think they're the right kind of things, they're not scientific.

John Taylor:

I don't understand that point of view. I mean, it is clear there are ways of analyzing in detail. We build systems, models, of the brain, in which there are emotional aspects and you can...

David Papineau:

We can have a science of what's going on in people when they make friends. It will leave out some truths about why friendship is important and worthwhile. I just think that's obvious.

Susan Greenfield:

I think the distinction is, let's take something like, you love your wife, right? Now I think your issue would be, you can't scientifically prove you love your wife, for example, yeah? Whereas John Taylor would say scientifically we can study love, we can study what's going on in your brain, we can manipulate your love. So I think you're both right and yet you're both ... *general talking...*

John Taylor:

David, if you were here yesterday you would have seen robots following each other around, and I would say that that development of group activity, this is quite an important area of research in robotics, is to work out through group dynamics of robots how humans may have group dynamics, at a very low level. So I think it is not quite the case - I see the thin end of the wedge in this sort of thing.

Peter Fenwick:

John, Descartians, I wish they'd give this up and look at the world as a unity. They keep splitting the mental from the physical. In answer to your question, you said two things which I liked very much. First of all, blundering about in the library and only knowing where the books are, I'm sure that's right. The second thing you said was culture. And I like that very much because our culture determines the questions we ask. And as John knows very well as a particle physicist, if you ask questions about a wave function nature answers with a wave answer. If you ask it about a particle question you get a particle answer. We do not have the right questions to ask of nature yet about the nature of consciousness and that's because of our culture and our science. We're asking a lot of questions about the brain but we're not asking questions yet, 'cause we don't have the proper structure for it, about the actual nature of consciousness.

John Taylor:

Thank you. Yes, a question at the back there, please.

From the Audience:

I'd like to discuss the question of pain. I think pain is actually rather more complicated than people would believe. Now I think it is true, and I think I've experienced it myself, that if one touches something hot, yes, one withdraws one's hand. But one experiences the pain after one's hand has actually left the hot object. This raises an awful lot of questions about what is pain for. But if anybody would care to explain pain in more detail, please do.

John Taylor:

Can I respond by saying that I would lay claim to be, I think, the only person who's walked on fire.... *laughter*.... Anybody else prepared to say that? I've walked on fire, experienced a certain degree of pain, in that case I was trying to test out how much pain would normally be experienced by people. But I would like to come back, and I think this was a point which was made possibly by Susan earlier, that pain is highly cultural. The thresholds change during the day, but they also change across cultures, which comes back to something that Peter was saying. Pain in fact in some societies has a very high threshold. There are even people who do not experience any pain throughout life and in fact they die at the age of between 35 to 40 or so, because they've worn out their joints, they've put their hands on hot plates or in fires, because they haven't realized the danger. And I would suggest therefore that pain is cultural. It is learnt. But it is learnt so that there can be survival value. And that survival is obviously extremely important. It hasn't yet got, it seems, into the genome, as far as I can understand.

Susan Greenfield:

I'd like to refine that, and I wouldn't say so much about culturally conditioned. I think a one day old baby will feel pain, without any culture.

John Taylor:

Not according to some child psychologists.

Susan Greenfield:

Well nonetheless, I think anesthetists on the whole would anesthetise a one day old baby if he had to operate on it. I think. So the corollary of that is they must be going to feel pain then.

John Taylor:

Well if you pinch a baby, one day old, it gives a reflex response but it doesn't seem to have any grimace and it doesn't seem to avoid, other than through straight reflexes.

Susan Greenfield:

Maybe reflexes are different because of the lack of integrity of the descending paths, but that doesn't mean to say they're not aware of pain.

John Taylor:

O.K. Please, the question at the back.

From the Audience:

I'd like to ask Igor Aleksander about his statement that any state can possess consciousness. I would say that consciousness is always a property of any system with a large number of independent particles.

Igor Aleksander:

I've done it again! I've made myself completely misunderstood. I think I made quite heavy weather of the business of an iconic transfer, which means if you take any old control system which is in a state, it certainly isn't conscious because that state doesn't relate to any of its outward experience, outward activity, and so on, so I think there's a long, long way to go, and I think we're only just beginning to scratch the surface in understanding this question as to what's the difference between any old state and an iconic state. Iconic states can be very complex indeed. They can have to do with beliefs. They can have to do with things you've never actually physically experienced but have been induced by language. So it's a special subset of states. I need a mechanism to create these iconic states before I can even start talking about consciousness.

John Taylor:

Thank you. Please.

From the Audience:

I agree that you can't split the mental from the physical in the normal case. But I would disagree with the statement that materialism is all there is, that there is only patterns of firing neurons.

John Taylor:

Are you saying that as a matter of principle or just as we understand at this point?

From the Audience:

You choose to ignore the stuff you don't believe in, such as near death experiences, out of body experiences. You say they must have been mistaken, there is something wrong with the data.

David Papineau:

The PVS[3] cases clearly don't support what you're saying, because there is a physical difference between them. That's how you tell they are in PVS. You do tests on their brain. So it's not an argument against materialism, it's an argument for materialism. The conscious difference is due to physical difference.

John Taylor:

But as far as the cases of out of the body experiences and similar ones, you can ask what is the information that is actually gained by these people? I, having been involved in tests in a number of cases, find there is nothing other than that which was already partly cultural, partly happenstance. And indeed we've done tests on people under hypnosis to see whether they, when they regressed to other times, we've done tests on out of body experiences, and distant viewing. And again, I'm afraid to say that I find that the only things that are discovered are either what's already in the brain attained under these changed mental states, in quite a remarkable way.

If I can just give you an example of a young lady who could spout classical Latin. It was discovered that at the age of four her mother was housekeeper to a professor of classical Latin, and he would recite back and forth in his study next door to her nursery his lessons before he gave them. Now the human brain is an amazing thing and she only under hypnosis would have been able to give this out. I would say that in fact you've got to be very careful to make the assumption the way you're going, that you've got to look into these cases. And in fact I did it myself for a number of years, investigating such phenomena, and

[3]Persistent Vegetative State.

I found nothing there. And if you do it yourself, you may find the same. If you find there is something there, come and tell me, please.

Brian Josephson:

History tells that Professor Taylor once wrote a book saying there was something there.

John Taylor:

Two books: one that there was something there, possibly. The second saying that there was nothing there, definitely. *laughter*.... Please, your question. I think you were going to have the last one.

From the Audience:

This is a question to Brian Josephson. Do you think there is any connection between conscious experience and the brain?

Brian Josephson:

Yes, I think there's plenty of evidence of correlates between conscious experience and the brain. There's so much evidence that we shouldn't ignore the brain. It's obviously a very important part of the problem. The issue is whether that is the only system connected with consciousness.

Peter Fenwick:

I think my answer is almost the same, and that is that being a neuropsychiatrist I deal with people with head injury. So I'm familiar with the slow disintegration of consciousness with trauma to the brain. But I'm also very aware that the models we use are totally inadequate, and most people who are materialists, in fact are negative dualists. What they've said is Descartes was wrong, there wasn't a split, but they don't add the qualities back into materialism. So I think we need both. I think we need to model the brain. I think we need to get as much data as we can about the brain, and then I think we need to look at new models of how we can incorporate subjective experience into our science, as it was removed by Galileo and Newton.

John Taylor:

Thank you. I think that's a lovely aspect to end on, because it is now 4 o'clock. I know some of you have to get away. I feel we've had a wonderful time here.

I hope all of your consciousnesses are raised. You go away and do some neural modelling and look for new science! Thank you.

Index

D

E

F

G

H

I

J

K

L

M

N

O

P

U

V

W

Y

Z